THE GREAT KANTŌ EARTHQUAKE AND THE CHIMERA OF NATIONAL RECONSTRUCTION IN JAPAN

THE GREAT Kantō EARTHQUAKE
and the CHIMERA of NATIONAL RECONSTRUCTION in JAPAN

J. CHARLES SCHENCKING

COLUMBIA UNIVERSITY PRESS
NEW YORK

香港大學出版社
HONG KONG UNIVERSITY PRESS

Columbia University Press
Publishers Since 1893
New York   Chichester, West Sussex
cup.columbia.edu

Hong Kong University Press
The University of Hong Kong
Pokfulam Road
Hong Kong
www.hkupress.org

Copyright © 2013 Columbia University Press
All rights reserved

This English-language reprint edition is specially authorized by the original Publisher, Columbia University Press, for publication and sale only in Australia and Asia, except for Japan.

Library of Congress Cataloging-in-Publication Data
Schencking, J. Charles.
  The great Kantō earthquake and the chimera of national reconstruction in Japan / J. Charles Schencking.
      pages cm
  Includes bibliographical references and index.
  ISBN 978-0-231-16218-0 (cloth : alk. paper) — ISBN 978-0-231-53506-9 (ebook)
  1. Kantō Earthquake, Japan, 1923.   2. Disaster relief—Government policy—Japan—History—20th century.   3. City planning—Social aspects—Japan—Tokyo—History—20th century.   4. Tokyo (Japan)—History—20th century.   I. Title.
  DS888.S45 2013
  952.03'2—dc23

2012048647

10 9 8 7 6 5 4 3 2 1

Printed and bound by Liang Yu Printing Factory Ltd. in Hong Kong, China

Cover image: Uchida Shigebumi, ed., Taishō daishin taika no kinen
Cover design: Jordan Wannemacher

References to websites (URLs) were accurate at the time of writing. Neither the author nor Columbia University Press is responsible for URLs that may have expired or changed since the manuscript was prepared.

For Janet

Chimera
|kīˈmi(ə)rə|
|ki-meer-uh|
noun

1 (Chimera) (in Greek mythology) a fire-breathing female monster with a lion's head, a goat's body, and a serpent's tail

2 a thing that is hoped or wished for but in fact is illusory or impossible to achieve

CONTENTS

List of Illustrations xi

Preface and Acknowledgments xv

INTRODUCTION 1

1 CATACLYSM: THE EARTHQUAKE DISASTER AS A LIVED AND REPORTED EXPERIENCE 13

2 AFTERMATH: THE ORDEAL OF RESTORATION AND RECOVERY 47

3 COMMUNICATION: CONSTRUCTING THE EARTHQUAKE AS A NATIONAL TRAGEDY 78

4 ADMONISHMENT: INTERPRETING CATASTROPHE AS DIVINE PUNISHMENT 116

5 OPTIMISM: DREAMS FOR A NEW METROPOLIS
   AMID A LANDSCAPE OF RUIN   153

6 CONTESTATION: THE FRACTIOUS POLITICS
   OF RECONSTRUCTION PLANNING   187

7 REGENERATION: FORGING A NEW JAPAN THROUGH
   SPIRITUAL RENEWAL AND FISCAL RETRENCHMENT   226

8 READJUSTMENT: REBUILDING TOKYO FROM THE ASHES   263

9 CONCLUSION   301

Notes 317

Bibliography 347

Index 363

ILLUSTRATIONS

FIGURES

1.1	Map of tectonic plates that produce earthquakes in Japan	17
1.2	People attempting to flee the ruined landscape of Tokyo	19
1.3	Evacuees and their belongings in front of the Imperial Palace	21
1.4	Nihonbashi Bridge and surrounding neighborhood before the earthquake	22
1.5	Nihonbashi Bridge packed with people as a firestorm approaches	22
1.6	Nihonbashi Bridge and surrounding neighborhood after the earthquake	23
1.7	Dead bodies at Mukōjima, Honjo Ward, in Tokyo	30
1.8	Manseibashi Train Station in Kanda Ward with statue of Commander Hirose Takeo	31
1.9	Bodies at the site of the Honjo Clothing Depot	33
1.10	Cremating the dead at the Honjo Clothing Depot	34
1.11	Map indicating burned areas of Tokyo	39
1.12	Ningyō-chō, a once bustling street in Nihonbashi, Tokyo	41
1.13	Men looking for work at a city employment-matching agency	42

XII - ILLUSTRATIONS

1.14	The remains of Kanda Ward in Tokyo	44
2.1	Fukagawa Ward burning, as seen from an army reconnaissance plane	60
2.2	Navy personnel repairing docks and army personnel clearing train lines	64
2.3	Army personnel unloading food at a distribution center	65
2.4	People surrounding a water truck	66
2.5	Refugees evacuating Tokyo by train	67
2.6	Refugees in Hibiya Park	69
2.7	*Jikeidan*: well-armed yet ill-trained vigilantes	75
3.1	The whirlwind firestorm that swept through the Honjo Clothing Depot	91
3.2	Neighborhood around Matsuzakaya Department Store in Ueno ravaged by fires	92
3.3	The imperial tour that inspected the ruins from a lookout at Ueno Park	94
3.4	The crown prince touring devastated Tokyo	96
3.5	Children selling maps and postcards in postdisaster Tokyo	98
3.6	Dead bodies in the ruins of Nihonbashi	99
3.7	The bodies of dead prostitutes from Yoshiwara	99
3.8	City official directing citizens to the site of the Honjo Clothing Depot	100
3.9	The site of the Honjo Clothing Depot on September 1	100
3.10	Dead bodies at the site of the Honjo Clothing Depot on September 2	101
3.11	Mound of ash from cremated victims at the Honjo Clothing Depot	106
3.12	Buddhist monks offering prayers to the spirits of the dead	108
3.13	Gotō Shinpei giving condolences at the Honjo Clothing Depot	109
3.14	Buddhist monks offering final prayers for the dead at the forty-ninth-day service	110
4.1	Mitsukoshi Department Store before the earthquake	131
4.2	Shell of Mitsukoshi Department Store following the earthquake and fires	131
4.3	The catfish rectifying evil trends in society	134
4.4	The twelve-story tower dominating the skyline of Asakusa	140
4.5	The spread of fires around the tower in Asakusa Park and Hanayashiki	142
4.6	Hanayashiki after September 1	143
4.7	The twelve-story tower before and after the earthquake and fires	144
5.1	Tenement houses in Ryūsenji, Shitaya Ward, Tokyo	158

5.2	The earthquake as opportunity	165
7.1	A discerning customer after passage of the luxury tariff in 1924	247
7.2	The catfish reappearing, two years after the earthquake	260
8.1	The aims and objectives of land readjustment	266
8.2	Land readjustment districts in Tokyo, 1924–1930	268
8.3	Readjustment area 12 before land readjustment	271
8.4	Readjustment area 12 after land readjustment	271
8.5	Shōwa Dōri after reconstruction	285
8.6	Number of bus and tram trips in Tokyo	286
8.7	Kiyosu Bridge spanning the Sumida River	287
8.8	Sumida Park in 1930	289
8.9	Interior of a public dining hall in reconstructed Tokyo	291
8.10	Children's day-care facility in Ryūsenji, Shitaya Ward	293
8.11	Maeda's winning entry for the earthquake memorial hall	296
8.12	The completed earthquake memorial hall in 1930	299
9.1	The emperor's motorcade as it left the earthquake memorial hall	302
9.2	Two exemplars of New Tokyo	304
9.3	World War II–era American propaganda image	307

TABLES

1.1	Number of killed, missing, injured, and homeless in the City of Tokyo	40
1.2	Number of unemployed on November 15, 1923, in Tokyo Prefecture	42
3.1	Number and gender of bodies at collection and cremation centers across Tokyo	104
4.1	Spending on phonographs and records in Japan, 1912–1925	149
4.2	Spending on cosmetics in Japan, 1912–1925	149
4.3	Spending on sake in Japan, 1912–1925	150
4.4	Spending on all tobacco products in Japan, 1912–1925	150
5.1	Number of factories, employees, population density, and saimin in Tokyo, 1919–1920	157
7.1	Spending on phonographs and records in Japan, 1912–1930	255
7.2	Spending on cameras and camera parts in Japan, 1912–1930	256
7.3	Spending on cosmetics in Japan, 1912–1930	257
9.1.	National government reconstruction expenditures	304
9.2	Tokyo Prefecture reconstruction expenditures	305
9.3	City of Tokyo reconstruction expenditures	306

PREFACE AND ACKNOWLEDGMENTS

"Oh! How I wish I could feel an earthquake!" is generally among the first exclamations of the newly-landed European. "What a paltry sort of thing it is, considering the fuss people make about it!" is generally his remark on the second earthquake (for the first one he invariably sleeps through). But after the fifth or sixth he never wants to experience another; and his terror of earthquakes grows with length of residence in an earthquake-shaken land, such as Japan has been from time immemorial.
—Basil Hall Chamberlain, 1905

In March 2011 the world was reminded of the extraordinary force that earthquakes and tsunamis unleash. In dramatic fashion the Tōhoku catastrophe revealed how vulnerable parts of our planet are to natural hazards. Disasters do more than destroy, however. They also compel reflection, inspire optimism, and lead people to believe that something better can and will emerge from the devastation. Some people suggest that disasters possess the potential to change everything.

Numerous individuals opined that the Great Tōhoku Earthquake would transform Japan. Some argued that rising to the challenge of recovery would instill citizens with a newfound confidence and make people once again proud to be Japanese. Many predicted that reconstruction spending would provide the economic stimulus necessary to end two lost decades of deflation. Still others posited notions that the Japanese people might lose their faith in science and demand a reorientation of the nation's economy, or that humanitarian aid from China might help resolve long-standing territorial disputes between both countries. Will these transformations ever materialize or will contestation and resistance limit policy outcomes? History suggests the latter.

In September 1923 Japan suffered a far more deadly natural calamity. Then, a magnitude 7.9 earthquake and resulting firestorms killed more than 120,000 people and turned roughly half of Tokyo and virtually all of Yokohama into blackened, corpse-strewn wastelands. Amid this desolate landscape, bureaucratic elites suggested that a once-in-a-lifetime opportunity to rebuild Tokyo as a modern metropolis had emerged. Others argued that the cataclysmic Great Kantō Earthquake could, if manipulated artfully, rouse urbanites from their increasingly consumer-oriented, hedonistic mindsets and enable the government to forge a more moderate, wholesome moral path for social regeneration. Even foreigners involved in humanitarian assistance succumbed to the postdisaster culture of optimism. Admiral Edwin Alexander Anderson, who oversaw the initial US relief effort in Tokyo, informed navy officials upon his return to US territory—at Pearl Harbor, Hawaii—that American aid and Japanese appreciation of such aid had so firmly cemented friendly relations between both countries that no possibility of war in the Pacific existed in his generation.[1]

We know that Anderson's dreams of a peaceful future did not survive. Did the ambitions of starry-eyed planners, politicians, and commentators who wished not only to rebuild Tokyo but also to reconstruct Japanese society suffer a similar fate? If so, why? Moreover, what did Japan's experience at this destructive, dislocating, introspective, and yet inspiring moment tell us about interwar Japan and the anxieties, ambivalence, and embrace of modernity? This book explores these questions. My findings suggest that the handmaidens of disaster opportunism—namely, contestation, resistance, and the desire for a quick return to routine and familiarity—tempered dreams of lasting, transformative reconstruction. In the aftermath of catastrophe, when everything in the physical world seems anomalous, the tugs of resilience and the attraction of normalcy often prove intense.

I can write with certainty, however, that the Great Kantō Earthquake transformed one thing fundamentally: me as a historian. I began this project in 2003 hoping to document the reconstruction of Tokyo from the elite, political, economic, and urban planning level. I wanted to trace the dreams of reconstruction opportunism and explore why Gotō Shinpei's grand ¥4 billion plan failed to materialize and change the built environment of Tokyo in radical and profound ways. I traced the opportunistic

PREFACE AND ACKNOWLEDGMENTS ~ XVII

dreams inspired by the calamity and the contested realities associated with reconstruction through parliamentary transcripts; the diaries, memoirs, and reflections of key political and urban planning elites; and the minutes of key reconstruction deliberative and consultative committees. Though I published these findings in *Modern Asian Studies* in 2006, I realized there was far more to this disaster than merely the reconstruction of Tokyo.

In looking at the postdisaster debates about the future of Tokyo, I learned that numerous political elites and social commentators saw the postdisaster period not only as a chance to build a new Tokyo so that it could reflect and reinforce new values but also as a time to reconstruct the nation on multiple levels. In looking at prescriptions for society, I asked the following question: what did people think needed to be changed or rectified? Here I tapped into a rich vein of material published in the late 1910s and 1920s, often in academic and popular journals, that described Japan in a state of moral decline and spiritual degeneration. This led me to investigate how people interpreted the 1923 calamity. Answering this question compelled me to first look at the writings of religious leaders and academics. In doing so, however, I found that numerous political, business, and military elites, as well as social commentators and journalists, also employed a divine punishment or heavenly warning interpretation of the Great Kantō Earthquake. I was struck by the fact that so many individuals from diverse classes, professions, and ideological backgrounds all found common ground in suggesting that the earthquake was a heaven-sent wake-up call or an act of divine punishment. In published journals and newspaper editorials, as well as in sermons and political speeches across Japan, virtually every commentator used the earthquake to admonish Japanese, especially urbanites, for leading lax, hedonistic, luxury-minded, sexually unrestrained, and material-driven consumer lifestyles. These individuals often expressed hope that the disaster would become an introspective event, akin to what the First World War had been to the countries of Europe, and encourage Japanese to change their thoughts, behaviors, and thus the trajectory of Japan's modern development. I published part of these findings in the *Journal of Japanese Studies* in 2008.

For this event to be a monumental turning point, however, various elites knew that people outside the disaster-stricken areas would have to accept a regional calamity as a national event. This inspired me to look at how bureaucratic elites, artists, songwriters, politicians, and members of the press constructed the Great Kantō Earthquake as a national tragedy.

My pursuit led me to explore a cache of vivid visual and textual materials. In doing so, it became clear that people across the nation became fixated on the disaster and embraced the humanitarian relief effort wholeheartedly. These findings were published in a special issue of *Japanese Studies* in 2009.

No such unity of purpose or support surrounded the physical reconstruction of Tokyo, however. As the disaster receded from the headlines, formal support for the mundane and expensive tasks of physical reconstruction diminished. Prescriptions made to encourage new, more wholesome behaviors as part of a larger project of national moral regeneration were met with ambivalence and often ignored by members of the expanding middle class who had embraced many aspects of Japan's urban modernity.

Despite finishing research on how the disaster was reported and constructed, I was still left with important questions unanswered: namely, how did Tokyoites experience the great earthquake calamity and, moreover, how did they respond to elite-level overtures for rebuilding the capital and reconstructing the nation? Exploring these questions directed me toward survivor accounts, diaries, and memoirs, often written while Tokyo was still a smoldering wreck. These harrowing accounts provided emotive insights into just how traumatic and confronting this catastrophe was for the inhabitants of Japan's capital. In the face of such destruction, I remained puzzled as to why countless Tokyoites responded to calls for radical reconstruction with such ambivalence and, in many instances, rational, well-calculated resistance. To answer these questions, I examined hundreds of petitions and letters of protest regarding everything from land readjustment in Tokyo to the design and construction of the earthquake memorial hall. Looking at these materials gave me added respect for the people of Tokyo who struggled to make their voices heard in order to shape the future of the city they called home.

In undertaking this study, I have examined all these questions in considerable detail and produced a book far larger and more holistic than I first imagined. The process of researching and writing this book has led me from my love of elite-level political and institutional history to embrace a more balanced historical approach that examines this calamity from the ground up as well as the top down. The process has convinced me that disasters, as Anthony Oliver-Smith has suggested, are totalizing phenomena that sweep across all aspects of society. Moreover, I have newfound respect for the editors of the *Japan Weekly Chronicle* who in 1923 suggested that "great calamities become an obsession."[2] Anyone who has known me over the past ten years understands that this is an accurate description not only of the 1923 Great Kantō Earthquake but also of my life.

Many people and organizations have not only shared my obsession but also enabled me to follow it with zeal. First and foremost I thank the Hong Kong Research Grants Council (GRF-750309), the National Endowment for the Humanities (FA-37067), and the Australian Research Council (DP-0208116) for providing generous financial support at various stages of the project. Without this funding I would not have been able to carry out the extensive research in Japan that was essential for this project. Assistance from these organizations also allowed me to share findings at the Association for Asian Studies Annual Conferences in 2005 and 2011 and the International Convention of Asia Scholars in 2003 and 2005. My research has also been supported by various small grants and periods of academic leave from the University of Melbourne and the University of Hong Kong.

Though funding organizations enabled me to carry out research, Anne Routon and her team at Columbia University Press made the production of this book possible. During the entire production process, Anne has been a star. Her enthusiasm and frank, sagacious advice has been a perfect complement to my passion for this project. Thank you, Anne, for believing in this project and for making the finished product stellar.

I completed this project at the University of Hong Kong, and a number of people at this institution deserve special recognition as supportive colleagues and wonderful friends. John Carroll tops this list. John read the entire manuscript and provided insightful comments and suggestions on how to improve the book. He also shared his thoughts and ideas about history, writing, and scholarship with me on countless hikes and walks in Hong Kong well before the manuscript was in its final stages. Frank Dikötter likewise proved to be a generous colleague and friend who shared his opinions on writing, organization, and research throughout the past three years. Drawing on his years of experience in academic publishing, Michael Duckworth provided me with a number of suggestions on how to make the book a publishing as well as an academic success. Michael and Frank, moreover, deserve a world of praise for introducing me to Anne

Routon and for encouraging me to proceed with Columbia University Press. I am indebted to Wilhelmina Ko, who scanned all the images used in the book and designed the companion website. I also thank the following people at the University of Hong Kong: Daniel Chua, Yoshiko Nakano, Kendall Johnson, Robert Peckham, Xu Guoqi, Peter Cunich, Maureen Sabine, Kam Louie, Priscilla Roberts, Carol Tsang, and Phoebe Tang. Other friends and family in Hong Kong whom I thank include David T. C. Lie, Daniel and Diane Daswani, Kumamaru Yūji, Matsunaga Daisuke, Lilian Wong, Eleanor and Reuben.

I would like to take this opportunity to thank David Pomfret for making all these Hong Kong connections possible. Had David not invited me to give a talk at the University of Hong Kong in 2007, I would never have moved to this university. It is hard to imagine not being at the University of Hong Kong surrounded by such terrific colleagues who are just eccentric enough to make my job as chairperson of the Department of History enjoyable, rewarding, and certainly never mundane. Thank you, David. Friends and colleagues I have known throughout this project whom I wish to thank include Melissa Walt, Roger Thompson, Naoko Shimazu, Michiko Yusa, Patricia Polansky, and John Stephan. In particular, John's unrelenting passion for history and the process of discovery has remained infectious and been a continual source of inspiration and awe for over two decades. Finally, I thank my sister, Leah, for taking me on my first international trip at the age of sixteen on a British Airways 747–100 and opening my eyes to all that the world had to offer. I feel like I have never stopped traveling since that time.

Considerable research was carried out in Japan during this project. I thank the staff at Waseda University Library, the National Diet Library, the Tokyo Metropolitan Archives, the Tokyo Metropolitan Library, and the Tokyo Institute of Municipal Research for their generous assistance. Katō Yōko from Tokyo University, Matsuda Kōichirō from Rikkyō University, Sakamoto Kazuto from Kokugakuin University, and Asada Sadao from Dōshisha deserve recognition for their encouragement of my research in Japan. Hayashi Yōko from Japan Publication Trading Company in Tokyo deserves credit for helping me locate so many of the rare books, pamphlets, and postcards that I have used for this project.

During my time at the University of Melbourne, a number of Japanese history and studies colleagues in Australia proved themselves to be good friends and generous colleagues. Sandra Wilson tops this list. Sandra read the entire manuscript and offered a wealth of insightful suggestions on how to make it better. Through a series of Japan Foundation–sponsored modern Japanese history workshops that she initiated, Sandra enabled me to meet others scholars of Japan including Stewart Lone, David Kelly, Matthew Stavros, Bea Trefault, Takeshi Moriyama, and Elise Tipton. Each has provided encouragement and offered constructive comments on my work. Stewart in particular deserves credit for introducing me to the satirical visual culture of Kitazawa Rakuten, first in relation to my previous project on the Japanese Navy and then with the 1923 earthquake. Through Sandra's conferences and successive conferences organized by Bea, Matthew, and Takeshi, I was able to share my ideas on this project with Sheldon Garon, Matsuda Kōichirō, Kerry Smith, Alexis Dudden, Morgan Pitelka, David Howell, and Michael Lewis. Greg Clancey, Timothy Tsu, and Greg Bankoff also shared their time and opinions about my work on a number of occasions. Thank you.

At the University of Melbourne, no one provided more assistance than Michelle Hall, the Japan librarian. Utilizing a generous bequest given to the Department of History by Kathleen Fitzpatrick as well as other library funds, Michelle ordered every book, journal, pamphlet, published speech, postcard, map, and image that I ever wanted for this project. On more than one occasion she took it upon herself to order material that she thought I might need, without asking. Virtually everything she ordered was useful. Dean of the Arts Faculty Mark Considine and Provost and fellow historian Peter McPhee, provided valuable support while I was at Melbourne. Others who deserve special recognition include Bill Coaldrake, Hisako Fukasawa, Pat Grimshaw, Miki Ikeda, John Lack, Michael Leigh, Andy May, Jun Ohashi, Merle Ricklefs, and David Runia. I also thank former students who became good friends over the course of this project, including Rachel Saunders, Jordan Winfield, Nicholas Gillard, Chris Mullis, Shir Lee Teh, Julia Madden, and Brett Holman. I owe Nick special thanks for trolling through RG38 at the National Archives in Washington on a break from his own research. Doing so, he found materials on the US relief effort in postdisaster Tokyo. Finally, friends and family in Melbourne who deserve special thanks include Bruce Borland, Lorraine Borland, Lawrence Mitchell, Linda Poskitt, and Vannie Winfield.

I first began to conceptualize this project in the summer of 1999 as a British Academy Postdoctoral Fellow. Looking for a break from the tedium in revising my thesis on the Japanese Navy for publication as a book, I became interested in the Great Kantō Earthquake. I remember distinctly how my former supervisor, Stephen Large, beamed with excitement on a

Thursday afternoon at the Boathouse Pub in Cambridge when I informed him of my interest in the 1923 disaster. From that discussion, I knew this had the potential to be a rewarding topic. I thank Stephen for being a wonderful champion of this project and my development as a historian of modern Japan. Gordon Johnson has remained an enthusiastic supporter of this project and a wonderful career mentor since 1996. Thank you for all your advice and friendship, Gordon. In a similar way, Dick Smethurst has been instrumental in my career. Since I received my Ph.D. degree in 1998, Dick has been the ultimate scholar-gentleman-mentor in my life. Thank you, Dick.

    I have saved the single most important person to thank for last: Janet Borland. As a fellow historian of Japan, you have enriched this study more than anyone else through your candid thoughts, astute suggestions, and unflagging support. It has been such a genuine and rewarding pleasure to share my research findings, ideas, and many research trips with you all along the way. As my wife and partner in all pursuits, you have added more to my life than I ever imagined possible, and I had extraordinarily great expectations. You have brought beauty, radiance, wisdom, rationality, passion, and love to all aspects of my life and, like an earthquake, turned my world upside down, but in all the very best ways. I am extremely grateful that our individual passions for Japan and Japanese history brought us together and that we have continued to share this and countless other passions since. Janet, I dedicate this book to you with all my love, admiration, and gratitude.

THE GREAT KANTŌ EARTHQUAKE AND THE CHIMERA OF NATIONAL RECONSTRUCTION IN JAPAN

INTRODUCTION

We do not have to become pessimistic or disappointed. In a sense, what struck us was a baptism by fire. If the whole nation sees it this way and moves forward, there is no doubt that a new life will be born.
—Abe Isoo, 1923

Prior to the earthquake . . . people expected nothing from the nation and the mutual help across generations and the truest in local communities was beginning to crumble. But maybe the Japanese people could use the experience of this catastrophe to rebuild a society bound together by renewed trust.
—Azuma Hiroki, 2011

The first seismic wave hit eastern Japan at two minutes to noon on Saturday, September 1, 1923. It toppled structures, crushed people, and unsettled everyone who survived. Within minutes, a second intense wave battered the already suffering region. This tremor killed scores more and triggered panic not seen before in Japan's imperial capital. Over the next seventy-two hours, roughly two hundred major aftershocks and a series of diabolical conflagrations unleashed pandemonium, killed tens of thousands, and incinerated large swaths of Tokyo and Yokohama. Both cities had been transformed into scorched, broken, and almost unrecognizable wrecks. The smell of death and the groans of the seriously wounded, half-dead survivors amid the vanquished landscape led one anonymous chronicler to ask, "If this were not Hell, where would Hell be?"[1]

When the fires extinguished themselves and people surveyed the landscape, even the earth looked wounded. The earthquake had ripped open great fissures in the land. At many points where seismic waves met human constructs, nature showed no mercy. Apart from destroying buildings, the earthquake buckled roads, collapsed bridges, twisted train tracks and tramlines, snapped water and sewer pipes, and severed telegraph lines. Hidden beneath the waters of Sagami Bay, 60 kilometers south-southwest of Tokyo and at the earthquake's epicenter, the sea floor fell by over 400 meters and triggered a series of tsunamis that inundated low-lying seaside communities. "Nature," Buddhist lay spokesman Takashima Beihō wrote, "raged all at once, collapsing the pillars of the sky and snapping the axis of the earth."[2] "The big city of Tokyo," he lamented, "the largest in the Orient, at the zenith of its prosperity burned down and melted away over two days and three nights."[3] To those who had experienced Japan's earthquake calamity, this was perhaps only a slight exaggeration.

The human tragedy was every bit as horrific to survivors as the desolate sights of the devastated capital. In less than three days, more than 100,000 people perished, often in violent, tragic ways. Almost immediately after the quake hit, thousands were crushed by falling objects or died in collapsed buildings. People were also trampled to death by the scores of panicked residents who clogged Tokyo's streets attempting to flee. Individuals who escaped the initial chaos found themselves confronted—often trapped mercilessly—by encroaching fires that turned Tokyo into an inferno. Some victims suffocated as fires consumed vast quantities of oxygen from the air. More succumbed to burns produced by the intense heat and flames. Others drowned in rivers or were boiled alive in small ponds and canals that provided virtually no protection from the approach of Hell. However their lives were extinguished, all who died experienced a terror that was almost unimaginable just hours before.

Referred to initially as the Taishō *shinsai* (earthquake disaster of the Taishō era), or the Tokyo *shinsai* (Tokyo earthquake disaster), its name grew in stature as the totality of its devastation emerged. The 1923 disaster became, and would forever be known as, the Kantō daishinsai (Great Kantō Earthquake Disaster), one of the most deadly, costly, and destructive natural disasters of the twentieth century. This cataclysmic event, moral philosopher Shimamoto Ainosuke declared, "overturned" Japan's "culture from its very foundation."[4] Owing to the government's decision in 1960 to designate September 1 as Disaster Prevention Day, the anniversary of the Great Kantō Earthquake has become a day that all Japanese, not just Tokyoites, associate with natural disasters and preparedness.

Songwriter, street performer, and political activist Soeda Azenbō felt compelled to preserve the sights and sounds of the disaster he experienced through a ballad composed in 1923 entitled "Taishō daishinsai no uta" (A song of the Taishō Great Earthquake). Soeda's evocative lyrics relayed the horrors of Japan's apocalyptic experience to countless people in the years after 1923. He wrote in part:

> We heard roars near and far.
> Fires spread to burn the sky.
> The streets turned instantly into a veritable hell.
> It was hell on earth, filled with cries and screams . . .
>
> People died helplessly,
> burned in the fire, going mad.
> Parents called to their children; children called to their parents.
> Looking for them in the fire and in the water . . .
>
> Falling after being tormented by both fire and water;
> Thrown to the river from a collapsing bridge;
> And drowning, falling from a burning boat—
> The number of people who perished was too numerous to count.
>
> A whirlwind fanned the fierce fire.
> There was nowhere to escape but a muddy pond.
> They immersed their bodies, avoiding the flying sparks.
> But it was no use.
>
> So many are steamed to death,
> Licked by the tongue of approaching flames.
> Stepping over countless corpses,
> People ran around, under the smoke and flames.
>
> Those who barely escaped with their lives
> had wounds too terrible to look at.
> Poor souls, they are more dead than alive,
> Breathing faintly in misery.
>
> The survivors have neither food nor water
> They sleep in the open, with only the clothes they happened to be wearing
> Day after day, night after night.
> They feel more dead than alive . . .

> What a terrible force of evil!
> With only one shock,
> It destroyed the great city of Tokyo, Yokohama,
> The Bōsō Peninsula, Izu, Sagami.
>
> Humans took pride in their civilization
> And enjoyed the luxurious dream life.
> But it has been destroyed completely,
> Ah, it has been destroyed with no trace left behind.[5]

In the course of three days, as Soeda described, Tokyo had become a fire-blackened, rubble-strewn, corpse-filled, stinking wasteland of a once vibrant city.

In that fateful autumn of 1923, economist Fukuda Tokuzō shared much of Soeda's anguish at the destruction of Japan's imperial capital. As a self-professed "proud Tokyoite," Fukuda was distraught over the state of the metropolis he called home.[6] Was anything, he asked, comparable to the tragedy Tokyo experienced? Initial comparisons between Tokyo's catastrophe and the one that had struck San Francisco in April 1906—the last major urban earthquake disaster on the Pacific Rim of Fire—ended abruptly when the true scale of Japan's calamity emerged.[7] Fukuda believed that only the war that had ravaged the European continent less than ten years earlier provided an apt comparison with what Japan had recently experienced.

Fukuda felt more than just anguish at the destruction of large parts of Tokyo. He also harbored a sense of sanguinity. The German-trained academician knew that however great Tokyo had become, it was far from perfect on the eve of calamity. In the latter half of his life, Fukuda believed that Tokyo had failed to become a modern imperial capital (*teito*) as it grew but rather expanded in an almost arbitrary fashion as a "mere extension of Edo, the city built by the Tokugawa regime (1600–1868)." Fukuda complained that the narrow, winding streets and mazelike urban plan that had served the Tokugawa government well—a purpose-built defensive construction designed to restrict the movement of potential enemy armies into Edo—had done nothing but hinder the free and easy flow of people and commerce in modern Japan. Yet it had not been redesigned even as Tokyo's population soared. Few parks or open spaces existed despite awareness that they could improve the quality of life for Tokyoites and serve as potential firebreaks in times of conflagration. Worst of all, however, successive governments had done little to combat the proliferation of slum neighborhoods in eastern Tokyo or deal effectively with the causes of urban poverty. "Those cursed aspects of Tokyo," Fukuda wrote, "were the center and origin of all cursed things about Japan."[8] Now they were gone, and in their place he hoped something better could be planned, created, and managed.

With the fiery destruction of much of eastern Tokyo, Fukuda thus felt a degree of jubilation at the prospect of a true, modern, imperial capital rising from the ashes of dead Tokyo. More than simply feeling ecstatic over the prospect of reconstruction, Fukuda proselytized to all Japanese that they must embrace the opportunity presented by the earthquake and reconstruct the nation. "Tokyo should not grieve that it has lost the remains of its old self," Fukuda wrote in the popular journal *Kaizō*, but rather "establish it [reconstructed Tokyo] as the base of a restored Japan." However deadly and destructive the disaster had been, it had done Japan a favor by purifying the capital through fire. He opined that the cleansing that Tokyo had experienced was similar to what the fire of London had accomplished in eradicating the black plague.[9] Both were blessings in disguise rather than something to bemoan.

Fukuda hoped that the fires that burned Tokyo would serve as the beginning of a "mass purification ceremony for the nation." It would only be a true blessing, however, if people were astute enough to comprehend it as a unique opportunity and to seize it. "The first task for the restoration of Japan," Fukuda therefore argued, was to "help turn the fires that burnt Tokyo, Yokohama, Odawara, Yokosuka, and other cities into a fire that will burn the wreckage of old Japan and the old mentalities of the feudal era that prevails." "Build a new Tokyo," he wrote, "that could serve as a pathfinder and leader of a restored Japan." More than this, Fukuda urged his readers to plan and construct "a city that could promote the reformation of the entire world." "Tokyo," he argued, had been made a "sacrifice of a Japan that had been following the wrong trajectory." A new Tokyo could shepherd Japan on a course of national renewal. If Japan failed in this undertaking, however, the "deaths of those people will be [remembered] as a waste."[10]

Fukuda was not alone in seeing opportunity embedded within Japan's most deadly natural disaster in recorded history. He was joined by a constellation of political elites, social commentators, journalists, academics, and bureaucrats who saw an unparalleled chance to fashion the 1923 earthquake as Japan's great national calamity. Once packaged as a national tragedy, it could be used, these actors hoped, not only to advance a project for rebuilding Tokyo as a modern, imperial capital but also to implement a much larger and more complex program of national reconstruction. Individuals from across the political spectrum, from all class backgrounds, and from a gamut of professions embraced the notion that the civilization-upturning calamity could be manipulated artfully to secure a transformative Taishō restoration. "When on earth," urban planner Anan Jō'ichi rhetorically asked, "will we have another great opportunity to construct a modern imperial capital to our heart's content if we miss this one? Never!"[11] Others argued that the social, political, ideological, and economic problems that had manifested themselves with intensity following the First World War—hedonism and decadence, extravagance and luxury mindedness, flippancy, frivolity, and laxness, political extremism, disunity, social agitation, inflation, and excess—could be not only arrested but also reversed through a project of national renewal. The earthquake "presented an unprecedented opportunity," Privy Councillor Ichiki Kitokurō suggested, to embrace a true spiritual restoration and thus improve "all aspects of life."[12]

From the wrecked landscape of Tokyo, many assumed that the government faced a monumental set of tasks. Virtually all commentators knew that both the reconstruction of Tokyo and the project of national regeneration would require significant funds, united leadership, and a sympathetic populace willing to accept sacrifice. Leading light of the social reform movement Abe Isoo suggested that people across the nation would have to endure hardship, willingly accept personal sacrifices, and unite behind a Taishō restoration much like he suggested Japan rallied in 1868–1869 to support the new Meiji state.[13] If reconstruction and restoration were successful, he and others predicted, the rewards would be enumerable. Shimamoto went so far as to predict that the catastrophe would forever be remembered as "the mother of all happiness" if a thorough reconstruction were carried out.[14]

Could elites construct the regional disaster into a "national calamity," and would people across Japan see it as such? How did people interpret, comprehend, or attempt to make sense of this disaster, and what did they believe it meant for Japan? Could Japan afford a revolutionary project that aimed to turn Tokyo into a modern imperial capital? More important, would people outside the capital support such an expensive undertaking? What voice would local citizens have in planning Tokyo's rebirth, and would they embrace a radical reconstruction of the city they called home? What did commentators mean by national reconstruction, and who would devise and implement this program? Could this disaster really be manipulated and turned into "*the* event" that compelled people to change their social behaviors and thus alter the overall trajectory of Japan's future development? This book explores these and other questions related to the 1923 earthquake calamity. In doing so, it provides the first study in English or Japanese that details how elites interpreted, constructed, and packaged the 1923 Great Kantō Earthquake and attempted to use it for larger political, ideological, social, and economic aims. It also explores how elites with competing visions for reconstruction and average citizens responded to the seemingly endless overtures for reconstruction and regeneration that emanated from the disaster zone between 1923 and 1930. Responses proved far more varied than many anticipated.

Rarely did the actuality of reconstruction match the starry-eyed dreams of the disaster opportunists. Reality, in fact, proved far less accommodating than most people imagined in September 1923. Plans for an expensive refashioning of Tokyo as an infrastructure-rich, high modernist dreamscape and the prescriptions made for reconstructing society ignited intense contestation from elites with competing visions. Likewise, plans for a radical makeover of Tokyo triggered resistance from landowners and spurred calls for a rapid return to normalcy. Both factors quelled grandiose reconstruction dreams conjured up by disaster opportunists.

Though Tokyo was rebuilt and the government celebrated its rebirth with a well-orchestrated series of commemorative services in 1930, the transformative urban, ideological, political, and social changes that many hoped the earthquake would facilitate remained as elusive and illusory as a mythical chimera. Amid the rubble and dislocation of Tokyo's postdisaster landscape, it was easier and perhaps more comforting for people to remember what life was like before the calamity and to yearn for a return to normalcy than to affirm a radically different future, however inspiring, new, and modern it was made to seem. Rather than alter Japan's trajectory in deep-seated ways, the Great Kantō Earthquake amplified many social, political, and economic trends that had already begun to define the increasingly contoured landscape of interwar Japan. Moreover, debates about reconstruction exacerbated many preexisting tensions within and between agencies and among various political actors in Japan. These two factors, along with the increasingly pluralistic nature of Japan's polity, meant that elites were unable to turn postdisaster reconstruction into a panacea for the afflictions, real or imagined, that they believed Japan suffered.

My findings have direct relevance for understanding today's world as much as they are relevant for understanding interwar Japan. Politicians and bureaucratic elites across cultures and from diverse political systems regularly make opportunistic and idealistic calls for reconstruction and renewal following major natural disasters. Frequently they attempt to use these events for larger political and ideological ends. Postdisaster pronouncements and policies following the Indian Ocean tsunami (2004), Hurricane Katrina (2005), the Sichuan earthquake (2008), the Haitian earthquake (2010), and the Tōhoku earthquake (2011) are just a few examples that support this assertion. Because bold reconstruction plans are often made without consulting citizens who are most affected by such undertakings, disaster victims frequently respond to such policies with ambivalence or, when they can in democratic societies, creative and sustained resistance. Moreover, competing bureaucratic elites with differing agendas likewise challenge and often limit public policy outcomes following catastrophic occurrences. Rarely is the optimism and opportunism that is unleashed by natural disasters translated into fundamental, lasting, and transformative changes in society. Japan's experience calls into question the universality of claims that postdisaster utopias invariably emerge after calamitous occurrences, endure, and contribute to lasting transformations within society.[15]

In a broad scholarly context, my study complements the growing number of works that examine natural disasters and the reconstruction projects that follow. Anthropologists Anthony Oliver-Smith and Susanna Hoffman have been instrumental in establishing a number of theoretical frameworks that have enabled countless academics, students, and a wider popular audience to better understand natural disasters and their importance as objects of study.[16] Disasters, they suggest, rest at the intersection where natural phenomena with destructive potential meet humans in their natural and built environments. The degree of environmental, physical, structural, social, political, and economic vulnerability that exists in a society struck by a natural hazard determines the level of destruction, dislocation, and loss that result. Truly understanding disasters thus requires us to have specialist knowledge about the "dramatic event" or natural hazard with destructive potential as well as the "total system" in which it took place. Disasters are thus demanding objects of study. "As events and processes," Oliver-Smith and Hoffman conclude, "disasters are totalizing phenomena, subsuming culture, society, and environment together."[17]

Any detailed study of disasters and the environments in which they take place thus uncovers much about a disaster-struck society. In short, disasters reveal.[18] Educator Miura Tōsaku understood this concept in 1923 Japan. Writing in the education journal *Kyōiku jiron*, Miura suggested, "Disasters take away the falsehood and ostentation of human life and expose conspicuously the strengths and weaknesses of human nature."[19] Calamitous events do more than just reveal the best and worst characteristics of society, however. Disasters, as Oliver-Smith has suggested, "disclose . . . power structures, social arrangements, cultural values, and belief systems."[20] They provide an extraordinary window into society that gives students and scholars unique insights into human nature, cultural and social constructions, and the many key relationships—between humans, nature, the built environment, the state, and even the cosmos—that help define our existence. Historian Marc Bloch articulated this point nearly seventy years ago. "Just as the progress of a disease shows a doctor the secret life of a body," he suggested, "the progress of a great calamity yields valuable information about the nature of a society."[21]

It is not just a catastrophic disaster "as event" and the immediate relief and recovery efforts that follow, however, that illuminate. Hoffman has demonstrated that longer-term reconstruction processes are just as revealing. She describes the process of reconstruction as a time when the complex webs of the world are spun again "thread by thread."[22] Disasters are often perceived of as providers of a new start and postdisaster landscapes are often characterized as blank slates awaiting rebirth. These desolate environments empower people—perhaps as a coping or survival mechanism triggered by incomprehensible devastation and loss—to assume that sweeping changes and the construction of something new, modern, and better is not only possible but destined.[23] Blank slates, however, rarely exist. Survivors return to where their homes and businesses once stood and often seek to rebuild their lives as they once were. Discussions about what the "new society" will resemble, how it will be achieved, and who will create it often ignite arenas of contestation, foster resilience, and promote resistance to change. Each phenomenon was apparent in Tokyo between 1923 and 1930.

Given that disasters emanate from the nexus where a natural phenomenon with destructive potential intersects with humans, often in violent, confronting, and reflection-compelling ways, they are valuable objects of study for people outside the disciplines of anthropology or sociology. Specifically, they are compelling topics for historians. This is particularly true since disasters are constructed "catastrophe events" and often result in the large amounts of public interest, media coverage, written records, reports, and visual materials for later consultation.[24] Likewise, discussions and debates about reconstruction, organized resistance against plans forwarded by elites, remembrance and memorial activities, and commemoration services—all components of natural disasters and their aftermaths—create a rich and varied cache of source materials to examine.

Fortunately, numerous historians with diverse geographical expertise and chronological interest have begun to embrace the analysis of disasters. They have used catastrophic disasters and humans' responses to these events not only to illuminate the disasters as events but also to analyze the societies in which they occurred. As a result, our knowledge of hurricanes in the Caribbean and Atlantic world; cyclones in the Philippines; earthquakes in South and North America, Europe, and Asia; floods on virtually every continent; and volcanic eruptions along the Pacific Rim of Fire has increased markedly.[25] Historians of Japan have contributed to this burgeoning field with works by Alex Bates, Janet Borland, Greg Clancey, Greg Smits, Timothy Tsu, Gennifer Weisenfeld, and myself expanding our knowledge of earthquakes, typhoons, and floods.[26] These have added to a number of works in Japanese that have highlighted the important role of natural disasters and reconstruction in Japan's modern and early modern history. Remarkably, however, no one before now has written an in-depth, book-length analytical study of the 1923 earthquake in English.[27]

While providing new insight on the Great Kantō Earthquake as a calamitous event, my work also adds to the growing number of excellent studies on interwar Japan. Sheldon Garon has been instrumental in shaping the way a generation of historians have approached and understood state-society relations in Japan. His works, but most important *Molding Japanese Minds: The State in Everyday Life*, illustrate how state officials nurtured a spirit of guided self-management or self-cultivation for the betterment of peoples' lives and for the improvement of Japan's overall economic, social, and political position.[28] His findings, along with those from David Ambaras, Janet Borland, Sabine Früstrück, Sally Hastings,

Gregory Kasza, Mark Metzler, Kenneth Pyle, Dick Smethurst, Elise Tipton, and Sandra Wilson, have demonstrated how bureaucratic elites, working with citizen groups, neighborhood associations, and other government agencies, managed people—rather than browbeat or coerce them into compliance—for larger political, social, ideological, and economic aims and objectives.

Though I suggest that Tokyo and Japan were not transformed in revolutionary ways as many disaster opportunists had hoped, the Great Kantō Earthquake remains an event of extraordinary significance to historians. Foremost, the disaster and people's responses to it reveal much about interwar Japanese society. On one level the disaster illustrated the abysmal state of disaster preparedness that existed in Tokyo prior to 1923, especially in relation to the vulnerabilities that defined this metropolis. Various levels of government proved themselves ill equipped to deal with the postdisaster chaos, confusion, panic, and anarchy that erupted. In stark contrast to events following the March 2011 tragedy, countless individuals exhibited a clear absence of rational calm mindedness, resolve, discipline, and humanity in 1923. This suggests that stoicism and courage in the face of catastrophic natural disasters are not innate characteristics of Japanese people but rather learned responses nurtured through years of training and education.

On another level the Great Kantō Earthquake illuminated the intense, almost palpable anxieties that a multitude of commentators and bureaucratic elites felt toward modernity and the state of society in 1923. From 1905 onward, but particularly after 1915, Japan underwent a series of profound changes that many elites perceived as disorienting, alienating, unhealthy, and ultimately threatening. Japan became more urban, densely populated, industrial, cosmopolitan, pluralistic, wealthy, and diverse. As these transformations took place, society became less cohesive, less well ordered, less stable, and more divisive. The process of modern development and the acceptance of urban, consumer-oriented modern culture posed unfamiliar challenges to Japanese people, their governing elites, and the nation's built and natural environment. By 1923 a series of transformative waves unleashed by modernity had swept over Japan well before the seismic waves of destruction amplified the sense of anxiety, foreboding, and dislocation. Given perceptions surrounding the seemingly fragile nature of post–World War I Japanese society, it is not surprising that many concerned commentators imbued the Great Kantō Earthquake with moral and political meaning. Still others attempted to use it to admonish Japanese society and implied that the thoughts and actions of Tokyoites had helped trigger a disaster of catastrophic proportions.

The earthquake disaster emerged as a unique, unprecedented, all-encompassing event that enabled people to critique Japan's state of modernity and its developmental trajectory. It likewise emboldened many elites to devise and articulate numerous prescriptions they believed could reorient Japan on a more responsible, wholesome, and ultimately satisfactory course. Ironically, many of the prescriptions advanced by bureaucratic elites to reorient society following the 1923 tragedy were permeated with the belief that the difficult task of national reconstruction could be achieved by better harnessing technocratic modernity to mobilize, persuade, and condition Japanese people. While some competing elites argued for the adoption of more heavy-handed approaches, none suggested that Japan should face, or could overcome, the challenges of modernity by returning to some idyllic premodern past. Modernity, like the earthquake itself, was viewed as both the scourge of society and the elixir for rebuilding Tokyo and reconstructing the nation.

Regardless of prescriptions, however, not everyone endorsed even the most moderate policies aimed at rebuilding Tokyo and reconstructing the nation. In fact, unity of purpose or action was virtually nonexistent. People responded to calls for rebuilding Tokyo and reconstructing the nation with varying degrees of acceptance, vacillation, and resistance. Such ambivalence further heightened elite-level anxieties and disquiet about society's resolve. Though certain actors from left, right, and center perspectives argued for elites to flex their muscles and employ the heavy hand of the state to capitalize on the catastrophe and "solve" the perceived problems of the day through dramatic interventionist policies, a sizable majority of governing elites in 1920s Japan resisted such calls and instead championed more moderate and nuanced strategies. Official longer-term responses to the Great Kantō Earthquake thus stand in stark contrast to policies employed in the late 1930s and early 1940s. During those tumultuous years, governing elites responded to multiple challenges and catastrophes, almost all of their own making, in a far more heavy-handed manner that ran counter to many of Japan's pragmatic and pluralistic political precedents.

One

CATACLYSM: THE EARTHQUAKE DISASTER AS A LIVED AND REPORTED EXPERIENCE

The rumbling sound of apocalyptic fires added further terror to it. Screams and shrieks were everywhere, fires turned into an inferno—it was Hell. . . . The big city of Tokyo, the largest in the Orient, at the zenith of its prosperity burned down and melted away within two days and three nights.
—Takashima Beihō, 1923

Over the ruined city on Monday (September 3) rose the never to be forgotten smell of burned human flesh, as everywhere charred bodies lay exposed to the sun. No one who walked through that city of the dead will ever forget those sights.
—Osaka mainichi shinbun, 1923

In the early days of September 1923, novelist Uno Kōji followed a well-trodden path. It was a course taken by many Tokyoites who were considered lucky because they had survived. The path eventually led Uno to the hills of Ueno Park. The writer did not enter the would-be sanctuary—as many had done—to place or read a missing persons notice on the iconic statue of Saigō Takamori. Rather, he was drawn to the place out of a desire to observe, record, and reflect. The object of his gaze was Tokyo, or what remained of the once vibrant imperial capital. "As far as the eye could see," he mused, "Tokyo had been reduced to ash."[1] All he could make sense of through the "smoggy grey air," still heavy with smoke from residual fires, "was the vague image of the twelve-story tower of Asakusa that had broken into two, the Kannon Temple, and a few other buildings that were still standing." Uno was, in many ways, fortunate in that he had not perished in the event that others described as a "burning Hell," an "apocalyptic revelation," or an unprecedented calamity that "overturned Tokyo from its very foundation."[2] Though Uno witnessed horrors and trauma on an intimately personal level, the burnt-out, haunting remains of Tokyo viewed from the vantage point of Ueno

Park remained his most confronting and lasting image from that fateful autumn. What just days before had been Japan's imperial capital, he confessed, resembled nothing more than "the wreckage of an extinct city."[3]

Other survivors would be haunted by different traumatic experiences and confronting images. The image of one charred female body remained emblazoned in the memory of child-welfare advocate and education specialist Takashima Heizaburō. Among the tens of thousands of blackened, swollen, and decaying bodies that Takashima experienced on one of his many walks through "dead Tokyo," the female body in question stood out in its ghastliness: extending nearly a quarter out of this woman's dead body was the partially exposed yet equally blackened head of a dead baby. The mother, Takashima concluded, had suffocated or burned to death immediately before or during delivery.[4] Funaki Yoshie's most memorable and traumatic experiences took place on board a "rescue ship" that the Nakasu Hospital in Nihonbashi had chartered to evacuate patients soon after the initial quake struck and the city's roads filled with more able-bodied people attempting to flee. Tokyo's waterways proved no less crowded. On the jam-packed Sumida River, which snaked through Tokyo, Funaki joined exhausted patients and hospital staff who fought to extinguish sparks from the burning city "that fell like rain" on the ship's deck throughout the night. Their task was made all the more challenging owing to the number of people with hair, clothing, and belongings alight who jumped from low-lying bridges, river banks, and nearby barges onto the vessel.[5] As Funaki later described, it was a night of pure terror.

For writer Tanaka Kōtarō, no one event stood out in horror. Rather, he was haunted by the total, overwhelming scale of human suffering and misery he experienced. No place, however, was more confronting to Tanaka than the site where the Honjo Clothing Depot had once stood. There, over 30,000 people suffered a terrifying death. Writing weeks after the calamity, Tanaka concluded that he would never—no matter how hard he endeavored—be free from the sight and smell of the tens of thousands of decaying bodies that littered the 67,000-square-meter (6.7-hectare) site.[6] The smell of Tokyo and Yokohama also embedded itself in the nostrils and memory of one correspondent for the *Osaka mainichi shinbun* newspaper known only to his readers as "Mr. Miyoshi." Remarkably, Miyoshi had not set foot in either city before writing his special column on "destroyed Tokyo" in which he described in graphic detail the smell that hung over Tokyo and Yokohama. Rather, he had been overcome by the stench of death as he flew over the city in an open-cockpit army reconnaissance plane that had departed from Kagami-ga-hara airfield near Nagoya. Even at a height of 1,000 meters the disagreeable and unmistakable odor of death overpowered the smell of engine exhaust. Not surprisingly, it made both the reporter and the pilot feel wretched.[7]

What binds all these narratives, images, and memories together are the calamitous events that began on September 1 when the Philippine Sea tectonic plate subducted beneath the Okhotsk tectonic plate under the waters of Sagami Bay, 60 km south-southwest of Tokyo. This violent yet entirely natural occurrence, which released energy equivalent to the detonation of approximately 400 Hiroshima-sized atomic bombs, triggered an earthquake-initiated natural disaster that was followed by extraordinary firestorms. Soon the catastrophe became a human-inspired calamity defined by the almost total breakdown of law, order, and authority in Japan's capital and surrounding areas. Taken together, the 1923 catastrophe as "event"—packaged or reconstructed in this chapter as a collection of remembered and reported experiences—is where we first turn our attention. Before it emerged as an event that compelled introspection, inspired opportunism and manipulation, and unleashed intense contestation, the Great Kantō Earthquake was a lived experience for many residents of eastern Japan. Quickly afterwards, it became the national story as newspapers reported in detail on virtually every aspect of Japan's earthquake calamity. It was a cataclysm that turned the world upside down in perception and reality and left the capital of Japan's empire as a desolate, seemingly war-torn wasteland that fell under military occupation.

THE TREMOR AND ITS AFTERSHOCKS

September 1, 1923, was not the first time that Tokyo and the Kantō region of eastern Japan had experienced a mega thrust earthquake. Nor will it be the last. Lurking deep below the waters of the Pacific Ocean roughly 275 km east of Tokyo and running north along Japan's east coast, the Pacific tectonic plate subducts beneath the North American-Okhotsk tectonic plate from Tokyo northward past Hokkaidō. Along this geological boundary the Pacific plate subducts at a rate between 2 and 12 centimeters each year. Over time this subduction has created the deep-sea Japan

Trench that connects to the Izu-Bonin Trench (also known as the Marianas Trench) in the South and the Kuril-Kamchatka Trench in the North. Sudden movement of the Pacific plate along this boundary, often caused by a release of built-up tension along the subduction zone, sends seismic waves in all directions from the epicenter of subduction. Earthquakes along this zone have rattled the eastern coast of Japan from Hokkaidō in the North to the island of Iwo Jima in the South since well before recorded or human history. Subduction along this plate triggered Japan's largest earthquake ever recorded and the fourth largest earthquake in recorded history, the magnitude 9.0 Tōhoku daishinsai that devastated parts of northern Japan on March 11, 2011.

The Kantō region, which includes the population centers of Tokyo, Yokohama, and Kawasaki, however, is vulnerable to the movements of two other tectonic plates that have triggered catastrophic earthquake damage in the past: the Philippine Sea tectonic plate and the Okhotsk tectonic plate. The subduction zone created by the interaction of these two plates sits roughly 100 km south of Tokyo, virtually bisecting Sagami Bay. Movements associated with these two plates triggered an 8.2 magnitude approximate (~8.2 M) Genroku earthquake of 1703 and the ~7.9 M Great Kantō Earthquake of 1923. Recent scholarship, moreover, has suggested that Tokyo is vulnerable to earthquakes triggered by movement of yet another plate or dislodged plate fragment located directly beneath the Kantō Plain upon which Tokyo and approximately 33 million people reside today.[8] Ross Stein and his research team suggest that movement of this plate fragment, sandwiched between the Pacific, Philippine, and Eurasian plates, is responsible for much of the regular seismic activity that strikes the Kantō region, including the ~7.3 M Ansei-Edō earthquake of 1855. Finally, much of the Kantō Plain, and in particular large parts of eastern Tokyo, rests on soft bay and river sediment, which, while not a cause of earthquakes, adds to the destructive impact of seismic events as these materials are far more prone to violent movement, sometimes even soil liquefaction, during a major earthquake.

Just after 11:58 a.m. on September 1, 1923, subduction of the Philippine Sea tectonic plate toward the north-northwest sent destructive seismic waves in all directions. The seaside communities of Odawara, Kamakura, and Yokosuka first fell victim to the shock waves, followed seconds later by the inhabitants of Yokohama and Tokyo. Though the Kantō region was the focal point of destruction and death, the waves radiated well beyond, with sensitive seismic instruments recording an earthquake of extraor-

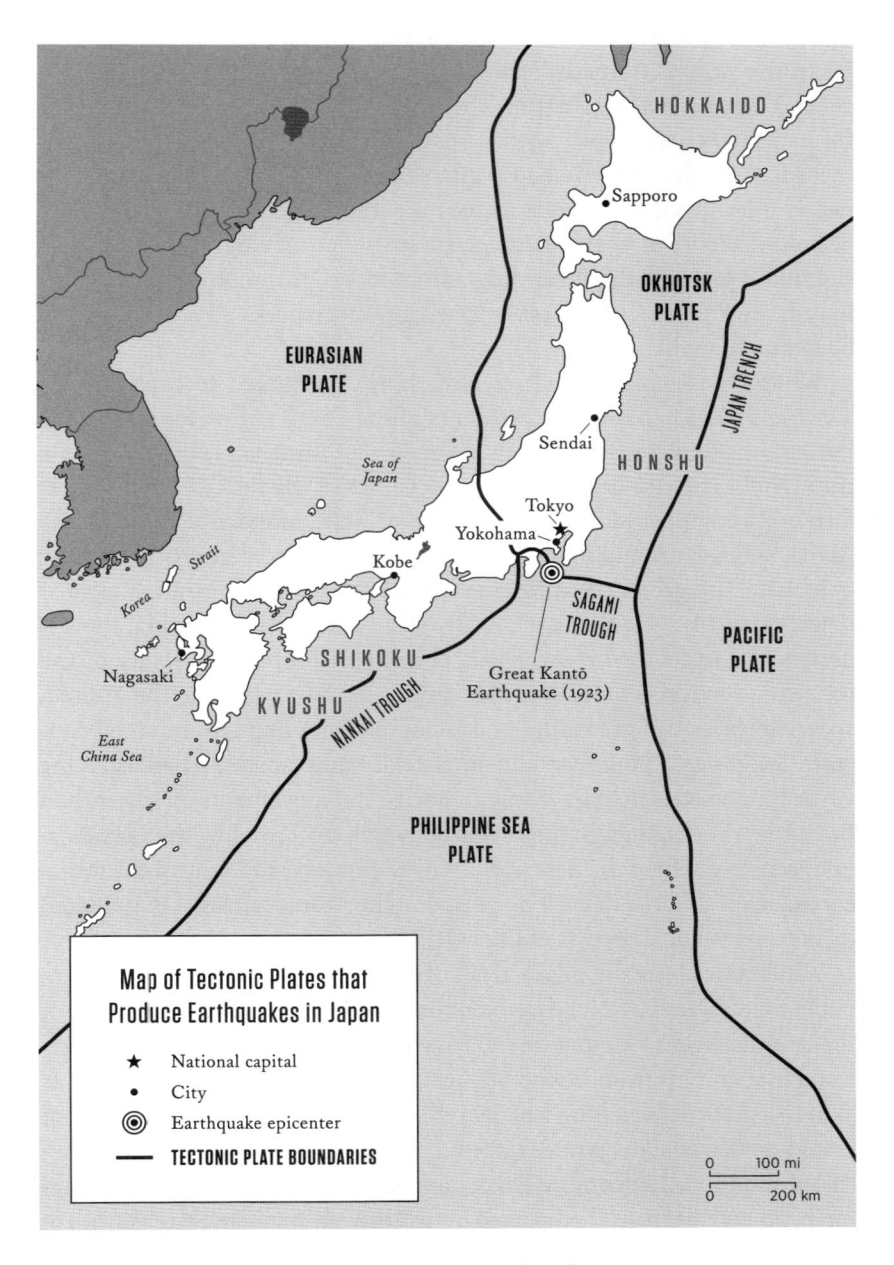

FIGURE 1.1 Map of tectonic plates that produce earthquakes in Japan

dinary strength as far to the east as San Francisco, California, and as far to the west as Granada, Spain. Early in the morning of September 1, central European summer time, the Vienna Institute of Meteorology informed the Reuters News Agency that "the most violent earthquake yet recorded" by the institute had occurred over 9,000 kilometers away.[9] For nearly four hours, technicians a quarter of the world away stood in a state of near disbelief as their seismograph just kept moving, recording both the initial jolt and the immediate aftershocks. In the ten days that followed, the Kantō region experienced 1,197 aftershocks strong enough to be felt by humans.

While the seemingly endless aftershocks continued to captivate far away scientists and unnerve residents of the Kantō region for weeks, the initial jolt shook Tokyoites from their normal Saturday routines. People throughout the city described the mechanics of the first jolt in a similar fashion to engineer Mononobe Nagao.[10] "At first," Mononobe recalled, the "earth shook back and forth for what seemed like 15 seconds," knocking objects to the ground and forcing people to steady themselves against objects that likewise shook violently. This continuous juddering, though destructive in its own right, paled in comparison to the ferocity of the following vertical convulsions that simply knocked people to the ground, snapped water and gas mains, and collapsed brick and unreinforced concrete buildings, chimneys, and industrial smokestacks. These striking vertical upheavals then gave way to horizontal movements stronger than the initial shocks, though they eventually tailed off to a low shudder. Seismic records confirm the descriptions by Mononobe and others: the initial earthquake comprised two long periods of horizontal shaking punctuated by massive vertical thrusts of the earth's crust.

While describing the physical act of shaking in a similar vein to Mononobe, writer Tanaka Kōtarō found himself awestruck first by the sound of the seismic waves rippling under Tokyo. To Tanaka, the noise that overcame him and his friends who had just sat down to smoke cigarettes was akin to a giant "blackening whirlwind" churning up the earth and all that rested on it from "deep underground."[11] The noise was followed by a series of shudders that conjured his tatami mats to life, undulating back and forth, up and down, as if possessed by an enraged spirit. In the eastern Tokyo ward of Honjo, academic Kawatake Shigetoshi likewise reflected that an eerie, disconcerting rumbling sound emanated from the earth as if a "violent storm was blowing" beneath the city, announcing its imminent destruction. For many others, the noise of household belongings crashing to the floor, the screams of terror, and the almost deafening clatter of heavy clay roof tiles falling to the ground and shattering were the defining sounds of the initial earthquake.

When the earth ceased shaking, hundreds of thousands of Tokyo residents scrambled out of homes, offices, theaters, shops, restaurants, and tramcars and absorbed the extent of the initial damage around them. Many observers commented that this was the last moment of silence they experienced that eventful day. If only momentarily, Tokyo's residents had been stunned into silent disbelief. The silence, however, did not last long. Once cognizant that a major earthquake had just struck the region, many individuals began a mad and panicked rush home to reunite with family members and to gather important possessions, money, and records. Many who were at home when the tremor struck scurried cautiously back into what was left of their abodes to collect valuables, personal affects, and loved ones who had not escaped their surroundings during the first shock. Soon after midday Tokyo turned into a disordered mass of motion, first geological, then human, and finally a combination of both as aftershocks struck the region with great frequency. Even in the best of normal conditions, the capital's infrastructure often proved wanting. Damaged

FIGURE 1.2 People attempting to flee the ruined landscape of Tokyo

Source: Postcard, author's private collection.

and faced with the panic-stricken movement of millions of residents all at once, Tokyo became a clogged, congested landscape of chaos, terror, and panic. Worse, it was about to become an inferno.

A CITY IN PANDEMONIUM AND A BURNING HELL ON EARTH

As disconcerting as the quake and its aftershocks were to survivors on a physical as well as psychological level, a far more deadly phenomenon erupted shortly after the initial seismic upheaval: fire. Conflagrations or the fear of them pushed people to flee with great urgency. Within thirty minutes of the first tremor more than 130 major fires broke out across Tokyo alone, many clustered in the densely populated eastern and northeast sections of the capital. Fueled by leaking gas from ruptured lines, overturned charcoal fires lit to cook midday meals, and debris from collapsed buildings, fires burned large sections of Asakusa, Kanda, Nihonbashi, Kyōbashi, Honjo, Fukagawa, and the Ginza districts. As panicked people fled Tokyo's eastern districts, the confusing web of roads, alleys, and bridges became packed with people. Congestion was compounded considerably because people chose to flee with many of their most valuable or useful household belongings.

Tanaka Kōtarō found himself both an observer of and a participant in the disorder that quickly descended over postearthquake Tokyo. After watching his neighbors in the ward of Hongō evacuate their home and gather food, water, and belongings on their street front, Tanaka was overcome by a sense of urgency and rushed out to buy supplies. He walked up Andōzaka (Andō Hill) and was first struck by the astonishing number of people who had begun to amass their belongings on railroad tracks in his neighborhood, as if they were preparing to abandon the city on foot where trains just minutes before had moved paying passengers. As he walked toward Koishikawa, two developments unnerved Tanaka: explosions originating from a fire at the Koishikawa Army Arsenal, and the panicked terror of people overcrowding the city's streets. He described a scene near Denzūinmae tram stop as follows:

> Among the whirling waves of people existed a group of several men and women pushing a cart and carrying a tall pile of items, including futons; a man carrying a chest of drawers and bundles wrapped in cloths on a

FIGURE 1.3 Evacuees and their belongings filling the open space in front of the Imperial Palace

Source: Tokyoshi, ed., *Tokyo shinsairoku*, 5:14. Reproduced with permission of Tokyo Metropolitan Archives.

two-wheeled cart; a woman in her undergarments with a newborn baby on her back and a young child holding her hand; a man carrying a chest of drawers filled with clothes over his shoulder; a man in his singlet carrying a safe; a man carrying a tatami mat over his shoulder; a man carrying a catlike figure of an old lady on his back; a man carrying under his arm a young man whose head was covered with blood; and so forth. All these people were walking around in a chaotic, dazed manner.[12]

After pushing through the crowds, Tanaka noted more and more buildings beginning to smoke and catch fire outright. Over the noisy and disorganized crowds, he reflected, heavy, smoke-laden air of a "dull yellowish color" enveloped the city. When Tanaka reached the Tōbu railway line he stood amazed as he looked out and attempted to cool his flushed face. What he saw, he later confessed, was difficult to comprehend. From the district of Kanda to Iidachō eastward Tokyo was engulfed in flames. With hot winds blowing across his face, frozen in momentary disbelief, Tanaka

FIGURE 1.4 Nihonbashi Bridge and surrounding neighborhood before the earthquake, Mitsukoshi Department Store resting majestically in the background

Source: Postcard, author's private collection.

FIGURE 1.5 Lithograph print of Nihonbashi Bridge packed with people as a firestorm approaches

Source: Lithograph print, University of Melbourne Japanese language collection.

FIGURE 1.6 Nihonbashi Bridge and surrounding neighborhood after the earthquake

Source: Uchida Shigebumi, ed., *Taishō daishin taika no kinen*, plate 6.

decided to hurry home, disregarding the supplies he had ventured out to purchase. In Tanaka's recollection, he admitted that at that moment he felt as if Hell was approaching. He would soon be proven correct.

The distinguished playwright and theater historian Kawatake Shigetoshi likewise described scenes of frenzied panic and terror that eventually engulfed himself, his family, and many residents of eastern Tokyo. While initially refusing to abandon his house in Minami Futabachō, Honjo, Kawatake reluctantly assembled his most important manuscripts and personal documents as the smell of smoke intensified and the sun became an eerie color of blood. Roughly two hours after the quake, as the winds increased and began to fill his neighborhood and then his house with the smell of a burning city, Kawatake evacuated. Attempting to move south from Honjo, he first faced intense difficulty—as many residents did—crossing the heavily congested bridges that connected much of the paved and unpaved infrastructure of eastern Tokyo. In his path he found panicked and disorderly people, overturned cars, rickshaws full of heavy items, piles of discarded belongings, and everywhere more people. As he progressed at a snail's pace across the 5.4-m-wide Tatekawa Bridge, Kawatake shuddered with fear. Not only was the river below his feet packed with heavily laden and sometimes smouldering boats and barges, but, more depressingly, he became aware that the bridge's pilings and supports were also burning. Unlike some who decided to chance their luck by climbing over the bridge or jumping off onto the boats below, Kawatake pressed on. He was one of the last to cross that bridge before flames consumed it.

Having crossed the canal, the situation only became worse for Kawatake. South from Azumabashi to Etchūjima in eastern Tokyo, the roads were congested with evacuees and their household belongings. He quickly found himself stuck in a crowd that filled an 18-m-wide road for as far as the eye could see. "No matter how hard I tried," Kawatake remembered, "I could not move toward the direction I wanted to go." "I was stuck," he confessed, in a "wave of people" that acted like a single, panicked entity. It was a long, agonizing moment of pure terror. Kawatake recounted that he felt like a cornered piece in a real-life match of Japanese chess. There was seemingly no escape, and death was approaching. Strengthened by winds that increased in velocity and changed directions, fires spread from three directions, forcing Kawatake and much of the crowd that surrounded him south. By dusk he had reached a large, open field in Etchūjima, where other refugees had congregated. Though he and others continued to extinguish small fires started by "sparks that fell on us like rain" from the burning buildings of the nearby Merchant Marine College, what truly shook Kawatake was what he saw to the north. Put simply, everything in that direction seemed to be part of a "blazing inferno." "I was speechless," Kawatake wrote, "watching the whirlpools of fire and the undulating waves of flames" consume much of eastern Tokyo and its inhabitants. "It was utterly beyond description." In relaying this tragedy to his readers in 1924, the gifted storyteller prefaced his remarks with a chilling warning: "We often use the expression 'take the story with a grain of salt.' But in this case, the story should not be taken with a grain of salt. On the contrary, this story was too horrible to be exaggerated enough." Kawatake later admitted that after his experiences on September 1, he never again used the expression "a burning hell" in a casual way because he and so many other Tokyoites had experienced it.[13]

As alluded to by Kawatake, those residents who fled or attempted to flee burning Tokyo by boat also faced numerous difficulties. Funaki Yoshie, a patient at the Nakasu Hospital in Nihonbashi, recalled the terror-filled day and night spent on the Sumida River. Having forced her way onto a boat that the hospital staff chartered, Funaki and other evacuating patients were astonished that the river grew as crowded as the streets of Nihonbashi, with "innumerable boats of all kinds and sizes, all overloaded with people and their personal belongings." Virtually boxed in on all sides, patients were forced to pour bucket after bucket of river water onto the boat's deck and then to sweep it off to keep "sparks that fell upon us like rain" and debris from adjacent burning boats from igniting their vessel. "The noises," Funaki recalled vividly, "the roaring of the flames and people yelling for water," went on throughout the night and into the next morning.[14] It was a night of unremitting terror.

Before the morning broke, however, Funaki recalled the disbelief that both patients and hospital staff felt in seeing Tokyo burn along both sides of the Sumida River. Worse, they witnessed hundreds, if not thousands, of people on riverbanks, bridges, and "cursed" flaming vessels jump into the churning river to extinguish clothes and often hair that had caught fire. A number of riverboat captains later claimed that the most confronting images of that awful day and night were the thousands of people standing tall on vessels and on the riverbank with their hair alight, like an image of the fire god Fudo, before they jumped into the river, screaming in agony.[15] More haunting, Funaki confessed, was the sight of packed bridges—the Ryōgoku, Eitai, and Azumabashi to name three—burning from both ends, thus trapping thousands of screaming people in a position as frightening as anyone could imagine. The desperate people on those bridges, and many more on the nearby riverbanks, she concluded, stood no chance of survival. The only choice left to them was death by drowning or death by fire. Many, she lamented, selected the latter.

Writer Koizumi Tomi was not as lucky as Kawatake or Funaki, though he was luckier than many. Like Kawatake, Koizumi found himself in Honjo on the day of the calamity. Soon after the initial quakes, he and the friends he was visiting rushed out of the house and headed across the road to a "vast and empty" parcel of land (6.7 hectares, 20,430 *tsubo*) where they reassured each other, "we will be safe here no matter what happens."[16] They were some of the first of over 30,000 individuals who took shelter at the vacant site of what was once a large production and storage facility operated by the Clothing Department of the Imperial Japanese Army, referred to in 1923 as *Hifukushō ato* (site of the Honjo Clothing Depot).[17] By early afternoon, as the sky had become "heavy with smoke" and the sun the color of a "huge drop of blood," more refugees, with belongings ranging from futons to baggage, bicycles to horses, poured into the makeshift reserve. It would be the last place many of the evacuees would ever visit.

As the air temperature reached 46 degrees Celsius by late afternoon and the winds increased, five independent whirlwind firestorms erupted in the areas of Asakusa, Higashi-Ueno, and Honjo, with the Honjo fire spreading from the neighboring Yasuda estate to engulf the Honjo Clothing Depot. It began, as Koizumi described it, with "a ghastly noise, as if coming from deep underground in the direction of the Yasuda family's forest. At the same time, roof tiles, rush mats, corrugated iron sheets, and everything began to fall on top of us like a rain of stones."[18] Soon after, a conflagration in the form of "an enormous wall of fire . . . like a tidal wave" as if released from Hell itself turned the air "as hot as melting rock" and ignited everything in its path, including scores of trapped people. The cries from humans, Koizumi reflected, went from "where is my boy," to "there is no way to escape," to simple, calm, if not resigned prayers of "Namu Amidabutsu" as a sense of imminent death enveloped the Clothing Depot. Koizumi hoped that he would "fall unconscious as if forgetting both time and space forever," and indeed one of his last memories of that unforgettable night was of the whirlwind lifting him off the ground and depositing him on a pile of hot, smoldering, blackened personal belongings. It would be a very different city when Koizumi awoke the following morning.

In other parts of Tokyo and throughout much of the adjacent prefectures that comprised the Kantō region, rumors began to spread as quickly as the fires; by the next day, as Funaki recounted, they were "flying around like arrows."[19] Often relayed by refugees, some stories suggested that Mt. Fuji had erupted or was about to erupt, while others claimed that a large tsunami had washed away Yokohama or that all cities from Tokyo to Nagoya had been devastated. Many other rumors dealt with Japan's largest ethnic minority group, Koreans, who in 1921 totaled nearly 81,000 out of a total population of approximately 56 million. Often these rumors led to brutal murders.[20] These murders illustrate just how far Tokyo had slid into complete anarchy.

The first component of this tragedy originated with people who fled the fires and disaster zone on foot. As refugees spread into neighboring wards, counties, and even prefectures, rumors of Korean duplicity surrounding the fires that engulfed Tokyo and Yokohama emerged. Tales that bands of lawless Koreans had started fires, looted shops and homes, poisoned wells, murdered women and children, and even organized an assault on what remained of the capital all found adherents. Rumors of Korean uprisings and violence spread as far afield as Hokkaidō, some 800 km from Tokyo.[21] Such rumors, as Michael Weiner has illustrated, were given legitimacy by the actions of select government officials. A message broadcast by Gotō Fumio to every prefectural governor by the government's wireless transmitter at Funabashi is just one case in point. It read: "organized groups of Korean extremists have taken advantage of the disaster and attempted to commit acts of sedition."[22]

Home Ministry bureaucrat Suematsu Kai'ichirō recounted a rumor he and others had heard, which he found simply too "absurd" to believe. The rumor, Suematsu explained in the Home Ministry journal *Chihō gyōsei*, revolved around Koreans (not censored in this article) using well-coordinated organizations and an intricate "coding system" to wreak havoc and terror on an already frightened population throughout the disaster area.[23] First a small group of discreet Koreans would scout an entire neighborhood, placing codes virtually unrecognizable to the untrained eye on doors or posts, indicating whether the house in question was to be burned or robbed, the well poisoned, or the family—if they were still in the home—simply murdered. Shortly thereafter an "execution gang" would enter the area and act on the code.

The second component to this tragedy was the manner in which select groups of people responded to and used such extraordinary rumors. In this regard, *jikeidan* (neighborhood vigilance groups) gained particular notoriety. By the middle of September, 3,689 vigilance groups emerged and operated under the pretext of preventing fires, stopping looting, and undertaking night watch and protection activities.[24] Military personnel reported that posters and other announcements that encouraged the formation of neighborhood watch groups often warned of illegal and seditious activities being carried out by Koreans. Tanaka Kōtarō remembered that as early as September 2 he saw posters that further unnerved anxious residents to be on guard against "bands of lawless XXX [Koreans] who were throwing small bombs and starting fires" in many parts of Tokyo.[25] Tanaka admitted being moved to join an armed neighborhood brigade after reading a poster that claimed XXX (Koreans) had been caught scheming around the elementary school near his home.[26]

Usually these neighborhood vigilance groups armed themselves with makeshift weapons, including clubs, iron pipes, swords, and bamboo spears, to repel what they claimed to be anticipated attacks by Koreans. On more than a few occasions, however, neighborhood vigilance groups stopped—and murdered without reason or warning—Koreans or those mistaken as Koreans who attempted to enter an area. Motivated by fear, anger, hatred, or opportunism, other individuals and loosely organized bands of rabble sought out Koreans and murdered them under the pretext of either fire prevention or punishment for the fires that many erroneously claimed that Koreans had ignited. Though an estimated 6,000 Koreans were murdered, only 125 members of vigilance groups were ever prosecuted for crimes committed after the disaster. Of these, only 32 received formal sentences. Two others were acquitted, while 91 received suspended sentences.[27] It was not only members of neighborhood vigilance groups who harassed, beat, and murdered Koreans. Evidence—sometimes circumstantial, other times direct—suggests that police officers and soldiers likewise murdered Koreans in the days following the calamity.

It is virtually impossible to reach any general conclusion as to why select Japanese murdered Koreans following the disaster. Racism, hatred, resentment, fear, and criminal opportunism all contributed to this tragedy. Whether these criminals were frightened, deranged, or coldly calculating citizens, poorly trained vigilance groups, or ill-disciplined police and military personnel, they shared malevolent intentions and actions. What is much easier to analyze, however, is how a number of commentators interpreted this murderous episode. Writing in the October 1923 issue of *Chihō gyōsei*, Hoashi Ri'ichirō suggested that many Japanese had lost all sense of rationality and "spread wild rumors and injured and killed others." In no uncertain terms, Hoashi suggested that what was inflicted on the Koreans was "extremely disgusting and internationally shameful."[28] The murder of Koreans, Suematsu concluded, was the clearest example of how "fiercely hostile," "frantic," "emotion driven," and "incapable people were of exhibiting rational judgment" following a disaster.[29] Future minister of commerce and industry Tawara Magoichi suggested that the "Korean incident" exposed a "major defect of the national spirit."[30]

Educator Oku Hidesaburō was more explicit in his description and condemnation. Writing in a 1924 edition of the journal *Kyōiku*, Oku claimed that "we kill[ed] them because they were XX-people; we kill[ed] them because they looked like XX-people. . . . people said we must kill Koreans because they, exploiting the postdisaster situation, were doing wrong and threatening our people." It was, he concluded, a disgraceful act that exposed "a moral flaw that was common among ordinary Japanese people" and tarnished Japan's international image.[31] The massacres were, in parliamentarian Tabuchi Toyokichi's opinion as expressed on the floor of the Imperial Diet, a "gross act of inhumanity" for which Japan's government should "apologize to its Korean victims."[32] No formal apology was issued. Unfortunately, it took nearly a week before order and stability returned to the Kantō region, assisted, yet in a few isolated cases hindered, by the eventual deployment of nearly 52,000 soldiers and the declaration of martial law. Any semblance of normalcy, however, would take much longer to materialize.

REFLECTIONS AND POST MORTEM ON A DEAD CITY

In the absence of any reliable or precise information about what had transpired in the city over the previous twenty-four to forty-eight hours, many able-bodied people who had spent much of September 1 attempting to flee Tokyo found themselves drawn back. Most returned out of a desire for information about loved ones from whom they had become separated, extended family members, their property, or the neighborhood and city they called home. One individual who did not have far to venture was Koizumi Tomi, who had spent the calamitous night of September 1 at the epicenter of death, the Honjo Clothing Depot. Upon waking early the next morning, Koizumi was first struck by what he described as an eerie, dead silence that hung over Tokyo, a stillness that replaced the "noises of trains or factory sirens that had always woken" him. His vision was the next sense to be confronted. All around him he saw death and devastation. First he noticed that all adjacent buildings had become nothing more than piles of blackened rubble. Quickly thereafter he realized that every tree in sight had been transformed into a "charred stick." "The gloomy scenes of human bodies," however, "the naked truth of the hell of the previous night," left the strongest impression. Koizumi found himself surrounded by "endless rows of bodies: red, inflamed bodies; black, swollen bodies; bodies partially buried under ash and smoldering remains."[33] The ghastly scene of so many corpses compelled him to question if the few people walking among the death and destruction that had been Honjo were humans or "hungry ghosts." It was only a matter of hours before Koizumi's final sense, that of smell, would be overcome by the stench of dead bodies.

A pitiful condition a dead body of several number Mukojima dyke in Tokyo.
東京向島土手無敷の死骸惨状

FIGURE 1.7 Dead bodies at Mukōjima, Honjo Ward, in Tokyo

Source: Postcard, author's private collection.

Amid all of the dead bodies, Koizumi pondered how he had survived the inferno of the previous night in which the "sky and the ground had become a world of hellish flame." The answer made him uneasy. As he had drifted into unconsciousness, Koizumi last remembered "piles of bodies," some "groaning, others screaming, many crying, and all seemingly collapsing" on top of him like dead weights void of volition. Though the bodies had crushed his legs, trapping him in a ghastly pile of misery, they likewise protected him from almost certain death. Soon, as Koizumi was quick to learn, the bodies became a magnet for the living. Throughout the days following the disaster, as the smell of death became more pronounced, Koizumi remembered Honjo as a place of macabre pilgrimage. Scores of people arrived at the site of "hell on earth" to look for missing family members or friends, even as fires continued to burn parts of Tokyo. Occasionally Koizumi would hear relieved calls of "there you are," "I found you," or "please hold on." More often than not, however, he witnessed stunned, exhausted people—themselves barely participants in the world of the living—desperately turning over corpse after corpse in vain, relieved that they had not found their loved ones but equally distraught because failure necessitated continued scouring of the dead, decomposing bodies.[34]

Unable to leave the grounds of the Honjo Clothing Depot on his own accord because of his leg injuries, Koizumi recalled that news of Tokyo's utter destruction reached him bit by bit in a piece meal fashion. Each hour brought word of this neighborhood's destruction or that area's demise as people from further afield ventured to Honjo. Nothing, however, prepared him for his eventual evacuation to a first aid station cum field hospital near Ryōgoku Bridge. Taken only a short way by a horse-drawn cart through Honjo four days after the initial carnage, Koizumi found all of his senses again confronted with the agonized deaths of thousands upon thousands of people. The burned out, foul-smelling landscape was strewn with "burned, bubbling, decomposing bodies beaten by the rain" interspersed with "reddish brown electric wires that crawled along the ground like barbed wire" and "blackened rubble." It was not the landscape of a

FIGURE 1.8 View of Manseibashi Train Station in Kanda Ward with statue of Russo-Japanese War hero Commander Hirose Takeo still standing

Source: Uchida Shigebumi, ed., *Taishō daishin taika no kinen*, 37.

city but of a hellscape, an environment Koizumi described as "controlled whimsically by devils."[35]

Kawatake Shigetoshi found himself walking through Koizumi's land of death and devils the day after the disaster. Having spent the night of September 1 at the vacant field in Etchūjima, he left on the morning of September 2 in search of his family. Resigned to the fact that his house in Honjo had no doubt burned down, Kawatake first plodded to his in-laws' residence in Ushigome in hopes that his wife and two children had fled there from the inferno. They had not. Kawatake thus returned to Honjo. Walking from the districts of Kanda to Yanagihara and then from Yanagihara back to Honjo, he recognized little, except for the scores of burned bodies. There were so many dead bodies between Ryōgoku and Honjo, Kawatake confessed, that he quickly "lost his ability to feel anything." Death had numbed his senses. He admitted that over the course of this short but poignant journey, he had become "paralyzed" by the tragic sight of so many bodies. Eventually he felt nothing turning over body after body, unmoved as to whether the victim was young or old, male or female, or had drowned or been burned to death. It became a strangely impersonal exercise for Kawatake. After six weeks of searching, reunion was bittersweet. Though his wife and daughter had survived, his son had perished. Strapped to his wife's back, their son had drowned when his wife jumped into the Ōkawa River to avoid an approaching wall of fire.[36] It was a fate that many infants and children suffered.

Six days after the inferno, with the ground still being shaken by aftershocks, Tanaka Kōtarō likewise trekked to Honjo with a friend who had previously resided in this now desolate area of the capital. As they crossed what was left of the Ryōgoku Bridge, Tanaka was immediately confronted by the image of the water beneath his feet littered with bodies "swollen like sumō wrestlers" bobbing up and down, back and forth, in the gentle waves floating next to charred doors, discarded belongings, and the remains of urban civilization. After crossing the hastily repaired bridge, Tanaka stopped to take a drink and asked a passerby for directions to the site of the Clothing Depot. "Don't go," the man replied, "the place is too awful to visit." After asking again, the fellow wayfarer relented and informed Tanaka that it "was just a little bit in that direction." Before parting, however, the stranger again issued a stark warning: "You don't want to go there, it would be better if you never saw that place." This exchange and the man's direct warnings, however, still did not prepare

FIGURE 1.9 Bodies at the site of the Honjo Clothing Depot

Source: Postcard, author's private collection.

Tanaka. Like the others before him, he was overwhelmed by the sight of thousands upon thousands of bodies, each burned, blackened, swollen, and decomposing. Thereafter he was confronted by the "fatty stench" of bodies slowly being cremated in makeshift funeral pyres. Accompanied by the low murmur of chanted sutras that floated on the heavy air, Tanaka passed pile after pile of bodies "stacked just like piles of fish that fishermen would make on shore before fish brokers came to buy them."[37] Others commented that workers moving bodies at the site had, out of necessity, employed iron meat hooks used commonly at fish markets and slaughterhouses.[38]

Upon closer inspection of the Honjo Clothing Depot, Tanaka realized that the 6.7 hectare compound was filled with stacked bodies. In fact, he discerned that the large, open space was not expansive enough to accommodate all of Tokyo's dead. The streets bordering the depot, adjacent lots, and the ditches surrounding this site also overflowed with bodies that Tanaka recalled "looked like charred, dried, sardines." It was not, he concluded, an exaggeration to claim that more than 30,000 persons had perished at the Honjo Clothing Depot. In walking through the grounds, Tanaka lost all willingness to examine any body close enough for possible

FIGURE 1.10 Cremating the dead at the Honjo Clothing Depot

Source: Uchida Shigebumi, ed., *Taishō daishin taika no kinen*, 44.

identification. All he could do was to hold his hand towel pressed close enough to his face to keep the overwhelming stench of death from making him stop and wretch.

Leaving Honjo, Tanaka decided to walk through what was once the entertainment district of Asakusa. It too was a pathetic wreck, and its sight did nothing but compound his sense of dread. In the shadow of the damaged twelve-story tower of Asakusa—the Ryōunkaku (Cloud-Surpassing Pavilion), also known locally as the Jūnikai (twelve-story tower)—which to many had epitomized the frivolity and unadulterated pleasures of modern urban life, Tanaka caught sight of a caged monkey resting amid charred rubble where only days before a circus had performed. The monkey, Tanaka observed, stared off with a stunned, almost vacant look that so many survivors likewise displayed. Later that afternoon as he gazed out at Honjo from the hills of Ueno, just as Uno Kōji and others would soon do, Tanaka confessed that he "could still smell the stench of dead bodies deep in my nostrils."[39] To Tanaka and many others, the Tokyo region had become an area defined by death and destruction. It would remain so for months.

REPORTING CATASTROPHE THROUGH NEWSPAPERS

For many residents of eastern Japan who lived through the Great Kantō Earthquake, it was a defining experience—if not *the* defining experience—of their lives. As many of the personal narratives and stories illustrate, the quake, the fires, the chaos and panic, the inescapable sights and smells of death and destruction became an all-encompassing event. It was an occurrence matched in traumatic intensity and significance only by the ordeal of unlimited war waged against the residents of Tokyo and most of Japan's other urban centers in 1945. While the 1923 catastrophe did not shape the lives of those from other parts of Japan as the Second World War would, it nevertheless dominated popular and elite culture and the media in the autumn of 1923. It did so first through newspapers.

By and large, newspapers played a lead role in introducing the terrible tragedy that had befallen the nation's capital to Japanese who lived outside the disaster zone. In the early 1920s newspapers were the chief form by which people gained knowledge of current events in Japan. In 1923 the total circulation of all major daily newspapers stood at just over 5 million, thus averaging about one newspaper per every eleven or twelve citizens.[40] The *Osaka mainichi shinbun* and the *Osaka asahi shinbun* led all national papers in 1923 with a daily circulation of just over 700,000 each. Tokyo's largest daily, the *Tokyo nichi nichi shinbun*, published approximately 372,000 papers each day. By 1924, however, owing to the interest in the earthquake disaster and the temporary closure of many of its rivals, the circulation of the *Tokyo nichi nichi* grew to over 700,000, while the *Osaka asahi shinbun* surpassed the 1 million circulation mark on January 1, 1924.[41]

Increased circulation figures provide partial clues as to why newspapers across Japan reported extensively on the Great Kantō Earthquake. In capturing the attention of readers throughout Japan, news of the disaster increased sales. Apart from the purely economic motivation of larger circulations, journalists and editors relished the opportunity to publish on the calamity because it was novel and the devastation Japan had experienced was unprecedented. As big stories go, nothing else in 1923 came close to the earthquake. Reporting news, after all, was the job of newspaper employees, and certainly this story proved as captivating to report as it was for consumers to read about. More than this, however, reporting on the earthquake was a job that was made easy given the fact that this calamity could be covered by journalists of all professional backgrounds, including those who reported on politics, economics, science and technology, religion, society and everyday life, international affairs, and Tokyo. The earthquake was a news phenomenon that swept the nation. Shimaji Daito reflected on this, stating, "All our everyday issues, including natural science, humanities, politics, education, economics, and morals, relate to this disaster."[42]

Though Tokyo was home to over fourteen major daily newspapers, only the *Tokyo nichi nichi shinbun*, the *Hōchi shinbun*, and the *Miyako shinbun* survived the quake and fires with their buildings and printing presses intact. The *Hōchi shinbun*, which had an almost exclusively Tokyo-centered audience, exerted little effort to construct or define the calamity as a national event for its readers. To most individuals of eastern Japan, the disaster needed no construction or dramatization. Unlike many regional papers that wrote for those outside Tokyo, the *Hōchi shinbun* served to relay news of, and for, survivors. In its first postdisaster publication on September 5, the paper displayed a hand-drawn map of the burned out areas of Tokyo and followed this with detailed instructions on how Tokyo inhabitants could best leave the city to seek refuge. The writers concluded that walking along the railway tracks that led to Tōhoku was the safest route available.[43] The *Hōchi shinbun* also published numerous death notices and paid announcements issued by businesses to reassure customers. On September 13, for instance, it ran an ad placed by the Kōike ginkō, a small national bank with one branch in Tokyo.[44] The announcement assured depositors that the branch would reopen that day and that, unlike many other banks in Tokyo, the contents of its vault, including the account books, were safe.[45] The *Hōchi shinbun* also published notices drawn up by the city of Tokyo for broader citywide appeal. Of great interest to survivors no doubt, it published where and when food could be collected and the location and opening hours of employment-matching services run by metropolitan officials.[46]

Local and regional newspapers likewise fanned the flames of curiosity for those outside the disaster zone through traditional print. More than this, however, these papers helped craft the Great Kantō Earthquake as a dramatic, unparalleled calamity, an event unprecedented in Japanese history. Newspapers in virtually every region of Japan, as they had done in the opening and closing days of Japan's previous wars, unleashed a torrent of published extra editions crowded with emotive, provocative, and sensational headlines and stories about the catastrophe. Many editions were published even before most reporters and editors knew the full extent of destruction: the absence of specific facts or details did little to constrain reporters from constructing what was quickly becoming *the* national story.

On Japan's northernmost island, Hokkaidō, the *Otaru shinbun* published eleven extra earthquake editions between September 2 and 5. On September 2 the paper informed its readers that Japan's entire capital had become a "burning hell" (*shōnetsu jigoku*). The following day, having heard no reply from reporters in Nagoya, it asked a provocative question that further fueled fears of an even greater national calamity: "Has Nagoya too been annihilated" (*Nagoya mo zenmetsu*)? Returning its readers to Toyko, the *Otaru shinbun* suggested that for all intents and purposes Tokyo had become a dead city, one overflowing with bodies whose inhabitants had been burned to death, drowned, crushed, or suffocated. Tokyo was a city no more, but rather an environment defined by mountains of corpses around which rivers of corpses flowed. Japan's imperial capital, it summed up, no longer existed: Tokyo was a graveyard.[47]

In western and southwestern Japan, stories of the capital's complete and utter destruction likewise dominated newspapers. Readers in Japan's southernmost main island, Kyūshū, were told by the *Kyūshū nippō* that the entire city had been transformed into a vast "sea of fire" (*zenshi hi no umi*). The destruction in the districts of Honjo and Fukagawa had been so complete—not one structure was left standing—that the paper speculated as to whether an earthquake-inspired tidal bore or wave had rushed over these wards and washed everything to sea.[48] Soon thereafter, newspapers began a running total—updated daily—on the number of bodies that had been collected from Tokyo's streets, canals, and open spaces and deposited for identification and cremation at the former site of the Honjo Clothing Depot.[49]

Regular editions of newspapers replaced the earthquake extras in most regions by September 9. Thereafter one noticeable trend emerged in how many newspapers reported the disaster. To keep local interest in the calamity strong, most regional papers began to report on local connections to the disaster. Often papers published the names of prominent locals who perished in the earthquake and fires or listed the destruction of businesses in Tokyo or Yokohama with ties to regional Japan. Aware of the strong family ties many Kyūshū readers had to the Imperial Japanese Navy, for example, the *Kyūshū nippō* published names of navy personnel from Kyūshū who had died and been identified in Yokosuka, Yokohama, or Tokyo.[50] In a similar vein, the *Fukuoka nichi nichi shinbun* published accounts of local army soldiers dispatched to the disaster zone, both relaying confessions of shock at the desolate state of Tokyo and highlighting the importance of duty and obligation felt by those in uniform to those in need.[51] One unnamed soldier from Fukuoka confessed that he had been overwhelmed by the scale of destruction he witnessed upon entering what remained of Tokyo. It was far worse than he had imagined possible.

While survivor accounts and newspaper stories provide remarkable microlevel detail on the impact of the earthquake disaster on Tokyo and on the lives of its inhabitants, an overview reinforces the totality of destruction suffered. Tokyo resembled a cityscape destroyed by modern warfare. The word *zenmetsu* (annihilation) was often employed to describe what had befallen much of the Kantō region. Not surprisingly, it took months of investigation to record the totality of destruction and death meted out over the prefectures of Tokyo, Kanagawa, Chiba, Saitama, Shizuoka, Yamanashi, and Ibaraki. The human calamity, when viewed through sheer numbers, was immense. Over the seven damaged prefectures, 107,858 people perished while another 13,275 remained classified as missing twelve months after the earthquake. Table 1.1 breaks down the numbers by the wards of Tokyo.

Fatalities, however, comprised only one part of the immediate tragedy associated with the Kantō daishinsai. In Tokyo Prefecture alone, the homes of 397,119 families out of a total number of 829,900 (roughly 48 percent) standing on September 1 were either destroyed (349,134) or severely damaged (47,985).[52] If one examines the City of Tokyo, the numbers are even more appalling. Of the 483,000 families that resided there, the homes of 311,775 (65 percent) became uninhabitable.[53] Another 42,732 homes (9 percent) were subsequently classified as damaged but habitable. The destruction of so many residences left a large mass of homeless people. A total of 2.5 million individuals found themselves homeless across the seven prefectures damaged by the quake. Out of the City of Tokyo's 2.26 million inhabitants, 1.38 million (61 percent) were displaced.[54] Some 781,000 out of Kanagawa Prefecture's total population of 1.3 million (just over 60 percent) likewise found themselves homeless after the disaster.

The earthquake and fires did more than just destroy homes and kill people. They inflicted devastation across all aspects of the economy and

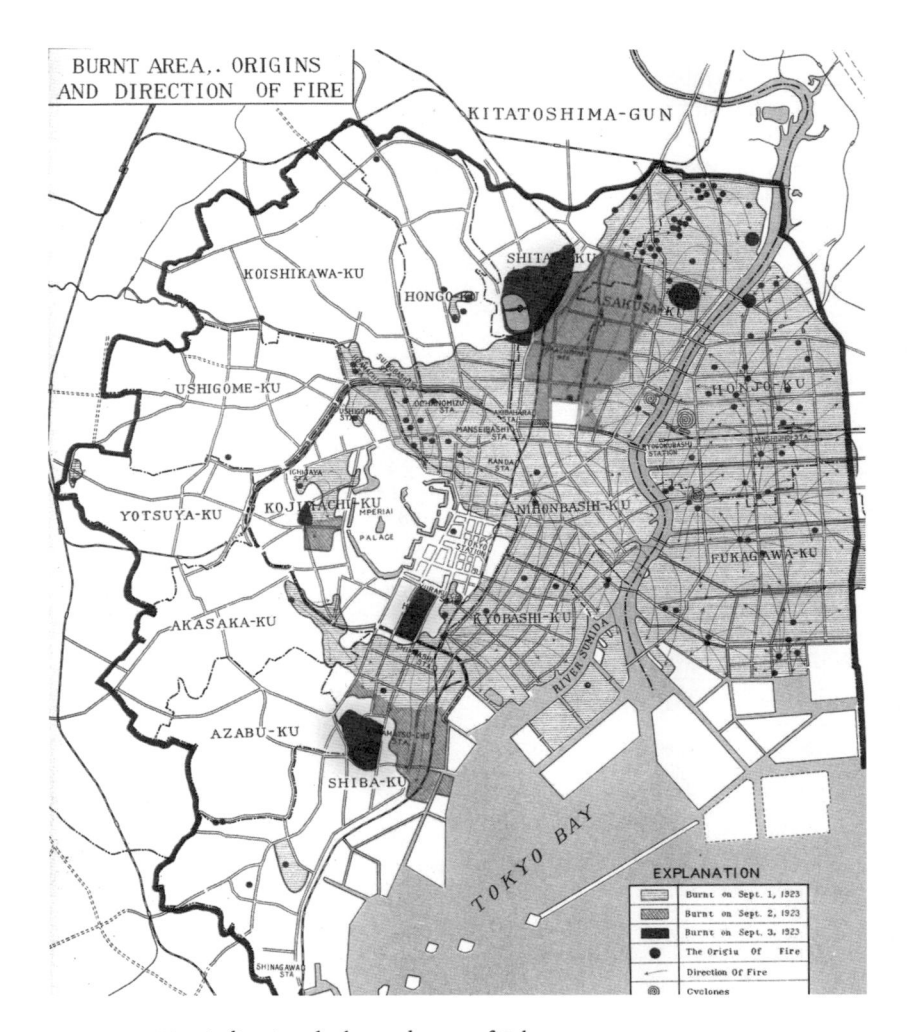

FIGURE 1.11 Map indicating the burned areas of Tokyo

Source: Tokyo Municipal Office, *Tokyo Capital of Japan*, 24. Reproduced with permission of Tokyo Metropolitan Archives.

society, making any quick return to normalcy virtually impossible. For one, the earthquake and fires destroyed the places of employment that so many Tokyo and Yokohama residents relied on to provide for themselves and their families. Roughly 7,000 factories met with destruction, including four major spinning factories, 264 dyeing factories, 445 tool manufacturers, 148 factories that produced ceramics, 37 pharmaceutical plants,

TABLE 1.1 Number of killed, missing, injured, or homeless in the City of Tokyo as a result of the Great Kantō Earthquake

LOCATION	KILLED	MISSING	SERIOUSLY INJURED	SLIGHTLY INJURED	THOSE WITH HOMES MADE UNINHABITABLE
Kōjimachi	95	42	73	334	32,832
Kanda	1,055	464	405	1,404	133,490
Nihonbashi	788	401	331	975	116,607
Kyōbashi	584	335	328	1,064	134,980
Shiba	361	133	275	898	81,930
Azabu	140	45	98	299	8,455
Akasaka	112	30	96	327	12,152
Yotsuya	68	35	61	189	5,019
Ushigome	150	53	115	355	7,366
Koishikawa	191	63	135	373	11,949
Hongō	218	102	121	427	38,354
Shitaya	577	314	388	1,078	147,211
Asakusa	2,597	1,070	809	2,576	254,692
Honjo	48,393	6,105	3,755	6,481	220,018
Fukagawa	2,775	1,364	886	2,152	178,794
Total	58,104	10,556	7,876	18,932	1,383,849

Source: Nihon tōkei fukyūkai, ed., *Teito fukkō jigyō taikan* [Survey of the imperial capital reconstruction project], 2 vols. (Tokyo: Nihon tōkei fukyūkai, 1930), vol. 1, part 5, p. 5.

28 breweries, 1,438 flour mills, 146 lumber yards, 114 paper manufacturers, and 23 paint and pigment factories. In the wards that comprised the City of Tokyo, the unemployment rate reached a staggering 45 percent (59 percent for men and 28 percent for women) in September 1923.[55] By November 15, six weeks after the catastrophe, the government calculated that the unemployment rate across all of Tokyo Prefecture stood at 37 percent.[56] Table 1.2 lists, by profession and gender, who found themselves unemployed on November 15.

FIGURE 1.12 Ningyō-chō, a once bustling street in Nihonbashi, Tokyo

Source: Uchida Shigebumi, ed., Taishō daishin taika no kinen, plate 7.

Displaced from their homes and their jobs, numerous individuals and families lost virtually all their assets, cash or otherwise, as well. In the City of Tokyo alone, 121 of 138 bank head offices and 222 of 310 branch offices were consumed by fire or reduced to rubble during the first three days of September.[57] Likewise, the four municipally managed pawnshops that gave loans and provided other financial services to the poorest of the poor working class of Tokyo burned to the ground. With the demise of these financial and banking institutions, many individuals' savings, financial records, title deeds, and other valuables also disappeared. For those who had the foresight and the means to purchase homeowners insurance, policies provided little to no comfort. All insurance policies throughout Japan contained a standard number of exemptions for which any insurance company would not be held liable. Along with war, invasion, volcanic eruptions, hurricanes, riots, and civil commotion, insurers found themselves under no legal obligation to pay policyholders for any "loss or damage occasioned by, or resulting from earthquakes."[58] For virtually all sufferers of Japan's worst natural disaster, insurance policies were, in essence, worthless. After intense and contentious political debates, however, the government agreed to provide long-term, low-interest loans of

FIGURE 1.13 Men looking for work at a city employment-matching agency

Source: Postcard, author's private collection.

TABLE 1.2 Number of unemployed on November 15, 1923, in Tokyo Prefecture

PROFESSION	MALE	FEMALE	TOTAL	PERCENT
Industry	51,965	18,784	70,749	39.5
Commerce	47,860	21,009	68,869	38.5
Public service	8,525	7,123	15,648	8.7
Transportation and communication	4,969	1,419	6,388	3.6
Agriculture	2,752	777	3,529	2.0
Domestic servants	977	171	1,148	0.7
Fisheries and marine products	215	33	246	0.1
Mining	212	29	241	0.1
Other	9,645	2,424	12,069	6.8
Total	127,120	51,767	178,887	100

Source: Nihon tōkei fukyūkai, ed., *Teito fukkō jigyō taikan*, vol. 1, part 5, p. 7.

¥64 million to Japanese insurance companies to settle claims. Under this agreement, however, no claimant received a figure higher than 10 percent of the original policy. In total, Japanese insurers paid out ¥78,181,446 from policies that totaled ¥1,382,772,556 in value, a figure that represented 5.65 percent of the value of all insurance policies held by those in the disaster zone.[59]

The miserable plight of the jobless, homeless, and impoverished disaster sufferers was compounded by the destruction of much of the physical, governmental, and social infrastructure that constituted Tokyo and Yokohama. Within the 33.4 million square meters of Tokyo devastated by the earthquake and fires, 362 bridges with a total surface area of 42,727 square meters were destroyed and another 70 with a combined surface area of 28,793 square meters were heavily damaged. Not only did the destruction of so many bridges make escape from the city difficult if not impossible for many refugees, it hindered delivery of relief supplies, including food and water, for weeks after the disaster. Compounding this problem was the fact that Tokyo's main aqueduct from the southwestern suburb of Wadabori collapsed for distances of over 18 meters in two places and was severely damaged and required extensive repairs in over two hundred other places along its path. In the districts of Fukagawa, Honjo, and Nihonbashi, moreover, 82 bridges that supported water-piping systems collapsed or were burned to the ground, making the reconnection of water supplies one of the most urgent yet most difficult tasks faced during recovery.[60]

In many ways the earthquake disaster also eliminated much of the physical presence of governmental administration in Tokyo, thus mirroring the decline of state authority in the disaster zone immediately after the initial quake and fires. The vast majority of police stations and municipal and ward offices and a sizable number of buildings that housed national government offices crumbled.[61] A large number of schools, often the most important sites of learning and spaces of state for many neighborhoods across Tokyo, were either burned or transformed into rubble.[62] Out of Tokyo's 196 primary schools, 117 were destroyed, along with 13 higher girls' schools, 13 business and trade schools, 10 technical schools, and 13 colleges and universities.[63] The meager social welfare facilities that existed to provide services to the poorest Tokyo residents were also annihilated by the disaster. They included both public dining halls (located in Asakusa and Honjo), both cheap lodging homes for male workers of "small means" (in Honjo and Fukagawa), both crèches that provided child-care services to families classified as the poorest of the

FIGURE 1.14 The remains of Kanda Ward in Tokyo

Source: Uchida Shigebumi, ed., Taishō daishin taika no kinen, plate 3.

poor where both parents worked similar shifts (in Honjo and Fukagawa), and the one women's work house that existed to train poor women for eventual employment (in Honjo).[64] Hospitals fared little better. Some 162 public and private hospitals in Tokyo were destroyed, leaving the vast majority of sufferers in need of medical attention destined to find care in the numerous field hospitals that opened during the weeks following the disaster.[65]

While numerous postdisaster sufferers may have held a strong desire and possessed plenty of reasons to pray or seek spiritual solace, finding a place for either proved difficult. By September 3, 633 Buddhist temples, 151 Shintō shrines, and 202 Christian churches had been destroyed.[66] Those who looked to comfort themselves with secular entertainment or jocularity in the face of widespread death and destruction likewise found this task equally difficult. Twenty out of twenty-two theatres, forty-three out of fifty-six cinemas, and sixty out of ninety-one dance halls collapsed or were burned to the ground.[67] The disaster, in short, left a large teeming mass of people—those who had not died in the carnage—hungry, thirsty, homeless, jobless, and often assetless: in a word, destitute. Worse, the survivors quickly found themselves inhabitants of a bleak, fractured, wretched, and stinking landscape of death, devastation, and want that had once been a vibrant imperial capital.

Aware of Tokyo's plight, Vice Home Minister Tsukamoto Seiji reflected on a conversation he overheard between a group of Japanese bureaucrats and an American official who visited Tokyo in November. The American had remained in Japan after bringing relief goods and commented on the extraordinary degree of destruction he had witnessed throughout Tokyo. Grateful for the dispatch of relief supplies, Tsukamoto however, like many Japanese, was sorry that foreign officials were exposed to Tokyo at its worst rather than as it stood on 31 August. In Tsukamoto's mind, Tokyo still resembled a city devastated by war. The American representative thought otherwise and stated emphatically that "even a war would not cause such a large scale of damage" as had been delivered by the earthquake.[68] The human and material loss Japan had suffered, the American representative claimed, was unprecedented in the annals of modern history. Tsukamoto and his colleagues agreed and understood that it would take a concerted national effort to restore normalcy, provide relief, rebuild Tokyo, and launch Japan on a course of recovery. To this end, the Imperial Japanese Army and Navy, two of Japan's most important and well-known national institutions, along with a host of public and private relief and aid organizations would play important roles.

In the days following September 1, 1923, survivors in Tokyo and Yokohama most assuredly realized that the earthquake and fires that destroyed much of these two cities was, and would forever remain, a defining moment in their lives. More than just "overturn[ing] Tokyo from its foundation" as socialist Yamakawa Hitoshi remarked, this disaster battered the social fabric that held together families, communities, the state, and society.[69] It wounded, killed, and displaced more civilians than any other event had previously done in Japan's modern history. It was the very definition of a catastrophe. Moreover, it annihilated large tracks of Tokyo and Yokohama's built environment, leaving scars of a physical as well as a psychological nature. Surviving it, moreover, pushed millions of victims to their physical and emotional limits. Given the sights, sounds, smells and almost indescribable suffering witnessed and experienced, it would be hard to imagine how victims of this disaster could see it in any other way but as an unprecedented calamity.

The Great Kantō Earthquake was an event that Takashima Beihō suggested all Tokyoites would forever remember.[70] Had Tokyo not been devastated on a similar scale in 1945 at the hands of American B-29 bombers loaded beyond normal operational capacity with incendiary bombs, the Great Kantō Earthquake would have remained the single greatest collective tragedy to befall Japan's capital. However deadly and confronting the earthquake calamity had been, though, the hardest tasks for government officials and for the citizens of Tokyo—restoration, relief, and recovery—were about to begin.

Two

AFTERMATH: THE ORDEAL OF RESTORATION AND RECOVERY

> The entire city was burnt to ash.... Tens of thousands of peoples scoured the ruins and searched the rubble for any food or water.
> —Yamanashi Hanzō, 1923

> The earthquake revealed not only the most virtuous quality that had been hidden inside our heart, but also the most evil side of us.... For what reason did so many creatures of Hell emerge among our people at that time? They damaged the pride of our empire considerably.
> —Fukasaku Yasubumi, 1924

Yamanashi Hanzō possessed many attributes among Japanese military men that made him unique. Born in 1864 in what would later become Kanagawa Prefecture, he rose steadily through the ranks of the military without clan or geographic ties to the powerful political domains of Chōshū or Satsuma, which had traditionally dominated Japan's military. What Yamanashi lacked in military pedigree, however, he more than made up for with experience. He was a seasoned soldier who saw combat during Japan's first three major wars: the Sino-Japanese War of 1894–1895, the Russo-Japanese War of 1904–1905, and the First World War. Upon promotion to lieutenant general in 1915, Yamanashi advanced within the military bureaucracy and served as army minister between 1921 and 1923. As the army's chief administrator, he emphasized quality over quantity.[1] In an era of military downsizing, the general stressed training, discipline, preparedness, and spiritual education as a means to compensate for reduced levels of soldiers. If such measures were extended to civilians, Yamanashi suggested that they might create a better-prepared and more harmonious society and counter the iniquitous trends of individualism that he saw in Japan. Given his experience and outlook, it is not surprising that Yamanashi was selected to serve as the chief of the martial law headquarters established in Tokyo following the Great Kantō Earthquake.[2]

Exposure to many different combat situations and experience with elite-level bureaucratic politics, however, did little to prepare Yamanashi for the tasks he faced in postdisaster Tokyo. He was the first to admit this fact. Specifically, the decorated general suggested that no previous experience placed a similar set of multiple, simultaneous demands on him, his staff, or the nation. Overseeing restoration, relief, and recovery was every bit as challenging as waging a war. Yamanashi described the tasks he faced not as a job but rather as an ordeal for which Tokyo and the nation had been ill prepared. Sufferers in eastern Japan were in want of everything imaginable, including food, water, medical assistance, shelter, clothing, and lavatories. Tokyo was a landscape of suffering. The hardships endured by Japan's survivors and the misery that Yamanashi saw every day were, in his own words, "beyond description."[3] Most troubling to Yamanashi, however, was the fact that Tokyo had descended into anarchy within hours of the earthquake. During this tumult, people had lost their sense of rationality, succumbed to exaggerated rumors, and acted like criminal thugs. This was especially true of people in vigilance groups, who in many instances had intimidated, beaten, and murdered. How, Yamanashi pondered, could this have happened in Japan's imperial capital, and why was Japan so ill prepared for such a disaster?

Shock and horror at the destruction of Tokyo and the violence and chaos that ensued fueled considerable elite-level anxieties about how people and the government responded to the extraordinary stresses associated with an unprecedented urban-based calamity. Both fears contributed to a burgeoning crisis anxiety that was summed up most effectively by Yamanashi. In a 1923 report that summarized his time as chief of the martial law headquarters, the general wrote that he had never been more "worried about the future of Japan's national security." Could Japan's citizens, he pondered, cope with the exigencies of a future conflict in which all "big cities would get bombed from the air and enemy propaganda [to confuse civilians and incite domestic unrest] would be more skillful?"[4] The experiences from the Great Kantō Earthquake did not give Yamanashi or others reason for optimism. Urban modernity nurtured manifold vulnerabilities.

CONFUSION, CHAOS, AND ANARCHY IN THE IMPERIAL CAPITAL

When the September 1 catastrophe struck, Japan was mired in a state of political confusion that amplified the chaotic situation created by the earthquake. Uncertainty began on August 24 when Prime Minister Katō Tomosaburō died of cancer. Though Admiral Yamamoto Gonnohyōe had been selected to form a cabinet shortly after Katō's death, he had made little progress in selecting ministers when the earthquake struck. Yamamoto, in fact, had spent the morning of September 1 at the navy reservist's club, the Suikōsha, discussing possible cabinet candidates with his son-in-law and future navy minister, Takarabe Takeshi. On paper, therefore, no formally appointed cabinet existed when the disaster unfolded.

Elite-level political uncertainty caused considerable confusion over who possessed authority to deploy police and military personnel as the disaster unfolded. While virtually everyone who witnessed the attempted mass exodus from Tokyo on September 1 understood the importance of reestablishing a semblance of political authority, no one knew precisely how to implement such a course of action. The deputy commander of the Imperial Guard Forces, General Ishimitsu Maomi (Sannomi), acted first, deploying troops to protect the imperial palace, the Akasaka detached palace, and other imperial locations within Tokyo.[5] He thereafter contacted the commander of the First Army Division stationed in Tokyo and encouraged him to ready his forces for immediate deployment.

Tokyo municipal employees and police officials faced a more complex task. As fires erupted and spread during the late afternoon, officials realized that their forces were outnumbered and woefully unprepared to respond to the human tragedy that was unfolding before their eyes. Though some survivor accounts mention heroic police, firefighters, or city employees who assisted with evacuations, police found themselves largely ineffectual at maintaining order. Aware of the immense task that the city faced, the inspector general of the Tokyo Metropolitan Police, Akaike Atsushi, believed that large numbers of military personnel were needed to gain control over an area in which all preexisting authority had vanished. But executing even this seemingly simple decision—to secure army troop deployments—proved problematic. First, no contingency plan or policy existed for responding to a catastrophic event of this magnitude. The police chief wondered if he possessed authority to request army assistance. Second, Akaike questioned, if he did not, who did?

At approximately 4:30 p.m., Akaike took matters into his own hands. Without seeking approval from the mayor, other municipal authorities, or any national politician, Tokyo's top police official contacted General Ishimoto and asked for all assistance that could be provided. When faced with this request, the general himself pondered whether he possessed authority to deploy troops in Tokyo. Hesitation lasted only momentarily as Ishimoto dispatched soldiers to key areas in the northern part of Tokyo. He then contacted the commander of the 1st Army Division and requested that their troops deploy to "vital points" in the southern half of the city.[6] These troops, however, like so many other Tokyoites, found movement difficult and order virtually nonexistent. "It was impossible," General Yamanashi admitted, "for Tokyo-based units alone to handle the disaster."[7]

As the extent of the previous day's calamity became apparent on September 2, virtually every surviving government bureaucrat realized that greater numbers of troops and police with increased powers were still needed to restore order before a nationwide undertaking of relief and recovery could begin. The chief of the Police Bureau of the Home Ministry, Gotō Fumio, recognized this point and used the powers of his office to request greater police assistance. Later informing his official biographer, Masuda Kaneshichi, that he felt as if he had "acted like the prime minister," Gotō dispatched messengers to ask—one might say plead with—police units from the prefectures of Nagano, Fukushima, Gumma, and Ibaraki to send as many officers as they could spare to Tokyo.[8] While prefectural officials dispatched 1,317 constables, it was still nowhere near enough manpower to return authority in Tokyo.[9]

The shape of Japan's elite-level government emerged by the afternoon of September 2 when the new cabinet was appointed. Given the extraordinary situation that hung over Tokyo, ministers focused their energies on charting policies that they hoped could reestablish political authority in the capital. If any official doubted just how dire the situation in Tokyo was, the circumstances surrounding the swearing-in ceremony reinforced the gravity of the challenges the new government faced. Cabinet ministers were sworn in on the lawn of the Akasaka Detached Palace, amid falling ash from the burning city, because all other ministerial office buildings within the vicinity of the imperial palace had been deemed unsafe. Moreover, palace officials informed those gathered to be ready to conclude proceedings at a moment's notice and evacuate if fires approached.

Not surprisingly, one of the first decisions reached by the cabinet was to mobilize troops from around Japan for deployment to Tokyo. Telling ministers that the government must deploy troops to secure "peace and order," Army Minister Tanaka Giichi dispatched infantry and engineering battalions from the 2nd Army Division (based in Sendai), the 3rd (Nagoya), the 8th (Hirosaki), the 9th (Kanazawa), the 13th (Takada), and the 14th (Utsunomiya).[10] This decision—followed by further deployments over the next week—ushered in the largest peacetime domestic mobilization and deployment of the army in Japan's pre–World War II history. The cabinet also agreed to declare martial law on September 2 over what was left of the capital in hopes that this would provide a pellucid chain of command for the divergent forces from various army divisions about to enter the ruins of Tokyo.[11] A formal state of martial law remained in effect until November 15.

The martial law headquarters possessed considerable power to regulate and govern activities within the disaster zone. The government granted martial law authorities power to administer relief and public safety activities, as well as the judicial power to prosecute lawbreakers. Beyond this, military authorities were empowered to prohibit public gatherings and to censor any public notice, telegram, newspaper article, or magazine story. Finally, martial law authorities were given the power to stop and search individuals, to prohibit passage of individuals within or through the disaster zone, and to enter private homes for any reason.[12] The subagencies created within the martial law headquarters best illustrate what areas of postdisaster life fell under the military's jurisdiction, namely, security, provisions, relief aid, transportation, intelligence, and general affairs. In short, Japan's army was about to occupy and administer all aspects of life and death in Tokyo.

If Prime Minister Yamamoto had any second thoughts as to the need to deploy so many army troops and police in Tokyo—in essence to give effective control over the disaster zone to the army until martial law was rescinded—they were dispelled as he journeyed from his first cabinet meeting in Akasaka to his home in Nagatachō on September 2. A number of events made this journey memorable. In the first instance, it was a tortuously slow ride. Repeatedly, Yamamoto's driver was forced to stop the car and, with the assistance of the prime minister's personal bodyguard, clear debris from the road. What made the journey most harrowing, however, were the people Yamamoto witnessed and encountered. Specifically, the distinguished elder statesman observed widespread chaos and disorder, few if any troops or police, and "gangs of vigilantes" who had constructed makeshift roadblocks and ignited numerous bonfires with leftover debris.

To Yamamoto, an unsettling, "riotlike" atmosphere hung over what remained of Tokyo. As his official car approached one roadblock, the chaos of postdisaster Tokyo took on a new and threatening personal dimension. Here Yamamoto experienced a glimpse of what a number of less fortunate Tokyoites confronted in the aftermath of the disaster: intimidation and the direct threat of violence. It began when his vehicle stopped near a roadblock and a group of vigilantes led by a man wielding a large club came to attack his car. In utter disbelief, the prime minister told his driver to confront the thugs. The would-be assailants stopped, according to Yamamoto's recollection, only when the cudgel-wielding leader saw a police captain sitting next to the driver, who informed them: "This is the new prime minister." "Oh, so a new cabinet has been formed," the chief brigand replied, before stepping back from the car and returning slowly toward his smoldering roadblock.[13]

Surviving this encounter unharmed but unnerved, Yamamoto was again confronted when he arrived at his home. The prime minister was astonished, though one might suspect also relieved, to see nearly thirty policemen guarding his residence. His family, however, was nowhere to be found. "Around 7 p.m.," one of the attentive guards stated, "we received word that Koreans were going to attack" and therefore moved your family to a "safe location."[14] Seven years after this incident, and no doubt influenced by the horrific events directed against Koreans that occurred over the following days, Yamamoto reflected just how painful it was for him to see the people of Japan fall victim to fear, lose control, and go mad over the outlandish rumors that had materialized.

Well aware of the dire situation that had descended upon the disaster zone, many municipal officials urged the martial law headquarters to take a more active role in restoring order and stopping violent acts. As part of the first martial law proclamation issued on September 2, authorities required members of all vigilance groups to visit their nearest security battalion, military police unit, or police station to receive orders. The martial law order also stipulated that no individual, apart from members of the military or police, had the power to stop, question, or restrict people in transit through the area under martial law.[15] Both laws, however, were virtually impossible to enforce because the first military deployments in the city were too small to enable the government to regain effective control. Moreover, security forces were given nebulous orders to "implement security measures" first at schools, government offices, and warehouses (or what remained of them) and at places where refugees had begun to congregate. Martial law officials also ordered these soldiers to distribute building materials and relief supplies held by the army in garrisons throughout the city.

The commander of the 3rd Army Division (Nagoya), who arrived in the city on the afternoon of September 2, was perplexed by these orders. He wondered if people giving orders had any idea of the situation on the ground. To him it appeared not. Providing relief and distributing assistance—let alone materials for shelters—he concluded, was impossible given the degree of anarchy and panic that defined the city. Moreover, he was stunned by the near total absence of reliable information about the state of affairs in Tokyo. Were fires still burning, and if so where? What bridges remained standing? How many sufferers remained trapped inside collapsed buildings? How were refugees leaving the city and where were they headed? These were just a few of the questions he pondered as he assembled his forces for work in the disaster zone.[16] Repeated attempts to contact municipal authorities for answers failed.

On his own accord, the commander therefore sent advance scouting teams into the epicenter of the disaster to gain information and locate, if possible, any other authority with which to consult and coordinate search, rescue, and restoration activities. Even this was problematic as he found there was no way to communicate with any of his troops once deployed: lines of communications were nonexistent. To assist placement and to get an overall picture of the destruction, he therefore ordered aerial reconnaissance over the city from flying battalions based in Nagoya.[17] To facilitate direct communications with troops in the field and to receive information from units already deployed around the Kantō region, the commander was forced to rely on approximately two thousand army-trained carrier pigeons.[18] The disaster zone, the commander concluded, was one of total chaos and pandemonium.

The implementation of martial law and its subsequent expansion in both geographic jurisdiction and range of legal powers provide clues as to the monumental task restoring order proved to be in postearthquake Tokyo. They likewise illustrate just how widespread chaos was and how difficult it became to contain. As initially implemented, martial law covered only Tokyo city and a few neighboring counties, including Ebara, Toyotama, Kita Toshima, Minami Adachi, and Minami Katsushika.[19] On September 3 military authorities expanded the zone of martial law to include all of Tokyo and Kanagawa Prefectures. In response to persistent rumors of Korean uprisings and substantiated reports of Koreans being killed in neighboring prefectures, martial law was extended for a third time to cover Chiba and Saitama Prefectures on September 4. All these initiatives, however, required more soldiers. Therefore, two further infantry regiments each from the 2nd, 8th, and 9th divisions, a telegraph regiment from the 5th division (Hiroshima), a heavy artillery regiment from the 15th division (Toyohashi), medical corps from every division, and engineer battalions from all the home island divisions that had not previously been mobilized except for the 4th (Osaka) and 6th (Kumamoto) divisions were ordered to Tokyo.[20]

In addition to these efforts, the government expanded the powers granted to the martial law authorities as greater numbers of troops streamed into Tokyo. Attempting, in part, to counter the proliferation of groups and written material that had erroneously implicated Koreans in seditious and criminal activities, martial law authorities gained the power to arrest individuals and disband any group suspected of engaging in violent actions. Anyone who authorities believed fabricated news, published material, or spread rumors with a view of disturbing public safety and order, moreover, was likewise subjected to arrest and, if convicted, imprisonment of up to ten years. Beyond this, martial law authorities also enabled troops to confiscate any weapon or article held by a civilian and deemed dangerous to public safety. Owing to the crimes committed by vigilance groups, all such organizations were thereafter requested, then ordered when few complied voluntarily, to send a representative to a local military or police command center and disarm themselves.[21] Informing the population of these laws and securing compliance, however, took time and effort.

Late in the afternoon of September 4, martial law headquarters received their first reports from field commanders sent to observe conditions in the various disaster zones of eastern Japan. The reports were somber reading.[22] From Tokyo, the unnamed commander wrote: "Relief aid has not yet been able to be delivered and there is a general state of anxiety among the population. Send more troops." The report from Yokohama was even more confronting: "The extent of damage is worse than that of Tokyo. The major institutions of the city and prefecture as well as the police have ceased to operate and wild rumors and crimes have largely unsettled the population. We wait with grateful anticipation for the arrival of more military forces." The observer who went to the Sagami Bay coastal area likewise reported on the "spread of groundless rumors" and the need for more troops to regain control of the panicked population. Neighboring Saitama Prefecture was described as in a "state of chaos due to the throngs of refugees and wild rumors that have accompanied them." The report stated that the large influx of refugees from Tokyo and Yokohama had also caused Chiba Prefecture to descend into a "state of chaos, rumor, and confusion."

Given the chaotic and violent environment that had developed, the commander of the martial law headquarters, General Fukuda Masatarō, with the backing of the cabinet, launched a bold policy unprecedented in Japan's modern history. From the evening of September 4 and continuing for over a week, military and police were ordered to collect and transport Koreans in the disaster zone to centers for "protective custody."[23] Koreans were also asked to voluntarily report to police stations, schools, and any other government center still in operation to facilitate their transportation out of the martial law zone. By the end of September, 23,715 Koreans had been taken to government-run detention centers: 9,980 had been taken into custody by police; 3,412 had been taken into custody by vigilance groups; 3,596 had voluntarily sought government protection; and 6,727 had been collected by military officials or had turned themselves in to other government agencies.

In conjunction with this policy, more troops were deployed in the capital, and the number of military checkpoints in Tokyo increased from forty-two to eighty-one. Travel restrictions were placed on anyone attempting to enter Tokyo or Yokohama, and stiff penalties were introduced for anyone who "started groundless rumors" or "instigated an assault, riot, or other crime that did harm to life, body or property of others."[24] Finally, the prime minister took the unusual step of communicating to people in the disaster zone through an official written pronouncement. Posted on September 5 at what remained of government buildings and places where refugees had congregated, and eventually handed out by military officials in person and dropped from the air, Yamamoto's proclamation urged all Japanese "not to lose their usual calmness of mind" and to "exercise strong self-control and live up to the principle of peace" in what had become an ocean of misery.[25] More than just Yamamoto's instructions were needed, however, to rein in the unprecedented acts of violence and mayhem unleashed in eastern Japan. No one knew this better than General Yamanashi, who assumed command of all military and police forces in the disaster zone on September 20. In General Yamanashi's opinion, it took approximately ten days for stability, peace, calm mindedness and public order to return to the disaster area. What it took in terms of manpower, he confessed, was equally extraordinary: nearly one in five members of

Japan's entire standing army—just over fifty-two thousand troops—had been deployed to Tokyo and Yokohama.[26]

There is no definitive answer as to why the government was so ill prepared to reexert control over eastern Japan and end the anarchy that emerged soon after the disaster struck. While virtually every Japanese person understood that Tokyo was fire and earthquake prone, no one in a position of authority had expected or drawn-up contingency plans for such a catastrophic series of events that began with a quake of extraordinary magnitude. The last destructive quake to strike a major urban center in Japan with such intensity was the 1891 Nobi earthquake, which killed just under eight thousand people. That quake had struck the much smaller city of Nagoya and had not resulted in massive conflagrations akin to what Tokyo experienced in 1923. While the 1854–1855 Ansei earthquakes killed over ten thousand in Edo (Tokyo), Japan's capital had become much bigger, more industrial, and more densely populated since the 1850s. Expansion in size and increased population density taxed the capital's infrastructure and made the city and its inhabitants far more vulnerable to seismic catastrophe and resulting fire than ever before. Given the extent of destruction to Tokyo's already inadequate infrastructure, there was little government officials could do to combat the fire or organize an effective, immediate evacuation of the capital. These facts became obvious in 1923.

Few government officials, moreover, believed that the acts of criminal behavior and murderous violence exhibited by residents of eastern Japan following the disaster were within the realm of possibility in 1923. Put simply, Japanese officials were unprepared and ill equipped to deal with the anarchy, pandemonium, and murder that transpired following the catastrophe. No detailed or coordinated plans had been put in place to deal with these contingencies. In this regard, Japanese leaders could be blamed for suffering from a failure of imagination. Japan's leaders also ignored the all too obvious social, economic, political, and geographic vulnerabilities that emerged in Tokyo from the turn of the century onward.

Imamura Akitsune, assistant professor of seismology at Tokyo University, understood the precarious precipice that Tokyo rested on. In the December 1905 edition of the popular journal *Taiyō* (The sun), Imamura warned that a potentially calamitous earthquake could strike the capital within the next fifty years.[27] He based his prediction on the fact that Tokyo had not been visited by a major quake since the 1850s. Imamura argued that since Tokyo had grown considerably since the last quake but had not developed any better system to respond to such an event, a "general conflagration" could sweep through the capital in the aftermath of a quake and kill upwards of 100,000 individuals. When Imamura published this article, he was not only lambasted by fellow seismologists but was also criticized by government authorities: seismologists ridiculed him for suggesting that earthquakes could be predicted, and government officials criticized him for publishing material that could incite social unrest. Imamura was quick to point out to newspapers in 1923 that his "theory had stood the test of fact" and that his opponents' had "fallen to the ground."[28] However prescient Imamura's prediction proved to be, the seismologist found himself amazed at the level of destruction meted out in 1923 and the chaos and confusion that ensued. It was worse than even he imagined possible. And the ordeal of recovery had yet begun.

PROVIDING RELIEF: EMERGENCY MEDICAL ASSISTANCE

Aware that the unprecedented disaster required an equally unparalleled response, the Yamamoto cabinet created the Emergency Earthquake Relief Bureau (*Rinji shinsai kyūgo jimukyoku*) to oversee disaster relief and recovery.[29] Relief activities were organized in the following areas: food, shelter and housing, materials and supply, communications, drinking water, medical aid and hygiene, relief funds and donations, financial accounts, general affairs, police, and intelligence. By the third week of September when relief activities were fully under way, the Relief Bureau employed over seven hundred officials. Employees worked in shifts around the clock. Given the demands of duty and the difficulty faced moving through the still devastated capital, many employees chose to sleep in tents and temporary barracks erected at the Relief Bureau's headquarters, the former Home Ministry building site.[30]

Though many disaster victims found themselves hungry, thirsty, and in need of immediate shelter, providing emergency medical assistance was one of the first tasks officials pursued. Few people, however, were rescued or successfully treated for serious injuries by government-led relief teams in the days immediately following the disaster. People either survived and coped with minor wounds and burns on their own or died. Many factors contributed to this failure. First, the challenges caused by a continuation of fires and the breakdown of public order proved difficult, if not impossible, to surmount in the short term. Unofficial checkpoints established by vigilance groups who stopped virtually all nonuniformed people not only unnerved rescue and relief workers but also slowed their progress. In part this kept many organized rescue and first aid squads from reaching places such as Honjo, Asakusa, and Kanda until September 4 or 5. Second, the transport of donated medical supplies and relief workers dispatched to Tokyo from neighboring prefectures was slowed because of the heavy damage sustained by Tokyo's infrastructure: it proved just as difficult for aid to enter Tokyo and be dispersed to sufferers as it was for victims and refugees to leave. Third, the medical and rescue teams that operated immediately after the earthquake also faced a dearth of supplies because so many hospitals and medical dispensaries in Tokyo and Yokohama had been destroyed.[31] Fourth, the congregation of tens of thousands of refugees—many of whom required minor medical assistance of some kind or another—in large open places such as Hibiya Park, Ueno Park, and the grounds surrounding the imperial palace compelled the Relief Bureau to concentrate their still scant medical resources at these locations rather than scattering to support smaller-scale search, rescue, and relief operations in many of Tokyo's most damaged areas.

A survivor account published in October 1923 provides a gruesome and harrowing glimpse of the medical treatment—or lack thereof—carried out over the first three days after the calamity. The anonymous chronicler of misery recounted how he found himself at a makeshift medical shelter near the Yasuda estate that bordered the site of the former Honjo Clothing Depot. In front of a sign that once read "Mikuraya Ferry," but which had been crudely painted over to read "Medical Center," he saw a "groaning boy around fifteen years old, lying on a half-burned board that had been used for drying washed kimono." "From the waist down," the account continued, "the boy was inflamed with burns and his left leg was crushed so badly that it was unrecognizable." As it began to rain, a soldier approached the author and asked him to help pick the boy up and take him inside the relief hut. Rather than being treated, however, the boy was taken into the tent to die with others, including "a girl whose whole body was covered with burns who looked already dead," "a man lying on his front side with his whole back burned," and "a woman whose body looked like it had stood under a shower of flames." "The center was filled," the raconteur concluded, "with about 150 severely wounded people who hovered between life and death," all waiting to "die one after another."[32] The sight of these people's suffering, reinforced by the constant sound of their groans and cries and the smell of death, convinced the author that the "lucky ones" had perished quickly the night before.

The government was more successful in providing medical relief over the mid to long term following the calamity. Given the dispersed nature of the postdisaster sufferers and the widespread devastation meted out over eastern Japan, mobile clinics and dispensaries that fanned out across Tokyo and Kanagawa Prefectures proved most effective at providing relief to those in minor medical need. Each unit comprised two medical doctors, two nurses, one assistant, and medical supplies secured from donations that eventually reached Tokyo from virtually every part of Japan. Initially the Relief Bureau created forty-one units, but their successes encouraged Tokyo Prefecture to create an additional twenty-eight, the Japan Red Cross Society to organize twenty-one, the Saiseikai, a relief organization created by the Imperial Household Ministry, to form seven, and the Mitsubishi Corporation to operate six.[33] By November 30, public and private mobile medical clinics had provided 447,111 sufferers with medical care. Along with these mobile units, the Relief Bureau also established temporary hospitals—some in no more than tents donated by the American and French governments—that accommodated just over 6,000 seriously wounded people.

PROVIDING RELIEF: FOOD AND WATER

Despite initial attempts to promote emergency medical care, providing food, water, and shelter became the three most pressing relief tasks that the Relief Bureau faced. Before the national government responded to what became a looming food crisis, Mayor Nagata contacted army officials from the Imperial Guards and asked how much rice the military held in Tokyo that could be readied for immediate public release, and where it was located.[34] At approximately 4 p.m. on September 1, army officials informed Mayor Nagata that the army supply depot in Fukagawa held 8,000 *koku* of rice that could be cooked and distributed to sufferers; 1 koku was equivalent to 278 liters and originally defined as enough rice to feed one person for one year. Theoretically then, 8,000 koku could feed 512,000 people for six days. When Nagata asked if the army could confirm whether the building had survived the quake, army officials gave no concrete answer but suggested that it had survived for two reasons: because

FIGURE 2.1 Fukagawa Ward burning, as seen from an army reconnaissance plane

Source: Uchida Shigebumi, ed., Taishō daishin taika no kinen, 47.

it was located away from residential housing and thus not as susceptible to fire as would normally be expected, and because it was constructed from brick rather than wood. Neither Nagata nor his army contacts were aware that an intense conflagration that spared almost no building was about to sweep through much of Fukagawa and neighboring Honjo.

The next day a "soot-covered" apparition who appeared at what remained of the mayor's office at 5 a.m. brought reality home to Nagata.[35] Nagata immediately asked who this ghost of a man was and pondered from where he had traveled. A closer inspection, which revealed numerous facial cuts and bruises and wounds to the exposed parts of the man's arms, legs, and feet, jarred Nagata's memory. The visitor was the head of the Fukagawa Ward, described in Nagata's memoir as Mr. Kawanabe. Kawanabe had fled burning Fukagawa late in the evening on September 1 and had taken all night to walk, crawl, and then stagger to central Tokyo. His first utterance was to apologize for his ramshackle appearance, explaining that his clothes had been shredded and soiled and his body bruised and battered as he journeyed through what could only be described as Hell. Relaying that the most treacherous part of his escape had been his crawl along a fiery hot water pipe on the underside of the still smoldering Eitai bridge, what he reported next caused "all color to drain" from the mayor's face. Without any exaggeration, Kawanabe told Nagata that not only had the army supply building been destroyed, but that Fukagawa no longer existed. When asked to explain, Kawanabe repeated that the entire ward of Fukagawa had been annihilated: no building remained. The gravity of the situation had now become apparent to Nagata.

The mayor immediately sent a messenger to the commander of the imperial guards with the message that the food situation for Tokyo was in "real trouble." Army officials needed no confirmation of the dire situation that now hung over eastern Japan. Military authorities had already begun to amass more than 120,000 combat rations for civilian use comprising 1,600 *kan* or 13,200 pounds of tinned beef and 60,000 rations of rice held at storehouses in and around Tokyo that had escaped destruction.[36] On September 4 the Imperial Guards and the 14th Army Division also directed army units in the Kantō region to distribute 7,200 kan (59,976 pounds) of dried noodles to sufferers. Realizing that this was still a stopgap measure, the army and navy likewise volunteered to transfer to Tokyo rice held at each army barrack and navy base throughout Japan. Military personnel thereafter transported 74,048 koku of rice from these military installations to distribution centers in Tokyo and Yokohama.

While the government assumed that drawing on these reserves would provide much needed relief in the short term, cabinet officials believed that far greater quantities of food would be required over the following weeks and months to satisfy demand. To secure food, the Relief Bureau therefore made direct appeals to regional elites. Using the navy's wireless broadcast station at Funabashi, Tokyo, the government sent requests to all prefectural governors for donations of rice and other materials, specifically medical supplies. The message sent by the Yokohama police chief illustrated the dire situation. It read: "We have neither water nor food . . . send relief at once."[37] Prefectural governments around Japan responded to requests for donations in a generous fashion. Governors pledged 61,490 koku of rice in the first week after the disaster with the prefectures of Aiichi (11,825), Mie (5,773), Osaka (5,304), Kyoto (5,234), and Hyogo (3,992) donating the largest sums.[38] On top of prefectural donations, the Provisions Bureau of the Ministry of Agriculture and Commerce also released rice stored at its main Osaka warehouse for immediate dispatch to eastern

Japan.[39] Eventually the government used 351,622 koku of rice from its primary warehouse in Osaka to feed those in need.[40]

But what would happen, Tokyo elites wondered, if those outside the capital failed to see the calamity as the nation's most pressing need or if initial donations were inadequate to feed all of the hungry sufferers? Concerned that people well removed from the immediate disaster area might not give enough to support relief, the cabinet also empowered the state to secure relief materials through a more interventionist approach if necessary. Imperial Ordinance No. 396 entitled "Emergency Requisition Ordinance," issued on September 2, allowed the national government to requisition "foodstuffs, material for building and medical care, vehicles, or personnel services" for the relief of disaster sufferers.[41] While guaranteeing that "compensation would be made according to average market price of the previous three years," within three months of acquisition, this ordinance also provided stiff penalties for those who refused to comply: individuals who did so faced imprisonment for up to three years and a fine not exceeding ¥3,000.[42] Rice purchased by prefectures or acquired under the national requisition program and then sent to the disaster area totaled 185,490 koku, with Tochigi (44,687), Niigata (30,381), and Ibaraki (20,818) supplying the greatest amounts. By December 1923 the government had requisitioned ¥2.5 million worth of food, ¥1.8 million worth of building materials, and transportation vehicles including cars, ships, and rail containers worth just over ¥1 million.[43] How and when the central government repaid prefectural authorities proved to be a continual source of contention. On a few occasions they did not. The governor of Hokkaidō, Miyao Shunji, who eventually played an important role in the reconstruction program, lamented the fact that regional Japan had been forgotten in the months and years after the disaster. Reflecting on this fact in 1930, he wrote: "When I heard Tokyo was destroyed, the first thing that came to my mind was that its people would need food. From Hokkaidō we sent rice, milk, dried squid, and salted fish.... For us, however, the most difficult thing about this disaster was that Tokyo never paid us [because of the moratorium]."[44] Others, particularly parliamentarians representing rural constituents, expressed similar concerns.

While securing donations and requisitioning rice from all corners of Japan did not prove excessively difficult, transporting rice and other foodstuffs to the disaster area and then to the sufferers was an immense logistical challenge. First, the seismic waves that rocked eastern Japan damaged virtually all rail lines into the Kantō region, particularly those feeding Tokyo from western Japan. To overcome this initial obstacle, prefectural authorities used merchant vessels to ship rice to Tokyo. Even this was fraught with difficulties, however, as the following story from Mayor Nagata illustrates. On September 3 Tokyo officials received word that the cargo ship *Senkai* had arrived in Tokyo Harbor loaded with 1,000 koku of emergency rice from Osaka. This was as much a goodwill gesture as it was a source of sustenance, but owing to the destruction of Tokyo's docks, rice from the *Senkai* could not be unloaded until September 10. It was disheartening, Nagata reflected, to have rice "right in front of us" with no means to collect it for distribution when so many Tokyoites needed food.[45]

It was not just the *Senkai* that experienced this fate. On September 2 Navy Minister Takarabe Takeshi ordered navy vessels to transport rice to Tokyo and Yokohama from Osaka and Kobe. Within a week of the disaster, the navy had employed 108 warships and 42 merchant vessels to transport food and relief supplies.[46] These ships, however, like the *Senkai*, were forced to wait to unload their precious cargo until Tokyo's docks had been repaired. Navy personnel therefore spent the better part of a week rebuilding and reconstructing eighty-six piers or docks at the Shibaura and the Ryōgoku distribution centers. Army forces likewise found their hands full in repairing the land-based infrastructure of eastern Japan. Soldiers removed the burned and damaged skeletons of over three thousand damaged railcars and trams left derelict on Tokyo's once modern mass transit network. They also rebuilt more than 85 kilometers of track upon which food and relief supplies could eventually flow. Twenty-seven bridges were replaced with temporary structures, and eleven bridges left partially intact following the disaster were repaired to enable transportation.[47]

After soldiers and sailors repaired the transportation infrastructure leading into and within the capital, a complex food distribution system opened. As food made its way to Tokyo from regional Japan, the Relief Bureau designated one minor and six major collection centers within Tokyo and Kanagawa Prefectures where military officials oversaw the dispatch of food. Three of these stations were supplied by navy vessels, including a large center that opened on September 5 at Shibaura, a large station created in Yokohama on September 9, and a small station opened on the banks of the Sumida River near the remains of Ryōgoku Station on September 5. Four other large collection centers administered by the army and supplied by rail opened on the grounds of Shinjuku, Tabata, and Kameido rail stations on September 5 and at Shinagawa Station on September 9. Military personnel worked around the clock unloading and

FIGURE 2.2 Navy personnel repairing docks in readiness for the arrival of supplies (*upper right*); army personnel clearing train and tram lines (*lower right*)

Source: Uchida Shigebumi, ed., Taishō daishin taika no kinen, 65.

redistributing relief food supplies to city officials until the last week of September. From these collection centers, city officials dispatched food aid to numerous distribution depots. To facilitate this dispatch, the Relief Bureau requisitioned over 300 automobiles, 2,000 rail containers, 800 hand carts, and 6,500 horse-drawn carts from as far away as Osaka. The government also mobilized 17,065 veterans, 4,749 members of youth groups, and 44,695 volunteers and requisitioned laborers from every prefecture except Okinawa to hand out food to disaster victims.[48]

Despite this herculean effort, survivor accounts published shortly after the disaster illustrated that acquiring food became a time-consuming preoccupation for at least one member of almost every needy family. In many cases families dispatched children to collect rice, a trend that eventually worried educators who feared that children sent foraging would be more prone to drop out of the temporary barrack schools that had been created throughout Tokyo by the end of September. "How sad it is," wrote one teacher soon after the disaster, "to have childing sobbing as

FIGURE 2.3 Army personnel unloading food at a distribution center in Tokyo (*upper left*); people waiting in line for food (*lower left*)

Source: Uchida Shigebumi, ed., *Taishō daishin taika no kinen*, 71.

they arrive late and say, 'Teacher, I am late because I had to collect rice.'"[49] Though some children might have been tempted to use food collection as an excuse for tardiness, newspaper reporters, relief workers, and observers noted that the numerous lines that fed into local distribution centers often reached, and maintained, a length of 700 meters until the last wave of hungry seekers had been allocated their daily allotment of rice.[50]

If detailed plans had been in place on August 31 for just such a post-disaster contingency, it is questionable—given the scale of destruction and eventually the degree of lasting want—whether results of food distribution activities would have been markedly different. While numerous inefficiencies existed with the distribution of food, the number of individuals receiving free foodstuffs was extraordinary.[51] Records up to September 16 were not kept because all available labor was utilized to construct distribution centers, clear transport links, and hand out food. Conservative estimates based on the sheer volume of rice distributed between September 6 and 10 suggest that 1.25 million people received food.⁵² From September 16 to 21, 2,097,170 people across Tokyo Prefecture each received 3 *go* (approximately 450 grams) of rice.⁵³ On October 15 and 16, six weeks after the earthquake, 670,532 sufferers lined up for free rice and other food—canned food from overseas had by then begun to arrive—from distribution centers. Some 51,014 individuals deemed "needy" by the Relief Bureau obtained food on December 20, and 1,300 received a handout on the last day city officials provided free rice: April 10, 1924.

The supply and distribution of water suitable for drinking proved to be an even more significant problem to rectify immediately following the earthquake. Though Tokyo's main water plant at Yodobashi in Wadabori had not been damaged heavily, water pipes and aquifers that supplied the city had been severed at over two hundred locations. Particularly vulnerable were the most heavily damaged wards of Fukagawa, Honjo, and Nihonbashi as over eighty-two bridges that supported water pipes into these locations had been destroyed. At 9 a.m. on 3 September, Relief Bureau officials agreed that all available manpower that could be spared should be directed toward transporting water to the disaster zone.⁵⁴ Apart from ordering warships to tank water from Yokosuka, Osaka, and Nagoya, officials also sent army personnel to requisition all available water barrels in undamaged parts of Tokyo for the transportation of clean water. Roughly 250 barrels, 4 *to* (72 liters) each, were placed in each damaged

FIGURE 2.4 People surrounding a water truck

Source: Postcard, author's private collection.

ward to be filled daily by water trucks that drew water from rail containers, warships, and eventually parts of the city where the water flow had been restored. Even toward the end of September, the daily distribution of water drew unanimous praise from refugees. One group of army reservists from Sendai in northern Japan who worked as volunteers in Tokyo wrote that citizens almost always cleared the road for the water trucks, raised their arms, and shouted *banzai* as they made their daily appearance.[55] Though by October 1 army engineers had repaired most of the damage caused to the city's aquifers, water continued to be shipped to many locations where temporary barracks had been constructed. By the end of December, vehicles, railcars, and vessels had transported nearly 40 million gallons (151 million liters) of drinking water.

PROVIDING RELIEF: RELOCATION AND SHELTER

While providing water, food, and basic medical care proved to be a monumental task, it was matched if not superseded in complexity by efforts undertaken to provide shelter to the throngs of displaced earthquake

FIGURE 2.5 Refugees evacuating Tokyo by train

Source: Postcard, author's private collection.

survivors. In the days immediately following the disaster, shelter options proved bleak for Tokyo Prefecture's 1.55 million homeless. The most fortunate attempted to leave the worst parts of the disaster zone. First on foot and then on a restricted rail service that opened on September 11, nearly 800,000 people left Tokyo or Yokohama to stay with relatives or friends in less damaged parts of the capital, its neighboring prefectures, or further afield.[56]

While a sizable number of refugees, 528,038, sought shelter in undamaged parts of the seven prefectures affected by the calamity, nearly 250,000 people dispersed to other parts of Japan. Many did not return. For example, 17,704 relocated to Kobe and cities and villages in Hyōgo Prefecture, 7,600 settled in Hokkaidō, 1,304 journeyed to Ehime Prefecture on the island of Shikoku, and 1,006 fled to Kagoshima on Japan's southern island of Kyūshyū. Some even escaped to Japan's formal empire: 513 disaster refugees resettled in Taiwan and 216 sojourned to Karafuto, Japan's colony that comprised the southern half of Sakhalin Island.[57] An anonymous survivor account published in October 1923 relates the extraordinary scene of thousands of people crowded into the damaged ruins of Tokyo Station in early September waiting for a train to take them anywhere beyond the disaster area. Evacuees, the author noted, had lost all empathy for the dead. Bodies—including those of children with their mothers—had been piled near the front of the station. At this indifference, the author asked, "How could we not be shocked?"[58] Few people, it seems were. The dead had become an accepted, if not normal, part of the landscape.

For those who either did not want or did not possess the ability to leave the disaster zone, life took on an even more surreal quality. Sufferers flooded virtually all the large open spaces in Tokyo. Tens of thousands of refugees congregated at Hibiya Park, Ueno Park, the outer grounds of the imperial palace, and eventually the main parks in Asakusa and Shiba wards.[59] Newspapers reported that the detached palaces at Hama, the imperial gardens at Shinjuku, and the destroyed imperial timber yard in Fukagawa had also been opened to refugees.[60]

Often packed person-to-person, evacuees waited for food, water, and news of family separated in the chaos that followed the catastrophe. Personal hygiene became such a problem that people began to swim in the imperial palace moat, prompting one newspaper reporter to suggest that this was the result of "an unprecedented disaster resulting in unprecedented scenes . . . things never dreamed of in ordinary days." Refugees,

FIGURE 2.6 Refugees in Hibiya Park

Source: Postcard, author's private collection.

the reporter continued, also exhibited no qualms about "drying their washing on pine trees in the open space of the imperial palace."[61]

Roughly one week after the disaster, Tokyo municipal authorities agreed on plans to accommodate the earthquake refugees in a more permanent fashion. On September 9 they started by constructing temporary barracks at Hibiya Park, Ueno Park, and the site of the Meiji Shrine.[62] Temporary housing built at the Meiji Shrine eventually housed nearly 6,000 refugees, while those at Ueno Park and Hibiya Park sheltered just over 9,500 and 7,000, respectively.[63] Many other large open spaces also gave rise to barrack communities, with Shiba Park housing just under 6,000, the Tokyo Botanic Garden in Koishikawa accommodating 2,500, and the grounds in front of the imperial palace serving as temporary homes for 2,400 sufferers. Many of these open spaces had filled quickly and, by the middle of September, no longer resembled parks but rather shantytowns. Hibiya Park, as one reporter for the *Osaka mainichi shinbun* suggested, looked like a makeshift "tin city" because of the more than 1,200 zinc-coated, corrugated iron sheets that served as roofs over the temporary barracks.[64] The extraordinary demand for galvanized sheet metal became so apparent that it prompted one Kobe-based newspaper that catered to foreign business interests, the *Japan Weekly Chronicle*, to ask if the disaster would compel Japan to end its exorbitant tariff on this product; the tariff was suspended until March 31, 1924.[65]

Pressed for space and worried that the throngs of refugees now housed in Tokyo's scant open spaces could increase the chance of disease and overtax the rudimentary sanitation facilities being constructed, Mayor Nagata also directed authorities to relocate refugees to smaller barracks at small parks, temples, shrines, and sixty-three primary schools that had been destroyed. Of the 39,977 individuals housed in these smaller barrack complexes on November 31, 1923, 29,780 found shelter in what were once sites of early childhood education. The eight barracks erected at the site of Honjo Primary School, a school that completely burned to the ground on September 1, housed more than almost any other primary school: 380 households, which comprised 1,138 people, found shelter at this location. Between them, the inhabitants shared 600 *tsubo* of total floor space, a figure that equated to 1,986 square meters. This translated to 1.74 square meters, or 0.53 tsubo, per person.[66] It was a tight fit, especially in comparison to the nearly 0.95 tsubo of floor space allocated per person at the most spacious barracks located at Nakasu Primary School in Nihonbashi ward. There, thirty-nine refugees from fifteen households found shelter from the autumn rains. Most barracks averaged about 0.6 tsubo of floor space per person.

Apart from being cramped, what else defined the barracks? In the initial weeks it would have been virtually impossible for inhabitants not to notice the smell, first of garbage and then of human waste. Both became acute problems. Government officials proscribed the burning of waste, fearful that unregulated burning amid the cramped refugee quarters could ignite further conflagrations in the disaster zone. Rubbish therefore accumulated for two weeks, not being removed until September 17. The disposal of human waste was fraught with just as many problems: it continued to fill and overflow makeshift latrines dug by city workers throughout the autumn of 1923. By the end of December, disposal crews had collected over 64 million liters of night soil and deposited it on nearby farmlands.[67] The barracks were also noisy. While a number of the barracks constructed at school sites were furnished with wood partitions, more than a few of the larger barracks used partitions made of rush matting (*anpera*).[68] Regardless of their exterior or interior walls, the roofs of many leaked when it rained, and virtually all the earliest barracks built in the wards of Honjo and Fukagawa found themselves flooded on September 24 by a massive, day-long downpour from the remnants of a tropical cyclone.⁶⁹ Reflecting in the November edition of *Chūō kōron* on the varied quality of the barracks that filled Tokyo, former Yomiuri journalist Yasunari Jirō wrote:

> So grand, as might be expected of Ginza.
> We envy the barrack,
> that is built with steel frames.
> Everyone is working,
> in order to escape the makeshift barracks,
> as soon as possible.⁷⁰

Few residents, however, escaped the barracks, at least for a while. On September 24, 27,429 barracks housed nearly 150,000 people.⁷¹ By the end of 1923, 122,261 still resided in barracks, a figure that would drop to 85,260 by July 1, 1924.

While only a modest number of refugees left barrack housing by December 31, 1923, to return to where their homes once stood—almost all of them owners, not renters—those who did were joined by a large number of returnees who had left the disaster zone to stay with family or friends across Japan in the days after the calamity. By September 17, 122,000 people had returned and erected 24,300 temporary homes in Tokyo.⁷² Just over a month later, these numbers had swelled to 539,450 people living in 111,791 makeshift abodes. While many of these private shacks were hastily thrown together with only the most rudimentary materials, the "reconstruction spirit" of these returnees struck many commentators. Bureaucrat Nagai Tōru commented that a "reconstruction competition" had overtaken many middle-class refugees who wanted to leave the cramped state-run barracks as quickly as possible to return to the site of their old homes.⁷³ Owing to both the larger than expected number of people reconstructing private residences and the shortage of building material, government officials decided to suspend most building regulations and allow all temporary housing in the disaster zone to stand until August 31, 1928.⁷⁴ Tokyo and Yokohama, as one foreign newspaper suggested, were no longer cities but shantytowns composed of "sorry hovels" built from "discarded sheet metal." "Tokyo," another writer concluded, was now a "city of the past." "It may take decades for it to rise from the ashes to its former prosperity and importance."⁷⁵

In virtually every way Tokyo resembled both a landscape physically devastated by war and one, if only temporarily, controlled by a military occupation force. Coupled with the existence of millions of destitute refugees who relied on state assistance and charity for food, water, shelter, medicine, and basic sanitation and hygiene needs, the environment of Tokyo served as an almost constant reminder that a totalizing, nearly incomprehensible tragedy had befallen the inhabitants of Japan's once vibrant capital. Not surprisingly, many commentators—foreign as well as Japanese—concluded that Japan had suffered a trauma even greater than what a modern war could inflict. As an editor for the *Japan Weekly Chronicle* wrote, "the Yokohama and Tokyo disaster did as much damage in twenty-four hours as the warring armies on the Western Front did in four years."[76] Ironically, even before Japanese commentators, elites, and military men reflected on what the disaster meant or agreed upon what it could be compared to, they had already begun to focus their attention on what Japan's 1923 experience might tell them about how their country would respond to a future calamity. These thoughts did nothing but amplify already heightened anxieties.

QUESTIONING AND PROJECTING: UNPREPAREDNESS AND FEAR OF THE NEXT URBAN CALAMITY

While countless individuals reflected on the trauma that Japan experienced in September 1923, others emerged from this calamity with an outright feeling of foreboding about how Japan might cope with a future calamity. In the first instance, the government's response to the 1923 earthquake illustrated to many concerned elites how ill prepared the city, the prefectures, and the nation had been to deal with calamity, to restore order, and to implement relief without significant intervention from the military. Far more troublesome, however, they believed it revealed how people were poorly trained to deal with a catastrophe and how violently some elements in society had responded during the postearthquake period of emergency. Commentators, citizens, and government officials were well aware that residents of eastern Japan had not only suffered at the hands of nature but been brutalized by their fellow citizens.

A few elites with direct experience of the September tragedy, such as General Yamanashi, wondered how people might respond to a future manmade calamity in which many of Japan's cities might be targeted by enemy airplanes intent on inciting panic and unleashing destruction. If Japan's earthquake experience was anything to go by, Yamanashi suggested, a future wartime calamity might result in widespread social, political, and economic upheaval and threaten Japan's ability to defend itself from attack. More to the point, Yamanashi argued that if Japan found itself engaged in an unlimited war as the countries of Europe had just experienced, it might not have the ability or the luxury to deploy 50,000 soldiers for domestic purposes. The tasks of restoring order, maintaining internal stability, and taking care of victims, he concluded, would have to fall on the shoulders of the police, local government, and the people.[77]

The postearthquake harassment and murder of innocent Koreans convinced Tawara Magoichi that Japan would again be headed for a human-inspired tragedy if a future emergency arose. Writing in the October issue of the journal *Chihō gyōsei*, the bureaucrat argued that unless the government became more assertive and prepared its people for a potential crisis, Japan would again find itself overwhelmed when the next disaster struck. Tawara suggested that the way in which people responded when government authority evaporated called into question not only the moral foundation of society but also the level of civilization that Japan had attained since 1905. Was the anarchy and murder exhibited by Japanese people, he asked, indicative of how civilized people behaved in the face of catastrophe? "No," he stated unambiguously. People's behavior following the calamity compelled him to question the efficacy of Japan's previous efforts to cultivate good subjects and citizens. Tawara suggested that the calamity exposed a "major defect in our national spirit" and urged government officials to rouse themselves from complacency, implement group training programs for all citizens across the nation, and cultivate a sound national spirit "for the sake of Japan's future."[78] People had to be taught and conditioned how to deal effectively with catastrophic situations.

Shimamoto Ainosuke suggested that however awful the earthquake had been, human responses to it had amplified Japan's tragedy. Violent responses to unfounded rumors and the general state of lawlessness that emerged illustrated how "pathetic" Japan's national character had become since the Russo-Japanese War. Shimamoto wrote that many individuals—even well-educated people—across seven prefectures believed groundless rumors, spread them as if they were truths, and murdered in response. The massacre of innocent people, in Shimamoto's words, was a "tragic comedy" that demonstrated the dangers of succumbing to a "mass psychology of fear." But, Shimamoto asked, was not this psychology of fear precisely what an enemy would attempt to cultivate in a future war with Japan? Yes, he concluded, it was, and unless the government educated and trained its people, tragedy would strike again. Shimamoto summed up his argument as follows: "If a foreign enemy attacks our imperial capital in the future, we will have no confidence as we used to have as a victorious nation. Fortunately, both the Sino-Japanese and Russo-Japanese Wars were fought outside our land; but it may not be the case with wars in the future. The thought of a future war makes me shudder. We must educate our people and remove this evil—the pathetic national character—before it is too late." Would Japanese people under the shadow of the bomber respond like "frightened cranes that panicked and took flight at the slightest disturbance," or would they face the challenges posed by an enemy with resolve and discipline? The 1923 disaster, Shimamoto concluded, illustrated that the introduction of group training revolving around catastrophe prevention must be a basis for the social and spiritual reconstruction of Japan.[79]

Rather than merely advocate future training regimes, some well-placed commentators also chastised officials who, they argued, had contributed to this human-inspired tragedy in no small way through gross negligence. Future privy councillor Ichiki Kitokurō lamented the fact that citizens had not been trained to deal effectively with emergency situations. "Those disgusting murders," he wrote in *Chihō gyōsei*, "would not have happened if people had been properly trained to deal with an emergency."[80] A longtime advocate of self-defense group training, Matsui Shigeru, argued that because neither local nor national government in Japan had taken concrete steps to organize and train local vigilance groups, members believed wild rumors about Koreans and "in a state of wild excitement, killed and harmed many people." Why, he asked, had government officials not tapped into the large numbers of ex-servicemen and retired police officers and used them to train and supervise self-defense groups?[81] "Many innocent Koreans were persecuted," bureaucrat Suematsu Kai'ichirō wrote, because people were incapable of making rational judgments. Would, he wondered, "the incidents of misconduct that were too numerous to enumerate" have occurred if government authority had not evaporated and self-defense groups had been trained properly by "experienced officials" on how to maintain order during a state of national emergency? Suematsu believed that Japan's experience in 1923 raised many questions that future governments would have to answer.[82]

Not surprisingly, many military men asked similar questions. This was particularly true of those who, in the aftermath of the European War, had encouraged Japan's government to embrace civil and air defense training

FIGURE 2.7 *Jikeidan*: well-armed yet ill-trained vigilantes

Source: Nihon Tōkei Fukyūkai, *Teito fukkō jigyō taikan*, 1:5. Reproduced with permission of Tokyo Metropolitan Archives.

programs. In fact, many were quick to use the breakdown of order following the disaster to call for more thorough mobilization procedures for society. Future army minister General Ugaki Kazushige is a perfect case in point. In the years after the European War, Ugaki advocated the adoption of civil and air defense training for all urban-based Japanese. Prior to 1923, however, his efforts had garnered little political support. Many factors contributed to his civil and air defense advocacy. First, the European War demonstrated to the distinguished officer that civilians would be targeted in future conflicts. Moreover, the recent conflict illustrated to Ugaki that warfare was no longer a competition between armed forces exclusively but also a struggle of economies. Based on what had happened in September 1923, Ugaki predicted that if Japan's workforce panicked under enemy attack, Japan's economy and thus the ability to wage war would collapse quickly.

Two other interrelated developments also triggered Ugaki's push toward air and civil defense advocacy. The advent of new aviation and naval technology was one factor that convinced Ugaki that Japan would, for the first time in history, find itself prey to enemy attack during the next war. Land-based bombers flown from Siberia, coastal China, or the Aleutian

Islands of Alaska, he suggested, would target Japanese cities, destroy industrial centers, and wreak havoc on its home islands. How would Japan's civilians respond to this pressure? If nothing was done, he concluded, people would respond poorly to the trials of a future war. On top of this, Ugaki correctly understood that future carrier-based aircraft from the United States Navy could threaten any part of coastal Japan. If Japan's leaders were unwilling to move Japan's capital from Tokyo, Ugaki argued that civil and air defense training would have to be a cornerstone of the nation's home island defense.

Second, Ugaki believed that Japan was arguably more vulnerable to enemy attack in more ways than it had been at any time in its past. Japan's modern industrial cities—almost all of which fronted a navigable ocean approach—were densely populated, fire-prone landscapes that would allow an enemy to concentrate its firepower at just a few locations to maximize destruction. Finally, Ugaki confessed in his diary that city dwellers had grown increasingly weak, both physically and mentally, as they lived what he considered to be a degenerate and hedonistic urban life of capitalist consumption and decadence. All these factors made Ugaki nervous about Japan's future. Writing in his diary just five days after the Great Kantō Earthquake, Ugaki summarized his feelings as follows: "Chills run down my spine when I think that the next time Tokyo suffers a catastrophic fire and tragedy on this scale, it could come at the hands of an enemy air attack."[83] As Ugaki informed his colleagues six months later in an official report on the earthquake and fires, Japan must not let the "precious experiences" of the disaster be lost. Rather, he urged his colleagues to draw up contingency plans and mobilization programs based on their experiences of 1923 to capitalize on this catastrophe and nurture a culture of preparedness in Japan. Such a policy, he suggested, would help reduce the chance of Japan ever experiencing a similar breakdown of order, violence, and anarchy following its next emergency.[84]

While Japan's elites, and no doubt its earthquake victims, could find solace in the fact that few people died from disease following the calamity and that relief supplies and shelter eventually became plentiful, the 1923 disaster exposed not only Tokyo's modern vulnerabilities but also Japan's unpreparedness for catastrophe. The general lawlessness of the immediate postearthquake period and the trials of restoration and recovery convinced many elites that along with reconstructing the capital, Japan's people would likewise need training to prepare for an uncertain future where tragedy might come from nature, an enemy combatant, or both. In this endeavor, many elites understood that nurturing a culture of catastrophe and preparedness out of a national crisis could be beneficial if the end result was to help foster an era of national reconstruction and spiritual renewal.

In a report compiled by General Yamanashi and submitted to his Army Ministry colleagues in 1924, the general suggested that the best way to inculcate preparedness was to use the 1923 earthquake as an example. No other event, he argued, "had introduced the citizens of Tokyo to the harsh realities of war" better than the recent earthquake. The real experience of the earthquake calamity, he suggested, was more effective than photo exhibits, documentary movies, and mock air raid drills in highlighting both the horrors that modern war could inflict on civilians and the importance of preparedness. "The Army Ministry," he concluded, "should not let this opportunity pass by," and he therefore urged it to "spread concepts of national security amongst the population."[85]

To emphasize the enormity of Japan's catastrophe and to facilitate the eventual culture of renewal and reconstruction, many elites realized that people across Japan would have to be exposed to the dreadful realities of this calamity and accept the regional disaster as a national tragedy. Not surprisingly, the government and the media—though for different reasons—undertook concerted efforts to make people aware that the Great Kantō Earthquake was a traumatic, national disaster befitting a coordinated national response. The response required would be complex and far reaching: it was not just to provide for disaster sufferers in the short term but also to foster a longer-term project of national renovation. Success in both undertakings, many elites believed, would hinge on how well the government constructed this catastrophe as a national tragedy and disseminated it to the public.

Three

COMMUNICATION: CONSTRUCTING THE EARTHQUAKE AS A NATIONAL TRAGEDY

> *Looking up we hear the spirits of the dead wail; looking down, we see the earth mourn... who could not feel heartbreaking grief when mourning the dead and consoling their families.*
> —Yanagisawa Yasue, 1923

> *Since the Meiji Restoration, the Japanese people have not gone through any large-scale national trial. There is no doubt that this [earthquake] is an unprecedented national disaster.*
> —Hoashi Ri'ichirō, 1926

On October 19, 1923, over 200,000 people gathered at a spot near the eastern banks of the Sumida River in central Tokyo. Tearful spectators and solemn participants filled an area that had once comprised the Honjo Clothing Depot and an open expanse of land owned by the Yasuda family. For over four hours, people listened to speeches, gave offerings to the spirits of the dead, and at times wailed uncontrollably at the epicenter of Japan's earthquake calamity. Forty-nine days earlier, nearly forty thousand Tokyoites had suffered a ghastly, terror-filled death at this now hallowed site. Survivors, relatives of those who perished, politicians, religious leaders, journalists, and military men reflected on Japan's tragedy and mourned its victims. However painful the remembrance process may have been, virtually every speaker—almost to an individual a politician—evoked the dead and spoke about the sacrifices the deceased had made. They did so for larger political aims and objectives. Through this emotive and meticulously choreographed memorial service, the dead were not only eulogized but also tied to the future of Tokyo and Japan.

Amid the star-studded cast of eloquent speakers, the leader of Japan's Imperial Diet, Kasuya Gizō, and Horie Shōzaburō, chairman of the Tokyo Prefectural Assembly, were standouts. "Who could attend this ceremony or visit this site," Kasuya asked, "without having his heart torn to pieces?" If one was to judge by his audience that day, few could. Kasuya claimed that the people of Japan must "not just stand staring up at the heavens with tears rolling down their cheeks." Rather, he exclaimed, "what we must do is evident: unite our efforts to reconstruct the nation." The Seiyūkai parliamentarian from Saitama ended his speech by asking those present to join with him in "mourning the dead from the bottom of our hearts." He urged all Japanese, moreover, to "do our best" in the reconstruction project because only by doing so, he concluded, can "we console the spirits of the dead." Horie, a Tokyo-based prefectural politician, articulated many of the same themes but went further in employing the dead and their memory for terrestrial uses. Horie beckoned the spirits of the dead "to come to us and accept our vow to reconstruct our imperial capital." "Protected by the spirits of the heroic victims," the assemblyman predicted that Japan would surmount all obstacles that might arise in the subsequent reconstruction process. Thus, he claimed, "the sacrifice victims made would be remunerated in the form of a new imperial capital." In accomplishing such a feat, Horie concluded that the "precious lives of our fellow citizens will not have been wasted." Both speakers concluded their orations to the sound of thunderous applause.[1]

Kasuya and Horie were just two of many who described the Great Kantō Earthquake as an unprecedented, civilization-upturning, national tragedy and employed it to secure larger political aims and objectives. Using speeches and ceremonial events, colorful lithograph prints and pictorial postcards, harrowing survivor accounts that often lionized selfless behavior, tangible objects taken from the epicenter of destruction, images and evocations of the dead, and documentary movies, government officials and other institutions constructed the Great Kantō Earthquake as an unprecedented national calamity. Often the analogy of war was employed in describing the carnage that was visited on Tokyo to relay both the degree of loss and the challenges that many elites believed the nation faced. Moreover, as with previous wartime experiences, themes of sacrifice—by those that had died and from people across the nation—were often emphasized to encourage a humanitarian response unrivaled in Japan's history. Leaders understood that people across Japan would have to see what was in effect a regional catastrophe as a national tragedy that would necessitate a long and expensive program of reconstruction and national renewal.

CONSTRUCTING CATASTROPHE THROUGH NEWSPAPERS

Even before government officials undertook a concerted effort to construct the earthquake calamity as a national tragedy to elicit humanitarian responses, newspapers had already introduced the disaster to people across Japan in time-tested and dramatic new ways. Often they did so to boost circulation. This was particularly true of the largest broadsheets, which aspired to the mantle of Japan's leading national paper. In this regard the *Tokyo nichi nichi shinbun*, one of the three surviving Tokyo papers, took a lead role. Throughout September its coverage was best known for sensationalism. Indeed, this newspaper helped construct the disaster as a traumatic experience of epic proportion through emotive headlines, stories of hardship and suffering, and vivid pictures of death and destruction. Eventually it also told a tale of great sacrifice and glorious kindness exhibited by the sheer volume of aid and relief supplies donated not only by Japanese but also by foreign governments. The paper went to great lengths to inform readers across Japan that the earthquake had killed many, destroyed much of Tokyo, and left survivors without food or water. In short, it had turned Tokyo into a wasteland in which inhabitants were in want of everything imaginable.[2]

The *Tokyo nichi nichi shinbun* also used the disaster to promote itself as Japan's leading national newspaper. With little modesty, its editors proclaimed that the newspaper possessed something that no other Tokyo paper did: motion pictures (*katsudō shashin*) of the calamitous September events. On September 8 the paper informed readers throughout Japan that its staff persevered despite seemingly insurmountable obstacles in order to give all Japanese the chance to see the disaster as it actually unfolded and to share in the event in a unique way through motion pictures. Just over a week after the disaster, it announced that movies recounting five memorable aspects were ready for national distribution, including *Prime Minister Yamamoto's Activities*, *Disturbance at the Suikōsha*, *Tokyo on Fire*, *Hardships Faced by Survivors*, and *The Relief Effort*. The *Tokyo nichi nichi shinbun*, the article concluded, was now taking applications from community groups and local associations throughout Japan to sponsor distribution of the documentaries.[3]

Japan's largest national newspapers, the *Osaka asahi shinbun* and *Osaka mainichi shinbun*, likewise framed the catastrophe through emotive media that many smaller papers could not afford, namely, the publication of extensive photographs and the distribution of documentary motion pictures throughout Japan. The *Osaka asahi shinbun* used the entire first page of its September 4 earthquake extra to invite readers to engage with what the paper described as the most engrossing depictions of the disaster yet published. With no little bravado, the paper published a detailed survivor account that also served to launch the paper's disaster photograph series. The survivor, Fukuma Genzō, was well placed to personalize the disaster, describe it in dramatic and skilful ways, and build excitement over the photos that were published the following day. Fukuma was an acclaimed *Tokyo asahi shinbun* journalist who, along with three photographers, recorded their struggles to escape Tokyo as it became hell on earth. Fukuma detailed in emotive prose the three-day struggle for survival that he, his colleagues, and virtually everyone in Tokyo faced, all without food or sleep. However powerful his descriptions were, Fukuma suggested that no words could compare to the images about to be published, images that the newspaper proudly proclaimed had been brought to Osaka by express courier. Introducing these images billed as the first representations of the disaster to be shown to the people of the nation, Fukuma issued a cautionary preamble. He warned readers that the images were not for the fainthearted. Japan's imperial capital no longer resembled a city but rather a battlefield (*senjō no yō na teito*).[4]

The publicity that the *Osaka asahi shinbun* gave to Fukuma's ordeal was not done entirely to increase sales but also to increase giving. Beginning in the September 5 edition and running for six days, the newspaper artfully intermixed charity and compassion with destruction and suffering. Each day it published a full-page pullout with Fukuma's pictures from the disaster zone on the front side. On the reverse the broadsheet listed the names of individuals, corporations, and associations that had contributed donations to the relief fund. The Tōyō bōseki kabushiki gaisha (Far Eastern Spinning Corporation), for example, gained top honors in the September 5 edition with a ¥20,000 donation, while an anonymous individual rounded off the donors with a solitary ¥1 offering.[5] In a symbolic act that demonstrated that the disaster must bring people together, both the *Osaka asahi shinbun* and the *Osaka mainichi shinbun* agreed on September 2 to work together to undertake the largest relief collection campaign in Japan.[6]

Though the owners of both Osaka newspapers agreed to work together for the cause of charity, the editorial team from the *Osaka mainichi shinbun* was determined not to be outdone by its rival in relation to disaster coverage. Beginning on September 3 and repeated on four different evenings until September 13, this paper brought the earthquake catastrophe to residents of central Japan through moving images.[7] The editors heralded their motion pictures as the most accurate, up-to-date, and heart-wrenching accounts of Japan's worst national calamity in history. Building anticipation, the broadsheet claimed that no viewer would be disappointed and that everyone who witnessed the films would demand a second screening. The full advertisement read:

> Quake Movie Shown Tonight
> Place: Okurayama Park, Kōbe
> Time: 7 p.m. September 12 & 13
> Films: *Burning of Tokyo*, *Tokyo after the Fire*, No. 2 Series, *Yokohama after the Earthquake*. These films, especially that of burning Tokyo, were taken by the Mainichi's movie corps at a risk of their lives and give vivid scenes of the catastrophe. None better yet shown on screen. Come early.[8]

Days earlier the newspaper reported that over twenty thousand people had crowded Nakanoshino Park in Osaka on September 10 to view the films in question. Demand was so strong, in fact, that the paper was forced to schedule additional showings just as its editors had predicted. Eventually eight different sessions from 7 p.m. to 11 p.m. were held. "Spectators," the newspaper reported in the September 13 edition, had "breathlessly watched the films," and more than a few had broken into tears during scenes showing devastated Yokohama.[9] When the series ended, moviegoers were given three simple instructions: (1) to leave the park from the Tenjin-bashi exit in an orderly fashion so that those waiting to view the films could enter; (2) to consider donating to the relief effort even if they had already done so; and (3) to keep abreast of the relief, recovery, and reconstruction efforts by reading the *Osaka mainichi shinbun*. Thereafter the films went to Kobe for a two-night showing before traveling to numerous regional locations in western Japan. The earthquake calamity had quickly become a national, visual phenomenon. Eventually its reach spread overseas. Owing to sponsorship by the Chinese Young

Men's Association and the Chinese Student's Federation, films from the *Osaka mainichi shinbun* even played to an audience of over two thousand in Shanghai on October 4.[10]

Virtually every paper across Japan encouraged people to give generously in response to the unfolding crisis. In its first postearthquake edition and in many others after, the *Tokyo nichi nichi shinbun* called on all brethren—*dōhō*—to give aid in response to the unprecedented national calamity.[11] Many smaller circulation papers implied that it was every citizen's obligation to give in support of disaster relief. The *Kyoto hinode shinbun* proudly informed its readers that it would join forces with its archrival, the *Kyoto nichi nichi shinbun*, and collect funds and donations for the relief effort.[12] On September 6 the Hokkaidō-based *Otaru shinbun* declared that supplying and distributing food was the number one issue not only for disaster sufferers but also for the nation of Japan.[13] The Kyushu-based *Fukuoka nichi nichi shinbun* asked its readers to give without "letting a single day pass." Moreover, it implored readers to consider the fact that ¥1 could provide 6 *go* (900 grams) of rice to sufferers in eastern Japan who had been left with nothing. To emphasize the issue of need, this paper referred to the nation's capital not as Tokyo but as a desolate landscape that had been reduced to its former bleak state of sixteenth-century Musashino. Everyone, the paper suggested, could give or spare something in the face of such clear and present need.[14]

Did such emotive coverage of the disaster have any impact on circulation figures or donations? As mentioned in chapter 1, major national newspapers received a significant boost in circulation numbers between 1923 and 1924.[15] Apart from increasing sales, it is nearly impossible to pinpoint a direct link between the way the disaster was packaged and the amount of money donated in response. A multitude of reasons shaped why people donated money and supplies or chose not to give. One fact cannot be denied, however: in the early 1920s newspapers were the chief form by which people gained knowledge of current events in Japan. Had newspapers not gone to such lengths to construct and relay this tragedy, far less aid would likely have been donated in the first weeks following the disaster.

It is worth documenting how much aid was donated, what forms aid took, and from where aid originated. In examining these aspects of the overall relief effort, it is possible to discern a number of patterns associated with giving. Aid took the form of relief goods, reconstruction supplies, and cash. Initially the Relief Bureau welcomed supplies such as blankets, clothing, bedding, food, and medical supplies. Though they continued to accept salted fish from Hokkaidō, pickled plums from Aiichi, and rice from any prefecture, officials in Tokyo asked regional elites on September 18 to send all future donations in cash. The total monetary value of supplies and cash given was sizable. Statistics compiled and published for the Imperial Capital Reconstruction Exhibition in March 1930 listed the total valuation of goods and cash donated at ¥74,471,360.[16] This equaled nearly 6 percent of Japan's overall national budget in 1922. The monetary valuation of goods donated totaled ¥20,011,000, while cash contributions totaled ¥43,874,160. The remaining ¥10,586,200 comprised a combination of relief supplies and cash donated by the imperial family.

Donations came from every prefecture in Japan and from every one of Japan's formal colonies. As with collections taken during Japan's previous wars, donations were accepted at all prefectural offices, schools, police stations, town halls, and offices of local and regional newspapers. In looking more closely at where donations came from within Japan, a couple of trends are obvious. First, people from the wealthiest urban centers were the most generous. Near the epicenter of destruction, people from the undamaged parts of Tokyo Prefecture donated most generously, with ¥16,891,545 contributed to the relief and recovery fund. Following Tokyo Prefecture were Osaka Prefecture (¥4,855,315), Hyōgo Prefecture (¥2,929,962), and Kyoto Prefecture (¥2,304,101). Not surprisingly, the poorest and often least urbanized regions gave far less: Akita Prefecture in northwestern Japan donated the third smallest amount at ¥245,657 while Okinawa provided ¥98,418 in relief. The prefecture that donated the smallest total amount was rural Yamanashi, which itself had suffered damage from the earthquake. Its residents gave a mere ¥37,840. In terms of Japan's colonial territories, a similar pattern existed. The wealthy, urban territory that comprised the Kwantung Leasehold in Manchuria provided the most generous assistance, with ¥2,266,870, while the sparsely populated tropical island territories of the Nan'yō provided ¥16,942.

One final trend associated with aid is apparent. Prefectures closest to the disaster area gave a much higher percentage of goods and supplies than those further afield. Ibaraki, Chiba, Saitama, and Yamanashi, which all bordered or were within two prefectures distance from Tokyo, gave the majority of their donations in goods. People from more distant prefectures of western Japan, such as Hiroshima, Yamaguchi, Fukuoka, Nagasaki, Kumamoto, and Miyazaki, donated more money than goods. The exceptions were the timber-rich prefectures of northern Japan, such as

Aomori, Iwate, Akita, Yamagata, and Miyagi, which donated more in goods and reconstruction supplies than money. These prefectures provided the lion's share of timber needed to rebuild temporary bridges, docks, and housing in disaster-torn Tokyo. The total amount of aid donated following the Great Kantō Earthquake was the largest amount given for any humanitarian undertaking in pre–World War II Japan.

Nagata Hidejirō, Tokyo's mayor, was cognizant of this fact. As a small token of gratitude on behalf of the city he represented, Nagata visited Japan's six largest cities to thank politicians and the people directly for their generosity. Beginning in Osaka, Nagata's trip proved to be a reflective experience that stirred different emotions for the exhausted mayor. Nagata informed his hosts that the aid given by Osaka was extraordinary and that it represented the best qualities of the Japanese spirit. Sheepishly, however, Nagata admitted that Osaka's vitality "somehow offended him."[17] It made Nagata angry "seeing all the houses in Osaka standing without a single crack," all the while knowing that Tokyo was still a wasteland. He recounted in 1930 that there was "no logic" to how he felt, especially since the people of Osaka "had taken such great care of us," but, he reiterated, that was how he felt. Nagata attempted to soothe his undiplomatic frankness by stating that Tokyo would stand with Osaka in solidarity if it ever found itself destroyed. Nothing else could ever come close, he concluded, to repaying Tokyo's debt of gratitude.

STORYTELLING: TALES OF HOPE AND HEROISM

Though popular and novel, newspapers and films were not the only techniques used by journalists or governing elites to engage the citizens of Japan and introduce them to the disaster. In describing and constructing the calamity, select government agencies and publishers went to great lengths to publicize ordinary Japanese who undertook heroic acts in the face of tragedy. The glorification of their actions was meant to encourage all Japanese to behave in similarly selfless ways. The stories thus emphasized virtues that the government expected from the people at this moment of crisis, including sacrifice, diligence, filial piety, and charity. The Tokyo municipal government exerted considerable effort to create and exalt everyday heroes amid the tragedy. In many regards, uplifting, beautiful stories from the areas of destruction transformed narratives that often focused on death, dislocation, and devastation into tales that celebrated the values and actions of ideal subjects acting in exemplary, if not scripted, ways for the betterment of society. If sufferers could exhibit charity, loyalty, courage, and civic-mindedness amid chaos and destruction, could not all Japanese behave in similarly glorious ways in ordinary times?

City officials believed so and published a major compilation of beautiful tales in a 774-page book entitled *Taishō shinsai biseki* (Beautiful outcomes from the Taishō earthquake).[18] Written or selected by seventeen primary school principals and teachers, the tales were classified under headings such as people who rescued others, those who prevented and extinguished fires, individuals who acted in a responsible manner, those who demonstrated compassion, and those who provided relief. Some individuals were lauded for escaping tragedy because they had taken precautions and followed simple practices of disaster preparedness. Others from outside Tokyo were lauded for providing postdisaster material as well as psychological support, thus acting out of a sense of national duty and civic responsibility. Apart from serving as a positive counterpart to many other survivor accounts that focused on death and destruction, these tales were meant to instill hope and compel recognition. Just as important, these heroic stories provide clues as to how many government officials hoped all Japanese would continue to act during the extended recovery and reconstruction phase: selflessly, responsibly, loyally, and up to society's heightened expectations.

Many of the tales mirrored beautiful tales published by newspapers throughout the autumn of 1923. By the middle part of October, the *Yomiuri shinbun* published stories of beautiful deeds (*bidan*) that its reporters had purportedly witnessed amid the ruins of Tokyo. One moving account documented an unnamed Korean couple who adopted a Japanese earthquake orphan and "nurtured the child like it was their own." As the paper pointed out, this action illustrated a selfless act of humanistic devotion and civic-mindedness.[19] Another detailed the heroic exploits of three nurses who, with little thought of their own individual suffering or lack of proper medical supplies, nevertheless assisted disaster victims who resided in makeshift abodes or temporary barracks in devastated Tokyo.[20] Their tale was one of self-sacrifice and civic duty. A final beautiful tale centered on a retired teacher who had been made destitute by the earthquake disaster. Rather than focus on his plight, however, the *bidan* documented the remarkable spirit of generosity and filial piety displayed by an anonymous former student who, upon hearing of his former master's plight, sent seven bags of rice to sustain him through the period of recovery.[21] As the editors of this paper suggested, this was a story that highlighted—far more than simple charity—loyalty, reverence, and filial piety.

Not to be outdone by newspapers or local government, Japan's Ministry of Education took even greater steps to construct heroes from the disaster. It did so to champion past actions and to foster future ideal subjects. Bureaucrats sent Boy Scouts into parts of Tokyo and Yokohama after the disaster to collect stories of exemplary behavior amid the tragic backdrop of destruction and loss.[22] Similar to the way in which previous governments constructed war heroes for acts of valor, sacrifice, or loyalty, Education Ministry officials lionized many nameless Japanese in official publications for selfless acts of which the nation could be proud. Stories praised individuals for assisting disaster sufferers and thus exhibiting the virtue of charity. Others gained recognition for saving portraits of the emperor from burning or collapsed buildings, thus demonstrating profound loyalty. Many stories revolved around civic-minded citizens who undertook acts of self-sacrifice for the betterment of others and society at large, such as adopting disaster orphans.

What made these stories unique was their targeted dissemination and use. These tales of ethical and ideal behavior were published in three volumes and sent to every school across Japan in November 1923 under the title *Shinsai ni kansuru kyōiku shiryō* (Education materials related to the earthquake).[23] Once at school, they were used as supplementary materials in existing moral education lessons. Students were told to reflect on and embody the thoughts and actions of Japan's earthquake heroes in daily life as models of human behavior.

When pressed on the floor of the Imperial Diet by representative Uehara Etsujirō over what initiatives the state had undertaken to use the disaster to educate the nation and to assist with its spiritual reconstruction, Minister of Education Okano Keijirō gave a spirited reply. The first policy he mentioned was the publication and dissemination of these written materials. Okano went further than this, however, arguing that not only had the government "published education materials related to the earthquake" for classroom use, it had also tapped into the extraordinary rich supply of visual images and material objects from the disaster and was now sharing these with the people of Japan. Moreover, Okano highlighted a number of public lectures, films, and traveling exhibits that the government had created to inform and educate the people of Japan.

"The outcome [of these initiatives] to date," he claimed, "has been good." He predicted that they would continue to "to produce worthy results well into the future."[24] Whether these undertakings produced "worthy results" is hard, if not impossible, to judge. Many efforts alluded to by Okano proved to stir interest throughout Japan, including a Ministry of Education–sponsored national tour that Kurushima Takehiko led between October 10 and December 1, 1923.

Kurushima's arrival in Kyoto on October 10, 1923, created a media sensation.[25] As one of Japan's best-known authors of children's literature, Kurushima was no stranger to popular stardom. The spirited reception he received in Japan's former capital was greater than even he had expected. The throngs of people who gathered at the first stop on his well-publicized tour, however, were not necessarily drawn to Kurushima for his literary accomplishments alone. Rather, they came to learn about the earthquake disaster that had befallen eastern Japan. More than just learn, they came to experience, in some small, safe way, the enormity of destruction and the severity of loss suffered. The road show traveled to thirty-nine other cities over the following two months. With assistance from government bureaucrats and aided by graphic documentary movies, colorful lithographic prints, emotive postcards, harrowing survivor accounts, tales of exemplary behavior in the face of disaster, and tangible objects from the epicenter, Kurushima crafted an unforgettable story of calamity. In doing so, he hoped to construct the 1923 disaster as Japan's greatest national calamity of the modern era.

While pretour publicity focused on the five reels of documentaries that comprised 3,700 feet of celluloid, audiences had a bevy of material from which to learn about the disaster. These included numerous "beautiful tales" of heroes that Kurushima narrated to spectators. Beyond this, however, the tour provided many citizens throughout Japan with the chance not only to hear stories and see films but also to view and hold objects from the earthquake's epicenter. Included in the government's display were a charred bicycle; a twisted piece of corrugated iron; fire-damaged watches, jewelry, and spectacle frames; fountain pens and glass bottles melted into shapes reminiscent of objects trapped in a Salvador Dali painting; and mounds of coins fused together by the intense heat from fires that took so many lives. Engaging with visual and material items brought the catastrophe to people of Japan in a way reminiscent of how the Imperial Japanese Navy relayed stories of and displayed objects from the Russo-Japanese War in 1905.[26] That many of the earthquake artifacts and images were eventually housed in the Taishō Earthquake Memorial Hall points to the status or importance government officials placed on the educational and memorial value of such items.[27] In a pretelevision age, Kurushima's well-orchestrated, visually impressive, and tactile exhibit served as a conduit of well-calibrated information about the disaster and what would be expected of all Japanese people in the larger task of reconstruction.

Not surprisingly, the skilled orator and his team of officials took great pains to thank Japanese well removed from the disaster zone for their acts of sympathy, generosity, and hospitality following Japan's unprecedented catastrophe. Lists, charts, and hastily bound books detailing the donations that individuals, businesses, charitable organizations, and local and prefectural governments had given to the relief effort were showcased and lauded as acts that revealed the true selfless spirit of Japanese. Officials, however, were quick to inform audiences that however generous past support had been, more would still be needed. They reminded listeners that securing donations for the task of reconstruction would be more difficult as stories from the rebuilding projects would be mundane in comparison to the sensational nature of the disaster itself. Moreover, they suggested that while the disaster had been a single, civilization-upturning event, reconstruction would be a drawn-out process that would recede as the central point of everyone's attention. Such frankness revealed much of what his aims and objectives had been in the autumn of 1923. Put simply, Kurushima and his retinue traveled around Japan to disseminate news and to construct the Great Kantō Earthquake and subsequent reconstruction process as events of grave national significance. They also took great pains to remind spectators that continued recovery, relief, and reconstruction would require the united efforts of all Japanese people. Organizers of this tour saw a unique opportunity to fashion a national catastrophe and employ it to secure larger political and economic goals. The Ministry of Education was just one organization that understood the malleability of calamity and thus the pregnant possibilities of its artful manipulation.

CRAFTING CATASTROPHE THROUGH VISUAL CULTURE

Individuals, organizations, and various governmental agencies used other media to relay many aspects of the Great Kantō Earthquake tragedy to people around the nation. Created and disseminated for a multitude of reasons, artwork, postcards, and other emotive images of the tragedy helped create a rich and varied visual culture of catastrophe in 1923. Earthquake-related visual material was produced both by recognized artists and by anonymous individuals who wished to capture a specific aspect of the unfolding tragedy or its aftermath. Some items were created purposely with the intent of enriching their creator monetarily, while other items were forged simply to record some part of the catastrophe for posterity. Though most visual images were created with a domestic audience in mind, postcards with captions translated into rudimentary English were crafted by individuals clearly cognizant of a larger market beyond those fluent in Japanese. Government agencies, newspapers, and other nongovernmental institutions created or sanctioned many visual representations of the disaster, while others, such as postcards of dead bodies, were classified as contraband. Regardless of who produced these materials or why, they provide a remarkable and often remarkably emotive picture of tragedy.

Lithographic prints (*sekiban*) depicting the catastrophe were a medium through which people across Japan learned of the disaster. Mass produced in quick fashion in the days and weeks following the disaster, these prints are striking both for the vivid colors employed and for the scenes and subjects portrayed.[28] Many compelled viewers to engage with the scale and severity of Japan's earthquake calamity in ways that published survivor accounts and beautiful tales could not: they stirred emotions through sight. One of the most famous lithograph prints, entitled *Honjo kōhan dai senpū no shinkei* (View of the great whirlwind approaching Honjo), from the series *Teito daishinsai gahō* (Images of the great imperial capital earthquake) portrays the whirlwind firestorms that converged on the grounds of the Honjo Clothing Depot on the night of September 1. The central, unmistakable feature of this print is a graphic representation of a golden yellow-orange, tornado-like whirlwind that helped ignite the grounds of the Clothing Depot. Closer inspection of the print reveals that it is a visual representation of numerous aspects of that fateful night. The tribulations of a night spent in a vessel on the crowded Sumida River, as Funaki Yoshie so harrowingly relayed to readers, are brought to the viewer. Immortalized too, in graphic detail, are the panic-stricken people trapped on the fiery bridge, likely Kuramae Bridge, as Funaki Yoshie and Kawatake Shigetoshi both recounted. While highlighting what was destroyed and lost in the fires, this print also shows, through two inserted picture boxes,

FIGURE 3.1 The whirlwind firestorm that swept through the Honjo Clothing Depot
Source: Lithograph print, author's private collection.

what was left; namely, the remains of humans, both ash and bone mounds of those who had been killed, who now must be mourned and remembered as tragic victims. This print reflects and reinforces many survivor accounts that would become part of the lasting history of this catastrophe.

Another striking lithograph print that documents the sheer pandemonium is entitled *Gekishin to mōka ni osowareshi Ueno Hirokōji Matsuzakaya fukin no shinkei* (Scene of the neighborhood around Matsuzakaya being ravaged by fires from the severe earthquake). In essence, this print captures the terror that many urbanites faced while attempting to flee the fires in the Ueno Hirokōji district near the department store, Matsuzakaya. One thing that makes this print novel and memorable is the emphasis the artist placed on how each major form of transportation available to Tokyoites is shown to be virtually useless. The picture depicts a stopped tramcar engulfed in flames; a motorcar partially overturned and suffering from a broken axel; a rickshaw likewise top-sized with the belongings of its owner in flames; a horse-drawn cart about to crash as its horse rears up on two legs with the approach of flames; and a solitary, black, abandoned bicycle with a bent wheel and frame. Only the most primitive

92 – COMMUNICATION

FIGURE 3.2 The neighborhood around Matsuzakaya Department Store in Ueno being ravaged by fires following the earthquake

Source: Lithograph print, author's private collection.

form of transport—walking—is shown to be anywhere near effective in enabling people to flee the burning capital.

Even fleeing on one's own volition, however, is an act portrayed as fraught with life-threatening danger. What most menaces those trying to escape is the encroaching fire, epitomized by the image of the large Matsuzakaya Department Store building consumed by fires and spreading flames to other parts of the neighborhood. People too, in their selfish acts of fleeing with their belongings—thus clogging already congested streets—are likewise portrayed as hindering the escape of panicked crowds. In short, this print relayed the trauma of escape, the perils of selfish attachment to material belongings, and the feebleness of Tokyo's transportation infrastructure in the face of catastrophe.

There could well be a deeper, more political message behind this print. The print depicts government officials dressed in white uniforms attempting to direct panicked crowds. They are clearly overmatched by the situation, and their efforts to impose any sense of order are futile. More striking, however, is the fact that people are ignoring these officials, implying that the government had lost complete control over the disaster zone. On one level, this picture is a manifestation of the fears that many elites held about 1920s society: people had lost their respect for authority. The burning backdrop of Matsuzakaya Department Store is also full of political symbolism. Many elites had become alarmed over the rapid increase in luxury-oriented consumption from 1917 onward. The burning store symbolizes in a graphic way a message that many commentators attempted to drum into the minds of all Japanese after the disaster: luxuries and high-end consumer goods that were often sold in leading department stores were ephemeral and ultimately harmful to individuals and the nation. Finally, the burning tramcar not only points to the ineffectiveness of Tokyo's modern transport system but takes viewers back to the tumultuous Hibiya Riots of 1905, the last time martial law was declared over the capital to suppress protests against the government. As Andrew Gordon has documented, tramcars became popular targets of urban protestors and rioters not only in 1905 but also in the years between 1903 and 1923.[29]

A final lithographic print worth discussing depicted an all too familiar sight that many Tokyo residents eventually witnessed: widespread devastation of the urban landscape. The print in question is entitled *Ueno kōen ōundai oite denka saigaichi o shisatsu no zu* (Illustration of the imperial tour that inspected the disastrous ruins from a lookout at Ueno Park). Mass produced from September 17 onward, two days after the Crown Prince and Regent Hirohito conducted one of three official inspection tours of the disaster zone, this lithograph compels readers to confront the totality of destruction brought about by the earthquake and fires. Though Hirohito and his horse-mounted entourage are depicted in vivid colors and excellent detail, all that is before them—the east-southeastern part of Tokyo—is a barren, still smoldering wasteland, resembling the emotive description provided by novelist Uno Kōji. Implying that the destruction went further than the eyes could see, as Uno indeed suggested, the crown prince holds a pair of binoculars with which to survey the magnitude of devastation. Little before the eyes of the imperial tour is recognizable, except the outline of the damaged twelve-story tower of Asakusa and the Kannon temple, now referred to as Sensōji. As with the Honjo whirlwind print, this print of destroyed Tokyo also reminded viewers of what, or who, was left as well as depicting in graphic detail what was lost. Behind Hirohito and his party stands the larger-than-life statue of

FIGURE 3.3 The imperial tour that inspected the ruins of Tokyo from a lookout at Ueno Park

Source: Lithograph print, author's private collection.

Saigō Takamori plastered with missing persons' notices that continued to adorn Saigō well into 1924.

Apart from conveying the totality of physical destruction Tokyo suffered, this print also illustrated something that government officials were keen to publicize: that a concerned monarch and his government ministers and military were aware of the destruction. Moreover, it implied that they were also involved in or engaged with the recovery process. Hirohito is portrayed as a benevolent representative of the royal family attentive to the trauma that has been meted out on the capital. The army reconnaissance airplane circling overhead alludes to the fact that more than just the crown prince and his entourage are aware of and responding to what has transpired. Finally, when compared with the previous print depicting the ineffectualness of government representatives, the print documents that order has been restored to the capital and that the military has asserted control over what was once a landscape of chaos and anarchy.

The image of the September 15 imperial inspection tour as well as various pictures and representations of the destruction delivered on Tokyo were also mass produced in smaller, cheaper, more easily distributable forms, most important as postcards. Mirroring postcards produced during and after the Russo-Japanese War, newspapers, private companies, individuals, and eventually city and national governmental agencies printed and distributed earthquake disaster cards between 1923 and 1930. These cards became a key medium through which government agencies and private interests disseminated news of the calamity and progress of the relief and reconstruction efforts throughout Japan up to 1930.

Many postcards were published in series that began with images of important buildings, landmarks, or famous scenes and places in Tokyo before the catastrophe. After these images, the postcard series often included one or two cards that portrayed scores of panic-stricken individuals attempting to flee an encroaching conflagration. Often producers enhanced pictures of approaching fires with brightly colored ink to dramatize their subject matter. Other cards focused on scenes of destruction, the plight of refugees, the construction of barrack housing, and other newsworthy events from postdisaster Tokyo. These included the torrential downpour of September 24 that flooded barracks throughout Honjo and Fukagawa, the opening of Tokyo City's first employment-matching agency, and the capture of the elephant that escaped the circus at Hanayashiki and survived the fires. Often designers artfully incorporated before-and-after scenes of buildings and important places in Tokyo in one impressive postcard. One could receive an almost complete pictorial history of the Great Kantō Earthquake and reconstruction of Tokyo from the postcard series produced.

Postcards also provided people with confronting glimpses into something that no lithograph print and only a few newspapers did, namely, images of the tens of thousands of bodies that littered Tokyo. There were so many different postcards published that featured the dead, one might ask whether they were really representative of the trauma Japan suffered. Or were they just a sensationalized, macabre medium that people sought, collected, and distributed because they were so unusual compared to any postcards printed before in Japan? Their publication and distribution across Japan became so widespread that government officials attempted, in a very selective way, to curb their sale and circulation, particularly to foreigners. The Kobe-based *Japan Weekly Chronicle* reported in its November 8 issue that "a pedlar [sic] named Kaibara Shotarō (39)" was arrested for selling postcards of dead bodies to passengers boarding the steamship *St. Albans* on October 25 shortly before it departed Japan.[30] The cards

FIGURE 3.4 The crown prince touring devastated Tokyo

Source: Postcard, author's private collection.

in question, the newspaper reported, illustrated "scenes of the holocaust at the Honjo Military Clothing Depot" and of "the corpses of Yoshiwara brothel girls." A week later the paper reported that police in Toriizaka had arrested "over forty persons... including a peer and a university student" and "charged [them] with selling forbidden photographs concerning the earthquake."[31] The editors questioned why police would waste their time with such pursuits given the fact that postcard and photo hawkers were everywhere and that a number of gruesome pictures had already been published in leading newspaper. Some of the first photos newspapers published of the catastrophe in fact were of dead bodies.[32] These arrests aside, serious attempts to confiscate the cards and jail or fine their sellers proved futile. They, like the bodies they depicted, littered the landscape of postdisaster Japan.

Graphic pictures of the dead reduced the tragedy of the Great Kantō Earthquake to the most basic, comprehensible level: human beings. Postcard images of the dead humanized the seemingly incomprehensible scale of devastation meted out across Tokyo in a brutally confronting way. Many of the cards captured all too common scenes across the disaster zone: blackened, swollen, and often crushed or mutilated bodies among the remains of buildings, bridges, or waterways.

Other postcards featured dead bodies at recognizable buildings or landmarks such as the statue of the Russo-Japanese War hero Hirose Takeo at Manseibashi Station or the Yokohama Specie Bank. Many different, yet all equally graphic, cards were produced that illustrated the horrific scene of hundreds of dead prostitutes from the Yoshiwara brothel district; many of these women were trapped and fated to a terrorizing death because their "employer-owners" refused them freedom to flee the approaching fires.

Few postcards were more powerful than those that depicted scenes of the Honjo Clothing Depot. While various writers, chroniclers, journalists, and political elites described the scene at this site as too ghastly and too horrendous for words to capture, picture postcards filled the void. By examining a number of postcards from different series, one can get an almost complete account of the dreadful occurrence and aftermath. One rare postcard still available today illustrates a city official directing fleeing refugees, many with belongings such as chests of drawers, bicycles, futons, and carts, to this vast expanse of open land. Another card depicts the packed site as the fires approached from three sides.

FIGURE 3.5 Children selling maps and postcards in postdisaster Tokyo

Source: Kaizōsha, ed., *Taishō daishinkasai shi*, plate, 85. Reproduced with permission of Tokyo Metropolitan Archives.

FIGURE 3.6 Dead bodies in the ruins of Nihonbashi

Source: Postcard, author's private collection.

FIGURE 3.7 The bodies of dead prostitutes from Yoshiwara

Source: Postcard, author's private collection.

FIGURE 3.8 City official directing citizens to the site of the Honjo Clothing Depot

Source: Postcard, author's private collection.

FIGURE 3.9 The site of the Honjo Clothing Depot packed with people, September 1

Source: Postcard, author's private collection.

COMMUNICATION – 101

(First) A number about three million a dead body of a clothing department in Tokyo.
(其一部) 東京被服廠跡々累死體其實數三万

FIGURE 3.10 Dead bodies at the site of the Honjo Clothing Depot on September 2

Source: Postcard, author's private collection.

Most, however, introduced people across Japan to the grim scene of misery and death that confronted photographers on the morning of September 2. An examination of these somber images lends strength to the assertion made by the *Japan Weekly Mail* that the Honjo Clothing Depot was the site of Japan's worst holocaust on record. These postcards demonstrated to all who viewed them that apart from devastating large tracts of Tokyo and Yokohama's urban landscape, the disaster also killed—often in brutal ways—humans on a scale larger than anything Japan had suffered in living memory.

THE HONJO CLOTHING DEPOT: EPICENTER OF DEATH AND REMEMBRANCE

Through sheer numbers and an unmistakable odor that many claimed blanketed the city, the dead became an inescapable feature of this devastated metropolis. Virtually every published survivor account reported in graphic detail on at least one aspect of the dead that remained in Tokyo. Tanaka Kōtarō recalled that bodies were strewn everywhere on both sides of what remained of Ryōgoku Bridge, along the banks of the Sumida River, and at the site of the burned down Kokugikan Sumō Arena.[33] Another survivor reported that Manseibashi Station resembled a morgue.[34] On a more personal level, Kawatake Shigetoshi shared his account of going from pile to pile, looking for the remains of his wife, her companion, and their two children. After he had located what he believed were their bodies in a group of charred corpses stacked alongside Tatekawa Street, he began to collect teeth from each of his loved ones to serve as a relic. Only when he opened the mouth of what he believed was his daughter did he realize that this was not his family: his daughter had lost her front teeth a week before the disaster. Experiencing both joy that these bodies were not those of his loved ones and revulsion for taking teeth from unknown victims, Kawatake decided that he would no longer search any other corpses.[35]

Initially Tokyo municipal and police authorities hoped that survivors, like Kawatake, would indeed search mounds of corpses near where their homes once stood in hopes of finding, identifying, and disposing of their family members' remains. This proved misguided on two counts. First, the chaotic movement of people had been so widespread on the night of the disaster that there was only a small probability that people would have died near where they once lived. Second, as Kawatake and other survivors reported, many of the bodies left behind were burned, crushed, or trampled beyond recognition, making identification impossible. The difficulties associated in dealing with so many dead bodies in poor condition under such trying circumstances was highlighted in Soeda's earthquake ballad. Three verses read:

> Look at the ruins of the fires.
> The hills of dead bodies of tens of thousands of our countrymen—
> The sight is too terrible to look at.
> It is more hellish than hell.

> It is impossible to tell from their faces,
> If they are brothers or parents and children,
> Or if they are men or women.
> Many of them are inflamed, and many are swollen.

> Heads are crushed; Bellies are torn;
> Bones are broken; Innards are pouring out.

The strong stench assails our nostrils.
The bodies are piled up as they are found.[36]

Before governing elites evoked the memory of the dead and their "sacrifice" in an attempt to secure political support for reconstruction, the bodies that existed in Tokyo's streets, canals, and open spaces emerged as a practical health concern that demanded attention. Beginning on the late afternoon of September 3, Mayor Nagata directed city authorities to join metropolitan police officials who had already begun to collect the dead and place them in piles near what remained of police offices, municipal government buildings, small parks, and open spaces created by the fires. Eventually fifteen main sites were selected as collection points. Using a combination of motorized carts, horse-drawn wagons, and push carts, over three hundred city employees joined all available policemen and began to clear bodies from Tokyo's streets, ditches, canals, and collapsed buildings. Officials also employed up to twenty-five boats per day to collect bodies from Tokyo's main waterways.[37] The sheer number of bodies collected, as recorded by city officials and reported in newspapers, gave considerable validity to survivor accounts that suggested dead bodies were indeed everywhere. On September 11 the *Ōsaka mainichi shinbun* reported that as of 9 p.m. on September 6, city officials had collected, moved, and stacked 47,200 bodies.[38] During the first two weeks of September, another 10,525 were removed from Tokyo's main waterways and rivers.[39] Of the 58,279 bodies collected by city officials only a small fraction were distinguishable by sex, let alone by features that would have enabled individual identification: 3,613 bodies were classified as male, 3,614 were classified as female, and 51,052 were listed as "uncertain" (*fumei*).[40]

As the bodies began to decompose and pose an even greater risk to public health and psychological well-being, city officials decided to gather as many as possible in one central location. Only one place was considered: the former site of the Honjo Clothing Depot. City officials selected this site for the simple reason that it already contained at least 40,000 dead. It would be far better, officials concluded, to have one central morgue and crematorium if possible, and one place where all those wishing to find and identify a body could go.

Corpses from the areas immediately surrounding the site of the clothing depot arrived first, with others streaming in for over ten days after the disaster. Though the city also created twelve other "corpse incineration

TABLE 3.1 Number and gender of bodies deposited at collection and cremation centers across Tokyo

LOCATION	MALES	FEMALES	UNCERTAIN	TOTAL
Kyōbashi	221	175	0	396
Shiba	221	175	0	396
Azabu	10	16	0	26
Akasaka	49	54	0	103
Kōjimachi	120	83	0	203
Yotsuya	2	0	0	2
Ushigome	24	28	0	52
Koshikawa	30	12	0	42
Hongō	54	30	9	93
Kanda	171	60	114	345
Shitaya	168	73	0	241
Asakusa	366	652	1,376	2,394
Nihonbashi	126	133	53	312
Honjo	767	970	48,807	50,544
Fukagawa	1,284	1,153	692	3,129
Total	3,613	3,614	51,052	58,279

Source: Nihon tōkei fukyūkai, ed., *Teito fukkō jigyō taikan*, vol. 1, part 13, pp. 26–27.

locations" (*shitai shōkyaku basho*), the vast majority of bodies cremated in Tokyo—roughly 85 percent—occurred at the site of the former Honjo Clothing Depot. When corpses arrived at the clothing depot, city officials, police authorities, Relief Bureau workers, and Buddhists from the Honganji Temple in Tsukiji recorded the locations at which the bodies were found. Officials also searched the bodies for any piece of personal identification or valuables. Rings, watches, and jewelry were regularly removed from the bodies. Cash, averaging about ¥10,000 per day, was also taken from the corpses and recorded.[41] If any thieves had thought of combing the dead for money or valuables, police, Buddhist priests, and eventually the press warned against such acts of brazen, ghoulish opportunism. As one byline in the *Osaka mainichi shinbun* stated: "The cash and valuables found on those bodies is said to retain a peculiar smell which ordinary people have no means of deodorizing. Many thieves who have run away with cash or gold watches, rings, or other valuables taken from those bodies have been traced and caught on account of the unforgettable smell."[42]

Soon the entire site of the Honjo Clothing Depot became, as Nakamura Shōichi described, "covered with the dead." "Some bodies," he continued, "were charred to dark brown; some were half-burned, with unrecognizable faces, looking like several-thousand-year-old mummies; and some were completely burned to the extent that the head had become a perfect skull." Beyond this, Nakamura believed that bodies held physical clues as to just how awful death through suffocation, burns, or exhaustion had been to such a large number of sufferers. "The clenched fists of each body," he concluded, "were telling of the agony of their last moment."[43]

Another chronicler described the scene at and around the site of the Honjo Clothing Depot as follows: "Most bodies around here were charred—some of them were exposing their guts, which were also charred and swollen, through their torn bellies. Some of them had popped-out eyeballs; and some of them had turned-up lips and noses. Each body was in an utterly miserable condition. It is so horrible that any human life should end like this—it was too awful to look at."[44]

Eventually, whether the victim was identified or not, Buddhist monks offered prayers for each and cremated the dead around the clock, leaving behind large mounds of ash and bones. From the vantage point of Ueno Park, novelist Uno Kōji reflected that the last fires of Tokyo he witnessed were those from the funeral pyres and makeshift crematoriums at Honjo. The haze, he confided, rested over the devastated city, adding an extra layer of ghastliness to the horizon.[45]

The shroud of cremation hung over the city for weeks, compelling Uno Kōji and many other survivors to reflect on the greatest collective tragedy experienced by the Japanese nation in his generation. Postcards, particularly those of the dead, contributed to the process of reflection across Japan. They also captured the solemn act of cremation carried out on makeshift funeral pyres and highlighted what remained of the process: mounds of ash and bone. Often they included pictures of sorrowful-looking survivors crying before these large mounds of ash and bone and the piles of adjacent personal belongings removed from the bodies prior to cremation. Important funeral rites carried out by Buddhist monks at the site of the Honjo Clothing Depot were also captured in cards and relayed across Japan.[46] Soeda described what he witnessed at the Honjo Clothing Depot as follows:

> Petrol is poured, and the bodies are burned.
> Choking on the smoke of cremation,
> The survivors say prayers in tears,
> For the repose of the dead.
>
> People cry silently, watching the cremation;
> People cry biting their sleeves, hearing about the cremation.
> Ah, what a terrible disaster,
> Ah, what a terrible disaster.[47]

Along with stirring emotions, the dead drew many people to the Honjo Clothing Depot to experience part of the human catastrophe firsthand. Postcards and stories of the dead provoked a voyeuristic response.

FIGURE 3.11 Mound of ash from the bodies of cremated victims at the site of the Honjo Clothing Depot

Source: Postcard, author's private collection.

Grade 3 student Kitō Shirō from Aoyagi Primary School in Koishikawa Ward recounted that after viewing postcards of the dead, his uncle felt compelled to see the scenes of destruction and death depicted firsthand. Like many, Kitō and his uncle began their pilgrimage in Honjo. Kitō recalled that "even just seeing the picture postcards made me feel horrible."[48] He soon felt much worse. He wrote that Honjo "was dotted with dead people's cremated remains piled in mountains like sand" and that he and his uncle "saw people's fat floating like tempura oil [on the water] in the ditches that surrounded the site." With a child's honesty and simplicity, Kitō concluded that he felt "sorry for all the people who had died."

Kitō was not alone. The remains of the Honjo Clothing Depot became a site of macabre pilgrimage. Some went there with the intent to find, identify, and claim the body of a family member or friend. Numerous others, including many government officials, toured the site to see if the location was really as horrid as had been spoken of in rumors, purported in newspaper accounts, or captured in photos. Almost everyone concluded, as Tawara Magoichi did, that "the spectacle of the ground covered with the bodies of men, women, and children of all ages was much worse" than described.[49] Still others, including Crown Prince Hirohito, visited this location to pay condolences—and to be seen doing so—as well as to publicly mourn the nation's loss. Few people left untouched. One anonymous sojourner exclaimed that "no word existed" to describe the grief he felt after seeing the bodies piled as high as mountains.[50]

In their final act, the dead, whether as charred, disfigured corpses or as mounds of ash and bone, compelled reflection and commanded remembrance. Four days after the first large-scale religious memorial service was carried out by Buddhist monks on September 21, Tokyo officials announced that the site of the Honjo Clothing Depot would be transformed into a park that would house a monument to those who had perished in the Great Kantō Earthquake.[51] The monument itself, Mayor Nagata proclaimed, would also serve as a reliquary, accommodating both the ashes of those who had been cremated and the personal effects taken off the bodies of the dead. The park and memorial monument would be open to everyone so that the relatives of the victims, the people of Tokyo, and any citizen who traveled to this site would never forget the calamity that took place or the loss it delivered.

Elites such as Nagata who ventured to the Honjo Clothing Depot realized that it could become a place of extraordinary political and

FIGURE 3.12 Buddhist monks offering prayers to the spirits of the dead in front of a large ash mound

Source: Postcard, author's private collection.

commemorative significance. Within weeks of the disaster, government officials used this hallowed site and the memory of the dead opportunistically to elicit responses from the living for the tasks of reconstruction and renewal. Evoking the memory of the dead for political objectives was commonplace at the large-scale memorial service held on October 19, forty-nine days after the calamity—a significant Buddhist memorial date to pray for the spirits of the dead. More than 200,000 citizens and virtually every major politician in Japan ventured to Honjo for this service. After the ceremony was opened with a song played by the Toyama Military Band, each cabinet minister, representatives from the imperial family, members of every noble family, and monks from every major religious sect in Japan paid official respects before a gable constructed on top of a mound of cremated remains. Decorated with flowers placed by government officials, including chrysanthemum wreaths made by the empress and crown prince from flowers grown on the grounds of the imperial garden in Shinjuku, the mound, surrounded by seventy large ash funeral urns, was used by dignitaries as a backdrop to deliver addresses rich in symbolism and meaning.[52]

Many invited dignitaries directed their comments first to the spirits of the dead, next to their surviving relatives, and finally to the people of Japan. Count Yanagisawa, chairman of the Tokyo City Assembly, urged those still "restless, wailing spirits" of the dead to release their worldly attachments, thus enabling them, their relatives, and the people of Tokyo to move forward from the terrible tragedy. Shinto Priest Kanzaki Issaku, representing members of the All Japan Shinto Association, pleaded with the souls of the departed to "hear our prayers and to rest with calm minds" so that the world may now become a peaceful place and that reconstruction could commence. Other speakers declared that while it was important to mourn the dead, it was now time for people across Japan to adopt positive actions and support farsighted policies so as to turn misfortune into a blessing. It could only be done, many suggested, if Japan completed an expansive reconstruction plan as a tribute to those who died. Kusunose Yukihiko, a retired general and former Army Minister, informed the crowd that it was solely up to the people of Japan to make certain that a great and glorious capital was reconstructed; otherwise the "sacrifice that the victims made" would go to waste. To secure this end, he implied that all Japanese must be willing to sacrifice in memory of those

FIGURE 3.13 Gotō Shinpei giving condolences to the spirits of the dead at the site of the Honjo Clothing Depot

Source: Postcard, author's private collection.

FIGURE 3.14 Buddhist monks offering final prayers for the dead at the forty-ninth-day service

Source: Postcard, author's private collection.

that died—a sacrifice that would be so little in comparison.[53] Such calls would be repeated almost uninterrupted for the duration of the reconstruction project.

SEARCHING FOR MEANING: USING WARTIME ANALOGIES

For those who lived in the disaster zone and experienced the great seismic upheaval and fires of early September 1923, Japan's earthquake calamity needed little construction or journalistic dramatization as a national tragedy to impart meaning. Everyday life for Tokyoites was surrounded by a degree of destruction, dislocation, and perdition. The blackened, rubble-strewn landscape and the hungry, homeless, destitute survivors were conspicuous reminders of the catastrophe that had befallen the capital. The charred, bloated, and traumatized bodies of the dead and later the mounds of ash that they became were confronting signifiers that the people of Tokyo had experienced unprecedented tragedy and suffered immeasurable loss. The artful use and construction of this disaster through new and well-established mediums, however, introduced scenes of this suffering and loss to people across Japan and helped define this disaster as a national calamity.

Newspapers described this event as an extraordinary tragedy in part because it was, and reporting it was their profession. It was also, however, an effective platform from which to sell newspapers. The largest papers were able to employ new technologies to bring the horrors of this disaster closer to people across the nation. This fueled greater interest in the tragedy and in turn sold more papers. Government officials took great care in making sure that as many people as possible within Japan saw the disaster as a national tragedy for calculated political reasons. First and foremost, they packaged the earthquake disaster as a national catastrophe to assist with the challenging task of securing relief and what they knew would be monumentally expensive tasks of reconstruction. Political elites understood that Japanese citizens from Hokkaidō to Satsuma were going to be called on to sacrifice, possibly for years following the disaster to assist sufferers and finance the reconstruction of the capital. Importantly, they wanted citizens to support such efforts with discernible unity of purpose as they had done during Japan's successful wars in the past.

Mayor Nagata held this opinion. For much of his predisaster tenure as Tokyo's chief representative, Nagata had sought repeatedly to secure greater financial and political autonomy for the capital. The earthquake changed all this overnight. Through speeches and publications and in private negotiations, the mayor argued that the 1923 catastrophe was not a local or even a regional disaster but rather "the most unprecedented calamity that our nation has ever had."[54] Nagata urged all people to continue to respond to the calamity as if Japan were still at war: give, think nothing of the self, and support the great national undertakings of relief, recovery, and reconstruction. Tokyo needed the people of Japan now more than at any time in its history. In fact, many of its citizens had become almost entirely dependent on the nation for virtually every aspect of their existence. The continued tasks of recovery and, most important, the expensive, long-term challenge of reconstruction, he realized, could be accomplished only by the continued mobilization of the nation's human and material resources.

Constructing the Great Kantō Earthquake as a national calamity and comparing the challenges of securing aid and undertaking reconstruction to that of mobilizing society for war thus made perfect sense. The challenges of war and what a war required from a nation's population removed from the battlefield were something that many commentators agreed all Japanese could comprehend. War and their related domestic mobilization campaigns had been not only experiences that most Japanese had shared at some point in their lives but also challenges that had been overcome.[55] Numerous elites understood this point in 1923. Had not, Takashima Beihō asked, Japan faced three previous national tests—the Meiji Restoration, the Sino-Japanese War, and the Russo-Japanese War—and survived only because of the hard work, spirit of sacrifice, and diligence of its people? Yes, he replied emphatically. In a similar vein, he claimed that "we will not be able to recover from it [the earthquake disaster], unless the whole nation again rises as one" as it had done in the past. Japan had fought a series of fierce battles with nature—if not an outright war—and lost. It was essential, Takashima concluded, for all Japanese to comprehend the Great Kantō Earthquake as a similar test of national resolve. He labeled it the test of Taishō Japan. The nation's leaders, he argued, must work tirelessly to make certain that Japanese saw it as a test as well.[56]

Takashima was not alone in holding this opinion. Gotō Shinpei, home minister and president of the Reconstruction Institute (Teito fukkōin), also understood the importance and potential impact of using the imagery of Japan's past wars to condition and mobilize the people behind a postdisaster project of reconstruction. He encouraged all state officials to emphasize this fact when dealing with the public, the press, and their fellow bureaucrats. Speaking to a group of local governors and officials in November, Gotō urged members of Japan's bureaucracy to invoke the memory of the Russo-Japanese War and in particular to emphasize the spirit of heroism and sacrifice displayed by Admiral Tōgō Heihachirō and the sailors under his command. Similar to the way in which sailors of Tōgō's battle fleet rallied for the nation—not for the navy or for thoughts of individual fame—when they saw the famous "Z" flag hoisted from the flagship *Mikasa*, Gotō suggested that all Japanese must be conditioned to respond to the challenge of their era: the reconstruction of Tokyo. Only by uniting the people of Japan behind the project of reconstruction, he argued, could Tokyo be rebuilt as the center of Japan's empire. If Japanese behaved like the heroes of the past wars or like the exemplars of virtue whose postearthquake stories were read by every child, Gotō argued, Japan could not only rebuild its capital, but that it could also reconstruct the nation.[57]

Numerous other elite commentators used a war analogy to describe the trauma Japan suffered and the challenges it still faced to compel reflection on the state of society. Initially this proved challenging. Professor Hoashi suggested that the Meiji Restoration of 1868, the Sino-Japanese War of 1894–1895, and the Russo-Japanese War of 1904–1905 were good starting points in trying to describe what the disaster meant to those in Japan who had not experienced it. He admitted, however, that the 1923 disaster was different in fundamental ways. While people from the armed forces had been most acutely affected by previous wars, the earthquake calamity was a civilian experience that had no precedent in Japan's past. Japan's past conflicts had been challenges, he concluded, but were they, he wondered, events that compelled solemn reflection?[58]

Kasuya Gizō, speaker of the lower house of the Imperial Diet, likewise cautioned against using such straightforward wartime comparisons. At the forty-ninth-day religious ceremony, he reminded listeners that a wartime analogy went only so far in explaining Japan's tragedy to those who had not suffered from it in person. What the people of eastern Japan had experienced, he suggested, was loss and devastation, and with such rapidity that few outside the disaster zone could comprehend, even if compared to an intense national struggle such as a war. At no single site on any major battlefield in any previous conflict, Kasuya claimed, had Japan or any nation lost forty thousand people in one night.[59]

Abe Isoo also admitted that it was difficult to employ Japan's previous wartime experiences in a comparative fashion with the 1923 disaster if the government's ultimate objective was to nurture introspection across society. Why? Because Japan, he argued, had never lost a previous conflict, and those wars had not led to a reflective examination of Japan's state of modern development. In the popular journal *Kaizō*, Abe wrote that more Japanese had died as a result of the 1923 disaster than in all the nation's previous wars. Moreover, he articulated that the monetary value of losses incurred and the costs of reconstruction would far exceed anything Japan had spent previously in war. Abe implied that the only analogy that could be successfully employed to compare to the earthquake calamity was the "Great European War." This conflict, he argued, had been a totalizing, traumatic event that compelled reflection even among the victorious nations.[60]

Tenrikyō Priest Okutani Fumitomo agreed with Abe's conclusions. Writing in the Tenrikyō journal *Michi no tomo*, Okutani suggested that the degree of destruction and scale of dislocation felt by civilians made this disaster comparable only to what virtually all people in continental Europe experienced ten years earlier. Why? In the first instance, it was because Japan had never before lost a war or experienced this level of devastation on its home islands. Victory in previous wars had therefore not compelled reflection or upended civilization in a way that the unmitigated, traumatic loss associated with the disaster had done. Second, Okutani suggested that Japan's earthquake ordeal was greater than a simple list of monetary losses, casualty figures, or destroyed square kilometers of physical space. Rather, he believed that there was something intangible about the full scale of Japan's loss that would forever be lodged in the minds of all those who had experienced it or witnessed scenes of the devastation. He predicted that the people of Japan would continue to reflect on the catastrophe for years, if not decades. He believed that only the European war had resulted in a similar state of reflection across all parts of society.[61]

House of Peers member Ōki Tōkichi wrote that because the European war had been so traumatic, it had resulted in extensive questioning among the combatants involved. The period of self-reflection following the war had led to "fundamental changes in the politics, economy, and society of every participant." He hoped that all Japanese would conduct a similar degree of soul searching so as to spur a similar transformation in Japan. No person who saw Tokyo in ruins, heard stories of the catastrophe, or saw the site of the Honjo Clothing Depot in person or in pictures, he suggested, could be immune to self-reflection. He believed, along with others such as social reform advocate Tomoeda Takahiko, that the people of Japan would turn over a new leaf in response to this calamity and embrace the idea of renewal and reconstruction under the mantle of a Taishō restoration.[62] Moral philosopher Shimamoto Ainosuke agreed with this assessment and suggested that "Japan's loss" in 1923 was spiritual and psychological and affected the "entire nation," not just soldiers or people from the Kantō region. He argued that it was the government's responsibility to convey the introspective meaning of this calamity across Japan.[63]

Was the Great Kantō Earthquake calamity really comparable to a major war? Or was this fundamentally a Tokyo-centered perspective that colored the way elites and commentators—almost all of whom were residents of the capital—believed people across the nation *should* view the calamity? Undeniably the economic loss meted out by the calamity was larger than anything Japan had suffered in war. In a physical sense, the catastrophe delivered on Tokyo and Yokohama far exceeded what had befallen any of Japan's major cities in living memory. Suggesting that the destruction and loss were anywhere close to what the major combatants of the First World War had experienced over the course of four traumatic years, however, was a convenient exaggeration.

The conflation of Europe's wartime suffering with what Japan experienced in 1923 illustrates the perceived malleability of catastrophe and war and the desire of elites to use the earthquake to foster acute introspection and national unity. Government leaders hoped that by constructing the Great Kantō Earthquake as Japan's national calamity and the reconstruction effort as a challenge akin to war, they could tie people—often well removed from the actual suffering and dislocation—to the epicenter of catastrophe to assist with recovery and reconstruction. Moreover, they hoped to foster a period of self-reflection that would, over time, encourage numerous Japanese from diverse backgrounds to ask penetrating questions about the disaster, its meaning, and its importance to the nation. Many did. Often they focused their intellectual energies on interpreting the catastrophe. In doing so, commentators and ordinary people asked why the earthquake struck Tokyo and what messages, heavenly or secular, lay behind the disaster. This marked an important phase in the recovery process and served as a critical factor behind the emergence of a culture of blame and admonishment in postdisaster Japan. In this environment, commentators used the disaster to criticize the people of Japan—and the residents of Tokyo in particular—for thoughts and behaviors that many commentators suggested had brought about a calamity of epic proportions.

Four
ADMONISHMENT: INTERPRETING CATASTROPHE AS DIVINE PUNISHMENT

Our people have been sinking into the abyss of corruption day by day, lapsing into flippancy; intoxicated with decadence and the principle of living only for the pleasure of the moment; and indulging in hedonism. There was an impression that all our efforts and the development of material civilization had been for the sake of this corrupted life.
—Hoashi Ri'ichirō, 1923

It is obvious that this calamity was punishment from the gods [kamisama]—the result of their disappointment and anger—at the sight of the capital, Tokyo. The gods have done a thorough cleanup for us. They said, "Come now! It is time to clean up every nook and cranny," and by the power of the heavens, it happened.
—Okutani Fumitomo, 1923

On Friday, August 31, 1923, Hanayashiki—the heart of the entertainment district of Asakusa, Tokyo—was a place of merriment. As on almost every evening, crowds of Tokyoites, often from the poorest neighboring wards of Fukagawa, Honjo, and Asakusa itself, flocked to this bustling locale. Joined by other pleasure seekers of all ages from near and far, nearly everyone who ventured to Asakusa entered a realm of and for the senses. To many, Asakusa embodied a state of being and a condition of gratification. Food and drink of every type could be purchased and consumed. Those wishing to be entertained by the exploits of other living creatures, be they human or wild beast, had a menagerie from which to select. Fortune tellers, street performers, freaks of nature, monkeys that began and ended each uniformed performance with a bow, a tiger that jumped through a flaming hoop, or an elephant trained to launch a double-jointed acrobat into the air by jumping on a garishly colored seesaw, allowed spectators to cross the threshold into a world well removed from the everyday tedium of modern life.[1] Patrons with more carnal urges could also satiate themselves through movies, dance and cabaret retinues, gambling, or prostitution. Asakusa was an escape. It was the place to lose one's worldly worries and everyday concerns amid an ocean of hedonistic decadence and excess. It offered, if only temporarily, many individuals the opportunity to leave behind not only the toils of the day but also, as novelist Saitō Ryokū suggested, the worries and responsibilities associated with tomorrow's work.[2]

Others, however, saw Asakusa differently. Many social commentators, moralists, journalists, and government officials viewed the entertainment district of Tokyo's "low quarters" as a cancer that weakened individuals and threatened the nation's spiritual well-being. It was the epicenter of modern vice that lured and entrapped vulnerable Japanese. Critics described Asakusa as a vast garbage dump where lower- and working-class laborers, juvenile delinquents, and people of questionable morals wasted time, money, and energy.[3] Tenrikyō priest Okutani Fumitomo suggested that Tokyo's entertainment districts were nothing more than "gathering places for sin" where all of the unhealthy desires of man could be satisfied through routine transactions.[4] Moral crusader Abe Isoo suggested that the pleasure and entertainment quarters of Tokyo were nothing less than the "dark side of civilization."[5] Defining and describing Asakusa as a place of either fun and release or decadence and menace was all a matter of perspective.

From whatever ideological vantage point one viewed Asakusa on August 31, by September 3 everyone shared an awareness that this district, many of the adjacent suburbs, and virtually all of "low" Tokyo no longer existed. They had vanished, shaken and burned beyond almost all recognition. Their complete and utter destruction, however, compelled people from all backgrounds, classes, perspectives, and religions to ask, reflect on, and eventually pontificate about one important question: Why? Why had this earthquake and conflagration struck Tokyo? What message or meaning lay behind it? Why had the poorest areas of the capital and the center of frivolity, decadence, and hedonism experienced the most thorough devastation? In exploring these questions, individuals looked beyond the bounds of science, urban planning, and secular reason and turned to the cosmos.

Numerous social commentators and critics, bureaucrats and governing elites, and members of the educated and chattering classes interpreted and explained the Great Kantō Earthquake as an act of divine warning and punishment (*tenken* or *tenbatsu*). In speeches, newspaper accounts, and journal articles, they invoked repeated references to divine intervention, warning, and heavenly punishment. Individuals implied that the general immorality associated with the post–First World War age of excess and frivolity had helped precipitate divine intervention. Others employed such interpretations for clear political and ideological reasons. Some admonished Japanese in hopes that this would encourage reflection on the general spiritual and ideological state of society. Others, however, hoped that this civilization-upending disaster would compel people to change their seemingly overindulgent ways and endorse moderation in thought and action. Virtually everyone classified the calamity as a moral wake-up call that placed Japan at an important crossroad or "watershed," as General Ugaki Kazushige described it, between decline and renovation.[6]

Bureaucratic elites also used the calamity to cosmologically bolster critiques leveled against people for their perceived enchantment with materialism, luxury, hedonism, and decadence.[7] Often these individuals employed the catastrophe to legitimate prescriptions for social, moral, and ideological reform. On other occasions they used it to admonish, berate, and blame Tokyoites for embracing suspect thoughts and behaviors that they claimed had triggered the calamity. Blame not the heavens, they argued, but the people who had forced the higher powers to act to save society from an even greater future apocalypse. The seemingly targeted destruction of large parts of Tokyo—namely, the entertainment quarters, the centers of consumer spending, and the slums—allowed a diverse range of commentators to coalesce around one extraordinary event and express concerns and project their fears about the state of Japan to a captivated audience. Analyzing interpretations of the Great Kantō Earthquake thus reveals the deep-seated fears and anxieties that many elites held about the state of urban Japan, if not modernity itself. It also highlights two important trends in interwar Japan that lasted long after the reconstruction project finished: First, it illustrates the fragility upon which many commentators believed that post–World War I Japan rested as it became increasingly urban, cosmopolitan, industrial, wealthy, consumer oriented, liberal, and pluralistic. Second, it points to the perceived malleability of crisis and catastrophe and the willingness of elites to employ these discourses to secure larger objectives.

COMPREHENDING JAPAN'S EARTHQUAKE CALAMITY AS DIVINE INTERVENTION

Across time and cultures, destructive natural phenomena commonly classified today as "natural disasters" have been described, explained, and interpreted as acts of heaven, the gods, or supernatural forces. Japan's myths—as well as those from other earthquake-prone regions—are studded with references to calamities originating from realms beyond the domain of everyday human beings.[8] In large part such metaphoric thinking about and supernatural representations of disasters stemmed from the fact that catastrophic seismic upheavals were far beyond anything people could make sense of through normal, everyday processes. Earthquakes could literally turn the world upside down by toppling human constructs. Moreover, they could open great fissures in the Earth and trigger massive waves of destruction from the sea. Humans could not comprehend, measure, or harness anything approaching the power unleashed by earthquakes until the middle of the twentieth century and then only just: the Tōhoku daishinsai of March 11, 2011, released energy equivalent to the almost simultaneous detonation of 25,000 Hiroshima-size "Little Boy" atomic bombs.

Early philosophers and naturalists in the West believed that the sudden release of trapped vapors, winds, gases, or "dried exhalations" from subterranean cavities caused earthquakes.[9] In East Asian scientific tradition, earthquakes were often associated with imbalances between the forces of yin and yang or positive and negatives poles above and below the surface of the Earth.[10] In Japan specifically, folklore suggested that the violent movements of subterranean catfish triggered tremors. Though renewed scientific interest in earthquakes triggered an avalanche of hypotheses among scholars and thinkers in continental Europe and England during the seventeenth and eighteenth centuries, no authoritative theory that explained earthquakes existed until the twentieth century. Geologist David Milne-Home argued this point in 1841, writing that "there is no department of physical science, over which, unfortunately, there hangs so deep a cloud of mystery."[11]

Only in the late nineteenth and early twentieth centuries, with significant input from scientists in Japan, did accurate theories on what causes earthquakes materialize. As Greg Clancey has documented, the modern science of seismology, with its own institutions, instruments, teachers, students, and diplomas, was largely founded in Tokyo.[12] By the dawn of the twentieth century, two of Japan's best-known seismologists, Ōmori Fusakichi and Imamura Akitsune, had traveled extensively around the world, particularly along what is today known as the Pacific Rim of Fire, studying earthquakes and gaining varying degrees of international acclaim and notoriety.[13] British seismologist Charles Davison described

Ōmori, who died weeks after the September 1, 1923, calamity, as "our leader in seismology."[14]

Regardless of the scientific advances within Japan or the West, earthquakes have rarely been divorced completely from the supposed forces of divine power or will.[15] Japan in 1923 was no exception. Rousing depictions of the catastrophe as an act of divine punishment flourished and aroused the curiosity of many researchers. Academics from the Office of Religious Studies at Tokyo University, including Harada Toshiaki and Takagi Toshiyuki, thus undertook an extensive study to document how people comprehended the 1923 calamity. In this project they collaborated with staff from the Kiitsu kyōkai, a research organization founded in 1911 whose members conducted research, held conferences, and published findings related to the interrelationship of religion, philosophy, ethics, sociology, education, and literature in Japan.[16]

Following the Great Kantō Earthquake, interpretations of the disaster and discussion related to the subsequent spiritual reconstruction of Japan became a key research focus for the Kiitsu kyōkai. Assisted by Anesaki Masaharu, professor of religion at Tokyo University, the research team examined academic journals, popular magazines, newspapers, literature, and song lyrics published between September 1923 and May 1924.[17] Their findings, published on the two-year anniversary of the calamity, concluded that the terminology of "divine punishment" (*tenken* or *tenbatsu*) was the most commonly employed interpretation of the disaster. Bureaucrat Shitennō Nobutaka agreed, writing in the Home Ministry journal *Chihō gyōsei* that the "divine punishment theory" was felt "acutely" across society immediately following the catastrophe.[18] Shitennō's conclusion, like those of the Kiitsu kyōkai, was not uninformed.

The Kiitsu kyōkai research team did far more than just examine published materials. To gauge the popular mindset of everyday citizens, researchers made inspection tours of the city, visited temporary shelters to interview residents, observed people, and took notes on conversations overheard while standing in line at food distribution centers throughout Tokyo. Researchers suggested that after the catastrophe the people of Tokyo developed and exhibited a heightened state of religious awareness. Rather than accept the calamity within what the Kiitsu kyōkai researchers concluded was a traditional Buddhist mindset of acceptance, resignation, and fatalism—as some in the organization hypothesized—researchers concluded that people embraced interpretations of the earthquake as a purposeful act of divine punishment and pondered the question of why the catastrophe had struck Tokyo. They also claimed that people's reading patterns changed to reflect these views. Kiitsu kyōkai members who tracked sales figures from both publishers and booksellers concluded that the sales of frivolous romance books decreased by roughly half after the earthquake while books on religion and ideology increased. Investigators concluded that "the bestsellers at the bookstores in Tokyo after the earthquake were religious books and books on war heroes."[19]

It was not just Tokyo disaster victims as described by the Kiitsu kyōkai who accepted the notion of the earthquake as divine intervention. Numerous elites embraced it and invoked the divine punishment paradigm in virtually every medium available. Okutani Fumitomo, a priest in the new religion of Tenrikyō, was one such person. As a priest within one of the thirteen officially recognized Shintō sects, Okutani possessed a well-established reputation for delivering energetic sermons. He did not let what he perceived to be the "great opportunity" of the disaster pass by quietly. Rather, he suggested in no uncertain terms that the catastrophe was not "accidental" but "*kamisama*'s [the gods'] punishment." Suggesting that the "sinful behavior" and "arrogance" exhibited by Japanese people "had surpassed all limits" in the years following the Great European War, Okutani informed readers of the religious journal *Michi no tomo* that "there was no doubt that the recent earthquake was part of heavenly retribution."[20] During a fire and brimstone laden lecture tour of Japan in October, Okutani told listeners that he was not alone in seeing this earthquake as a "disaster for the nation" that was brought about by the gods' understandable wrath.[21] As if to give further weight to his interpretation, Okutani quoted likeminded military leaders, literary figures, and government bureaucrats who described the disaster as divine punishment. He then concluded that there was no way for Japanese people to understand this calamity other than as an act of divine punishment. The question that Japanese people must grapple with now, over the coming weeks and months, he suggested, was "why." Why specifically, Okutani asked, had the gods destroyed a large part of Tokyo, and how should humans respond in the face of this manifestation of divine will?

Takashima Beihō, a lay leader in the New Buddhist movement and spokesman for the Buddhist lay community, also classified the disaster as an act of heavenly displeasure. With a background in political activism and religion, Takashima lashed out at those who could not or would not see the disaster as an act of purposeful, heavenly intervention.[22] Expressing his opinion to a different audience in *Chūō kōron*, Takashima wrote,

"Some people say that a natural disaster is only an accident and that there is no hidden meaning or intention." But, he asked, "is that really so?" "No," he concluded. The 1923 disaster was a clear act of divine punishment, specifically the "Buddha's punishment" leveled against a society that had become morally corrupt.[23] Tenrikyō relief worker Haruno Ki'ichi shared these views. Writing about his postdisaster humanitarian exploits in the journal *Michi no tomo*, Haruno admitted that he was struck by how many people in Tokyo saw and spoke of this disaster as an act of divine punishment. As if convinced by the people whom he assisted throughout the months of September and October 1923, Haruno acknowledged, "I too now think this may be the case."[24]

Such interpretations expressed by religious leaders are not surprising. History is full of examples in which religious elites explained natural disasters as dreadful visitations of divine punishment. Earthquakes that struck Europe and North America during the eighteenth and nineteenth centuries, including the devastating Lisbon earthquake of 1755, were often described as dislocating events "sent by a wrathful God to punish human transgressions."[25] Even minor earthquakes, such as the ones that shook New England in 1755, were interpreted as cosmological events. "What ailed the earth that it shook in such a dreadful manner?" asked Pastor Thomas Prentice. "Why," he replied, "it was because the Lord was angry," and thus it was "under the heavy Burden of our Sins" that "the earth groaned and shook and trembled in so amazing a manner" during that eventful autumn.[26] Beyond earthquakes, other natural disasters such as hurricanes, floods, and even great fires have been classified by religious elites as acts of divine intervention.[27]

What is most remarkable about Japan's experience in 1923, however, is that many nonreligious spokesmen likewise embraced the concept of divine intervention to explain why the disaster had struck Tokyo. Diverse individuals found a significant degree of common ground in describing the disaster as an act of heavenly warning and divine punishment, thus reflecting the unease that many elites held about the state of society. Suematsu Kai'ichirō is a perfect example. He was a Home Ministry official who worked his way up through the bureaucracy and gained recognition for introducing a female suffrage bill before the Imperial Diet in 1930. Neither known for exhibiting religious fervor in public or private nor known to possess a religious outlook, Suematsu wrote in the special earthquake edition of *Chihō gyōsei* that it was "not unreasonable to think that the recent disaster was a divine warning." "The people," he concluded, "badly need to wake up" and as such, "I have a feeling that it could have been a divine warning."[28] Army general and future army minister Ugaki Kazushige was less circumspect in his appraisal of what triggered the disaster and how people should view it. Writing in his diary, Ugaki expressed hope that the people of Japan would "accept this disaster, which seems like divine punishment," as "an awakening call." [29]

Some commentators confessed that while they did not consider themselves to be religious prior to the calamity, the scale of the disaster compelled them to think in new ways about the relationship between the heavens and the Earth. In the November issue of *Chūō kōron*, Horie Ki'ichi suggested as much. The professor of economics at the University of Tokyo specifically wrote that, "while I do not believe in divine providence [*ten no haizai*] or revelations of heaven's might [*shin'i*], I cannot help but think that there was some truth in theory that this was heaven's punishment."[30] Privy Councilor and future minister of imperial household Ichiki Kitokurō concurred. Ichiki admitted that while he did not give initial thought as to why the disaster occurred—instead focusing his attention on relief and proposals for reconstruction—"why" remained, to him and others, a nagging question. "It is often said that the recent earthquake disaster was a divine punishment," he wrote, "and indeed it does more or less give us the impression of it."[31]

Reflecting on the intellectual processes of disaster acceptance, interpretation, and the eventual construction of this calamity as an act of divine punishment, moral philosopher Fukasaku Yasubumi devoted two chapters of his 1924 book on ethics and society to the earthquake. It was common, he wrote in his chapter entitled "Impressions of the Earthquake Disaster" (*Shinsai shokan*), for people to find higher meaning, purpose, or intentions behind catastrophic, life altering events, particularly those of such a grand scale as the September 1 disaster. In this regard, he informed his readers, Japanese people were not unique:

> Generally speaking, humans have the tendency of personifying nature, interpreting its phenomena from moral points of view, even though nature seems to have no mind. They describe nature's inevitable phenomena, freely using figurative expression, and interpret nature's inevitability as moral inevitability. In so doing, they try to gain internal satisfaction both intellectually and emotionally. This is why they see divine plans in various natural phenomena and god's will in unavoidable incidents. This is why they find lessons in horrible reality and find materials for

self-reflection in tragic events. . . . This is why they call the recent earthquake disaster "divine punishment."[32]

Given Japan's experience in 1923, he concluded, it was not "absurd to interpret the disaster as a deliberate divine plan."

Following this matter-of-fact, rational appraisal, however, Fukasaku felt compelled to share his own personal thoughts and reflections on the catastrophe. Writing with the force of a true believer and with the zeal of a recent convert, he declared that it was directly because of the increasingly degenerate state of society that emerged following the First World War that "the gods cracked down a great hammer for the sake of our race, waking us up from idleness and urging us to reflect on our past deeds." The people of Tokyo, he suggested, had precipitated this disaster.[33]

Yamada Yoshio, scholar of philology and literature at Tōhoku University, was even bolder in his assertion that Japanese people, through immoral attitudes and questionable behavior, were somehow behind the disaster. Writing in 1923 and reiterating his conclusions ten years later in a second edition of his work, Yamada suggested in no uncertain terms that "this was a disaster they [the people of Japan] had brought on themselves because of frivolousness, self-indulgence, and greed."[34] The earthquake was proof, Yamada and many others suggested, that Japan had reached a stage of decline so severe that only a heaven-sent disaster and subsequent reconstruction of the people, the capital, and the nation could repair it.

What was it about 1923 Japan that led numerous people from diverse backgrounds to suggest that the earthquake was a targeted, vengeful strike from the heavens? What rested at the root of their anxieties and fears about the trajectory of modern Japan? It is important to first point out that deep-seated concerns over the economic, ideological, social, and political health of Japan did not begin in 1923 as Tokyo lay in ruins. Many historians, such as David Ambaras, Harry Harootunian, Sheldon Garon, Andrew Gordon, Barbara Sato, and Miriam Silverberg, have pointed out that elites held many well-defined anxieties about the perceived decline of Japanese society since the period just after the Russo-Japanese War. These concerns, however, grew in an almost exponential fashion from the beginning of the First World War to the eve of the earthquake. In many ways, Japan's First World War experience amplified elite anxieties. During the conflict, Japan further developed as an industrializing, urbanizing, and ultimately wealthy power, emerging as an international creditor for the first time in its history. Japan's newfound wealth, however, was not distributed equally, and, when savaged by severe inflation, wartime boom, and postwar bust, urban industrial workers experienced marked economic and social dislocation. More troubling to elites was the fact that people expressed their dissatisfaction, often in a confrontational or violent manner. The number of organized labor disputes is just one indication of this trend. In 1914 Japan experienced fifty organized strikes involving 7,904 workers; in 1919 over 60,000 workers had taken part in labor disputes.[35] Protests in Tokyo for universal male suffrage, moreover, attracted nearly 70,000 participants in 1921. While large, these figures paled in comparison to the roughly two million people who took part in the 1918 Rice Riots that erupted over the rapid increase in the price of rice.

Concurrent with displays of social unrest, commentators also witnessed a rapid increase in the consumption of items deemed luxuries by an emerging middle class and the new rich. Some elites claimed that the consumption or purchase of alcohol, tobacco products, perfume, and cosmetics had not just ballooned but skyrocketed between 1914 and 1923. Speaking in 1924 on the first anniversary of the calamity, Home Minister Wakatsuki Reijirō suggested that Japan had fallen prey to the "evil custom of overindulgence" and been tyrannized by a "restlessness of national thought." Wakatsuki, who served as finance minister (1914–1915), home minister (1924–1926), and later prime minister (1926–1927, 1931), claimed that alcohol consumption in Japan had jumped from 14.2 liters per person in 1914 to 22.7 liters by 1919. Moreover, he suggested that nearly 24 billion cigarettes (419 per person) had been smoked in 1921, up from the 7.5 billion consumed in 1914, a figure that equated to 142 per person. More astonishing to Wakatsuki was this simple fact: spending on alcohol, tobacco products, and precious gemstones reached ¥21.8 million in 1920. This figure equated to roughly 9 percent of what Japan's government spent on its army. Wakatsuki concluded that Japan had been cursed by twin evils since the outbreak of the European war.[36]

It was in this environment that the Great Kantō Earthquake struck. If wartime trends amplified elite anxieties, the 1923 disaster gave these fears tangible definition. The earthquake was a unique, apocalyptic, and in many ways galvanizing event. It allowed a diverse group of social commentators to consolidate around a meta narrative of catastrophe that drew on and reinforced their opinions that Japan was mired in a state of decline and degeneration. That such opinions were shared by a wide cross-section of people points to the fact that large numbers of elites were gravely concerned about the state of Japanese modernity. It was a fragile existence. In the course of becoming modern, they feared, many Japanese had lost vitality and Japan itself had grown weaker. By spurning the values of diligence, sacrifice, frugality, loyalty, moderation, obedience, and resolve, the people of Taishō Japan had threatened much of what their predecessors had accomplished.

The behavior of people had thus triggered divine intervention. With this emotive construction of calamity, elites possessed a vast narrative landscape on which to level specific critiques about the perceived state of Japan, its people, and their thoughts and actions. Heaven's punishment had not been indiscriminate, they argued, but rather targeted. The people whom commentators blamed for inviting heaven's wrath tell us much about elites' specific anxieties. The earthquake disaster became a tool for critics of all backgrounds to voice their concerns about the perceived degeneration of Japanese society. Individuals used the disaster to admonish, berate, and challenge Japanese subjects to confront and reverse "the decline of the popular mind" (*jinshin*) that many argued had manifested itself through excessive materialism and luxury-mindedness, individualism and selfishness, frivolity, decadence, and hedonism.

USING CATASTROPHE TO ADMONISH THE PEOPLE: MATERIALISM, GREED, AND LUXURY-MINDEDNESS

General Ugaki was one such individual who used Japan's catastrophic earthquake to express his displeasure with the course he believed Japan had taken prior to the calamity. Ugaki had, on many previous occasions, voiced concern over what he saw as the sinister spread of Western culture, ideologies, and particularly exploitative capitalism in Japan. Not surprisingly, he emphasized that the disaster had struck the center of Westernization in Japan, Tokyo. Ugaki argued that the earthquake was an act of "divine punishment" against "a nation that aspired to a culture of materialism and degrading thoughts." Who in particular did Ugaki believe had recklessly aspired to and furthered materialistic culture? He focused his ire on wealthy industrial and mercantile capitalists who had become singularly focused on making money to secure further material comforts at the expense of others. This, he concluded, led to a widening gap between rich and poor. As a result, most of Japan's cities had experienced an explosion of labor agitation and decreased social stability. Elected politicians, he argued, had done nothing to arrest this dangerous and unhealthy exploitative trend because they had become too closely tied to business interests. Heaven therefore took it upon itself to intervene and destroyed the capital.[37]

Other individuals also employed the catastrophe to rail against the evils of materialism. Future minister of commerce and industry Tawara Magoichi lamented that the "economic boom brought about by the European War had produced numerous war profiteers" who singularly pursued ever increasing profits. Their quest for capital gains with which to secure further material possessions not only "belittled those who worked hard and honestly" but led to resentment and growing incidents of labor unrest and social protest for those who had been exploited. More damning, Tawara suggested that the war rich—often referred to as nouveau riche or, in Japanese, *narikin*—indulged in a "luxurious loose life, swaggering about like big shots." In doing so, he argued, the *narikin* "forgot the true national character of the Japanese race" and turned their back on moderation, frugality, and diligence. To Tawara, the nouveau riche demonstrated a selfish, individualistic streak that was harmful to the nation. "This earthquake," he wrote, "has revealed how selfish today's society is." Tawara asked his readers, "Did behavior like this not deserve divine punishment?" "Yes," he concluded, it and the number of rich people of property who "became the property-less in a single morning" attested to Heaven's will. Tawara suggested that in destroying assets, Heaven clearly demonstrated the ephemeral quality of material possessions and wealth, even in its most basic and seemingly most secure form, property.[38]

Waseda University professor Hoashi Ri'ichirō expanded on the theme that the earthquake illustrated that material possessions were the essence of impermanence. Hoashi argued that it was no coincidence that Tokyo, "where the essence of our material civilization had gathered, was turned into ruins."[39] The disaster illustrated heaven's specific displeasure with human attachments to material constructions. Fukasaku Yasubumi was even more explicit, telling his readers that "this heaven-sent disaster has taught us" in no uncertain terms that material wealth, which he argued many people blindly associated with "nice garments and tall buildings," was ultimately illusorily. "Now that people are naked both materially and mentally," Fukasaku concluded, "perhaps people will see that wealth is not reliable and that humans must look elsewhere for something to rely on and be proud of."[40] In Fukasaku's opinion, the earthquake was a violent leveler of wealth inequality. He hoped that it could sever slavish attachments to material culture and possessions once and for all.

Other elites were even more forceful in tying heaven's displeasure—as manifested through the earthquake and fires—to the material-driven, consumption-oriented thoughts and actions of Japanese. One of Japan's strongest proponents of the divine punishment interpretation, Okutani Fumitomo, was also one of the nation's chief critics of what he described as the rise of frivolous, immoral, materialistic consumerism that, he argued, defined post–World War I Japan. Okutani focused his interpretative lens and criticism on the physical space and areas of Tokyo known for harboring places of materialism and consumer spending. "The areas that served to satisfy people's greed and desire," Okutani wrote, in particular the upscale shopping districts, had all been destroyed because of divine will.[41]

The Ginza was one such district singled out by Okutani. Since before the turn of the century, the Ginza district had become known as *the* upmarket, modern shopping area. This was by design. Following a fire that destroyed much of the Ginza on February 26, 1872, government leaders from the Ministry of Finance endorsed plans to construct a new, Western-looking precinct that embodied Japan's modernization and pursuit of civilization and enlightenment. Designed by Irish architect Thomas Waters, the Ginza looked modern in a Euro-American context. It made a strong impression on foreigners and Japanese alike with brick buildings, flashy businesses, provocative cafes, wide roads, paved sidewalks, gas lights, and eventually tramcars.[42] Moreover, it was a place for up-and-coming professional and business elites to be seen strolling down the paved sidewalks and looking smart, visiting European inspired cafes and restaurants, and, of course, shopping. The Ginza retained its charm and allure up until the afternoon of September 1, 1923. Fujimoto Taizō, one writer among many who attempted to capture the flavor of the Ginza, described it in 1915 as follows:

> The name Ginza recalls to all minds the most flourishing street in Tokyo. . . . Buildings on both sides of the street are all in European style, constructed of brick and stone. Large buildings are all occupied by big merchants and watchmakers, jewelers, dealers of foreign goods, and bazaars [where] . . . fine articles hung against the walls, such as hats, neckties, shawls, and umbrellas, stimulating the desires of purchasers.[43]

In describing the people who filled the Ginza, especially those he described as "the young of new tastes," Fujimoto wrote:

> Most of the men and women who wander about the street in the evening are rather of richer rank and come round here for shopping by the way of taking a walk. While you are looking, the people pass the cross-roads—young gentlemen in European dress of the latest style, beautiful daughters in gay garments, accompanied by their parents or maid-servants, happy couples in honeymoon, debauchees hand-in-hand with geisha.... Customers of the Ginza shops are proud of what they have purchased here and, all the goods and articles of Ginza being believed to be of the first rank, don't care about the higher prices. Articles of the same kind and quality can be bought at lower prices in the streets of Yotsuya or Kanda district, and yet ladies living in the Bluff quarters [high city] come down to Ginza in the evening by taking trams or by *rikisha* to satisfy their vanity for their neighbors and friends. If their clothes or articles are wrapped in the paper with signs of any Ginza shops, they think it an endorsement for the goods of first and best class. [After shopping] people enter the Café Lion. Here they can prefer any kind of wine, European and Japanese—beers, Masamune (genuine Japanese sake), whisky, liquor, vodka, and so on—and may pine after the evening of Paris or London, or dream of the pleasure at Berlin or St. Petersburg.

Eight years after Fujimoto penned his description, Okutani likewise drew attention to the consumption-oriented atmosphere and excesses of the Ginza. He also chastised the well-to-do, self-indulgent people who journeyed there. It was precisely because the unhealthy pulse of materialism, luxury, and immoderation flowed through the shops and streets of the Ginza that, in Okutani's opinion, the deities had destroyed it. "The streets of Ginza," he wrote, "where people strolled enjoying the quintessence of city atmosphere, are now mere ruins." More pointedly, Okutani suggested that Mitsukoshi, Japan's largest department store and a center of consumer spending "where crowds had flocked to satisfy their vanity, had been turned into a pathetic wreck." The gods' intentions, at least to Okutani, were clear. He destroyed the parts of the city where unhealthy extravagance had concentrated. To "prove this point," Okutani asked his readers to simply look at a map: the areas where profits and material culture were not pursued, "the residential areas of Koishikawa, Ushigome,

Kōjimachi, Akasaka, and Azabu," had been "hardly damaged." While admitting that he "felt sorry about the hundreds of thousands that died" in the catastrophe, Okutani also stated categorically that Japan was fortunate to experience "heaven's punishment." The earthquake provided unequivocal evidence of heaven's displeasure with the immoral state of society. Okutani hoped that the disaster would be recognized as such in order to complete what the calamity had started: the process of "sweeping the corruption of society away" forever through national reconstruction.[44]

The burned out shell of the once proud Mitsukoshi Department Store—an icon of consumerism in preearthquake Tokyo—also drew Horie Ki'ichi's attention. The Keio University economics professor reaffirmed Okutani's conclusion that the destruction of Mitsukoshi lent terrestrial evidence and "truth to the theory that this earthquake was heaven's punishment for the evil deeds the people of Tokyo conducted over the years." Perhaps due to his economic background, Horie often made a distinction that Okutani and others did not between consumerism and what he saw as the real evil of his time: the pursuit of luxury and extravagance, *shashi*. The wasteful "habit of luxury," in Horie's mind, was the true reason that the gods had punished Japan through the Great Kantō Earthquake. As a longtime champion of increasing tariffs on luxury goods and the introduction of laws, taxes, and disincentives to dissuade individuals from "wasting money on unproductive items" and pursuits, Horie took great pains to remind his readers that it was the center of luxury and extravagance that burned down after the quake. "The tall building of Shirokiya," a rival department store to Mitsukoshi, was "reduced to ashes within a second," while the extravagant "jewelry that decorated the windows of Hattori Tenshōdō, Mikimoto, and so on all disappeared."[45] This was no coincidence in Horie's view. Wasteful luxury, he claimed, also took the form of rich food and entertainment, and the gods were no less merciful on the districts and quarters renowned for upscale restaurants, dance halls, cinemas, and theaters.

"What clue should we glean from all this," Horie asked his readers in the November issue of *Chūō kōron*. The answer, he claimed, was obvious to all those who looked objectively at where the destruction was meted out in Tokyo. Japan's citizens were being punished for their pursuit of luxury and extravagance and, in more general terms, their self-indulgent and decadent behavior. Many business and industrial leaders, he claimed, "had made utterly unrealistic and excessive profits over the years of the

FIGURE 4.1 Mitsukoshi Department Store before the earthquake

Source: Postcard, author's private collection.

FIGURE 4.2 The shell of the once grand Mitsukoshi Department Store following the earthquake and fires

Source: Postcard, author's private collection.

war boom ... and had exhibited no end to their pursuit of luxury." The 1920s financial panic and economic downturn, in Horie's mind, had done little to rectify peoples' behavior. Even a recession, he suggested, could not break the insidious habit of luxury and reorient people back to a life of discretion and moderation. Heaven had therefore stopped—in the most ominous way possible—wasteful expenditures on luxuries by eliminating the items and the physical space in which they were sold. "The hard blow of the gods [shin'i]," Horie hoped, would highlight the immorality of the age and halt the hollow pursuit of luxury across the nation. "As the Heike Family showed us [at the end of the Heian period, 794–1185], pride goes before a fall, and after the height of pleasurable life comes the sorrow."[46]

People with no such economic background also employed the divine punishment interpretation to admonish Japanese people for exhibiting the traits of luxury-mindedness and extravagance and happily paying inflated prices for goods. Fukasaku highlighted this point by mentioning purchases he witnessed of "obis [a traditional sash or fabric belt worn with kimonos] that cost thousands of yen each and clothes of hundreds of yen that were displayed in the windows of big department stores in Tokyo."[47] It was the worst form of luxury—the very definition of wasteful excess—Fukasaku suggested, to pay excessive amounts of money for ornate, fashionable, or high-end names when cheaper and simpler alternatives existed. More galling to commentator Yamada Yoshio was the fact the people who frequented the upmarket department stores happily paid inflated prices for items in part because they were expensive, thus attaching value to the price as much as the item itself. It was not only the "new rich" he argued, who "devoured luxuries" for the sake of consumption, but increasing numbers of middle-class urban professionals:

> We witnessed with our own eyes that goods sold like hot cakes when they were advertised with a price several times as much as a fair price, while nobody paid attention to items that were priced fairly. At the department store that claimed to be the largest in the Orient, items were selling like hot cakes at ¥50 while the same items were five dollars in the USA (equivalent to ¥10 by the exchange rate).... The sellers had no moral principles; the buyers possessed a dangerous, loose tendency.[48]

The situation, Yamada concluded, was deplorable, and people's behavior had led to direct intervention from the heavens to correct this imbalance.

He wrote, "What a work of the universe—giving it a shake to correct the world that has lost the moderate path."[49]

Tawara Magoichi suggested that the excessive materialism present in postwar Japan was symptomatic of two terrible processes: a widening gap between the rich and poor that had resulted from the proliferation and blind acceptance of Western-style capitalism, and an overemphasis on individualism. Materialism, he argued, was the accumulation of money and possessions beyond what was needed for an honest, hard-working life. In the years prior to the disaster, he claimed, many Japanese had fallen into the trap of consumer-driven materialism and had accumulated money and possessions far in excess of what was needed. This led to resentment, social problems, and what he classified as the destruction of "the universal spirit of mutual support" that had made Japan so unique. Materialism had vanquished this wholesome spirit, and it had been replaced by the great "evil of modern civilization: the excessive emphasis on individualism." Moral philosopher Shimamoto Ainosuke agreed with Tawara's appraisal and suggested that the "harmful influence of Western culture" had grown steadily since the Meiji Restoration and had become increasingly acute since the turn of the century. Was it no wonder, he asked, that the heavens (*ten*) had "dealt a crushing blow to our civilization that had gone unnaturally too far?" "No," he replied, the earthquake tragedy heralded a need for fundamental rectification in Japan.[50]

Political satirist Kitazawa Rakuten, who had made his early career through the magazine *Tokyo Puck* (*Tokyo Pakku*), captured one of the dominant discourses related to the divine punishment interpretation of the disaster in a colorful cover image on the October 14 edition of *Jiji manga*. In this print Kitazawa placed an aggressive catfish with human arms thrusting a black cloak with the words "*shitsujitsu gōken*" (fortitude and earnestness) on a gaily dressed woman. At the same time, the catfish is tearing away from the woman a layer of clothing marked with the words "*kyoei kyoshoku*" (vanity and ostentation). Entitled "Catfish Rectifies the Evil Trends in Society," the print also alludes to the sexual dimension of degeneration that many elites articulated. Numerous commentators saw women not only as the primary consumers of expensive, frivolous, and ostentatious luxury items but also as the personification of lustful, carefree frivolity that many suggested defined the age. Both were responsible for evoking heaven's displeasure.

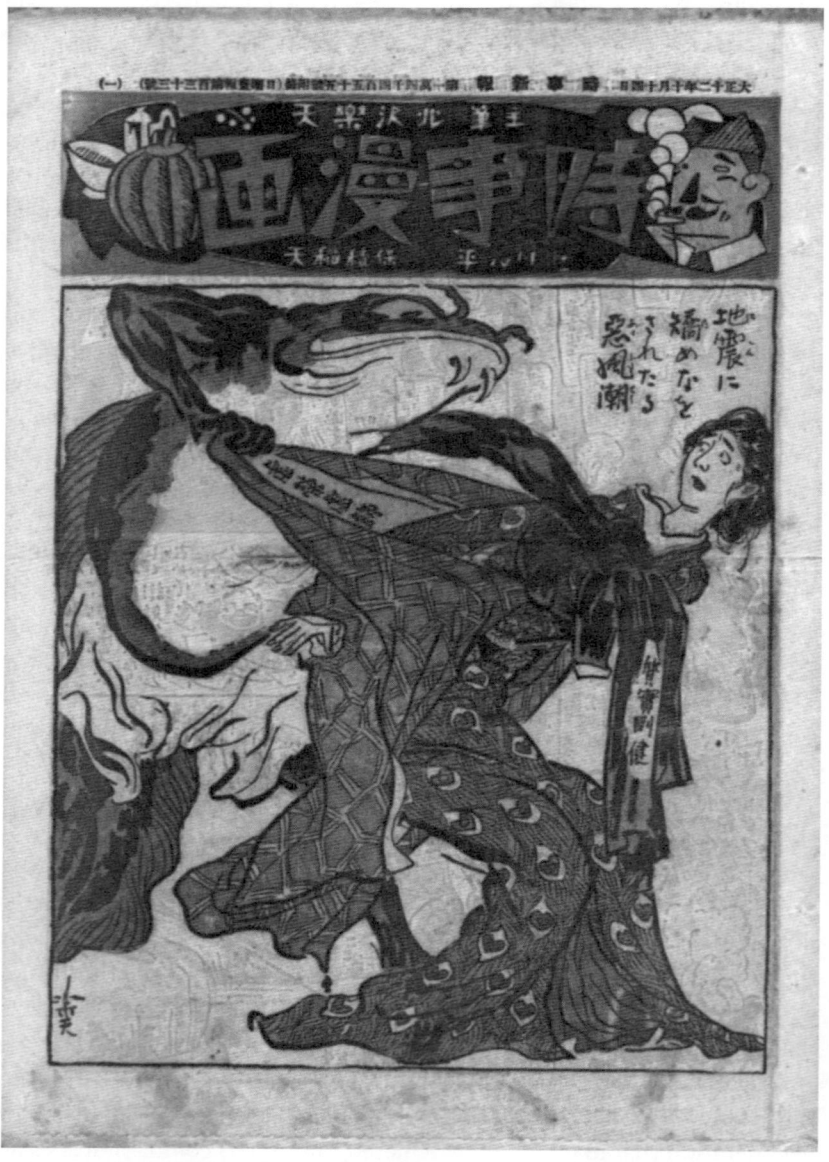

FIGURE 4.3 The catfish rectifying evil trends in society by replacing the cloak of vanity and ostentation with the cloak of fortitude and earnestness

Source: Jiji manga no. 133 (October 14, 1923), cover image. Ohio State University Billy Ireland Cartoon Library & Museum.

USING CATASTROPHE TO ADMONISH THE PEOPLE: DECADENCE, HEDONISM, FRIVOLITY, AND INDIVIDUALISM

In the months following the Great Kantō Earthquake, commentators found common ground in using the "heaven-sent" earthquake to admonish against far more than just perceived excesses in the economic sphere. They cast their gaze and moral aspersions toward Japanese for exhibiting hedonistic and decadent social practices and loose sexual morals, partaking in frivolous popular amusements, and supporting eccentric and radical political ideologies. Importantly, the perceived decline in morals was not limited to the *narikin* who possessed excess capital to spend on items deemed luxuries. In the minds of many elites, lower- and middle-class individuals also exhibited frivolous and harmful tendencies and wasted precious time and money on a multitude of insalubrious pursuits. While some individuals used the divine punishment interpretative framework to imbue and bolster their criticisms with cosmological legitimacy, others simply used the opportunity created by the destruction of Tokyo and the debates concerning its reconstruction to reiterate concerns about the declining moral underpinning of what many saw as modern, urban society. Commentators saw Tokyo as not just the imperial capital and the largest city of the empire but also the epicenter of immorality. It was a place that Tenrikyō relief volunteer Haruno suggested embodied, personified, and exemplified the "total spiritual corruption of mankind."[51]

The perceived slackening of proper sexual mores by urban Japanese was one target of condemnation. Fukasaku Yasubumi complained that modern city life, with all its temptations, helped foster promiscuity. In his mind, the large, informal, crowded urban space also created a greater number of opportunities for people to engage in anonymous acts of sexual depravity such as adultery. Adultery was easier to undertake in a densely populated city where men and women interacted freely, increasingly in the workplace, with fewer overt social controls governing behavior. Worst of all, Fukasaku believed that the increased frequency of adulterous relations made people more accepting of such immoral acts of lust and expressions of sexual desire.

To support this assertion, Fukasaku made reference to the Arishima Incident, a highly publicized affair and lovers' suicide that occurred just prior to the earthquake. That the perpetrators of this salacious event garnered "unreserved admiration" in many tabloid newspapers, Fukasaku lamented, illustrated how widespread the moral decay had become.[52]

This story broke in the summer of 1923 when the bodies of writer Arishima Takeo and his mistress, Hatano Akiko, were discovered at Arishima's holiday home in Karuizawa, Nagano Prefecture. Arishima was a writer who espoused socialism and other left-wing ideas and railed against bourgeois modern culture, capitalism, and consumerism. In many ways Hatano was Arishima's ideological, as well as physical, bedfellow. To many, Hatano was the epitome of the "new woman." Not only was she employed in what had been a male-dominated elite profession—writing, editing, and publishing—but she also worked for the women's journal *Fujin kōron* (Women's review). From this position she regularly espoused left-wing ideologies and championed women's rights. That she was also married, and thus an adulterer, added further sensationalism to the story.[53] Adultery was consensual and existed outside the bounds of arranged marriages and was thus described by an increasing number of social critics as the embodiment of harmful lust.

Newspapers, particularly the *Tokyo Asahi shinbun*, followed every aspect of this developing scandal and reported that the lovers had agreed to kill themselves when Hatano's husband gained knowledge of the affair. Newspapers reported that Hatano was adorned with expensive garments and surrounded by many accessories that were often derided as "luxuries." Arishima's body was also described as "well dressed," with over ¥200 and seven suicide notes found in his coat pocket.[54] While some newspapers emphasized the immorality of the whole affair, Fukasaku pointed out that the couple had also won considerable sympathy as star-crossed lovers, however hypocritical their ideas about bourgeois consumerism seemed given their appearance at death. To Fukasaku, such consideration was unfortunate. This affair, the double suicide, and the sympathetic response the deceased lovers achieved illustrated to him the "shameless praise for lust" associated with modern society. In his opinion, all three parts of this incident were "deplorable."[55]

Philosopher Nose Yoritoshi was even more explicit in critiquing what he described as the sexual degeneracy of 1920s Japan. In an article on the necessity of spiritual reconstruction following the 1923 earthquake, Nose emphasized the grave importance of combating the "obscenity and weakness" associated with ideologies, ideas, and actions based on the concept that he referred to as "freedom of lust." This freedom entailed the willful expression and attainment of sexual desire, sexual liberation, freedom of instinct, and the supremacy of love and lust. Why did people pursue this "new freedom"? The answer was simple: "People claimed the freedom of lust and supremacy of love only to satisfy their sexual desires." Whatever people might say, Nose argued, freedom of lust was about nothing more than selfish gratification. This was the most base and dangerous form of individualism and it threatened social stability. Nose wrote that lust weakened the moral foundation of society and dissipated people's spirit. Moreover, following the freedom of lust regardless of the social and legal consequences was the ultimate act of selfish, social irresponsibility. Such thinking challenged public morality, in Nose's opinion, and was increasingly difficult to combat because it had invaded literature, popular magazines, and the arts and also dominated popular entertainment in venues such as dance halls and theaters. "The supremacy of lust" was becoming so widespread, Nose suggested, that proponents of this ideology were out to establish new manners and customs and form a new moral framework for accepted social behavior. They also sought, Nose concluded, to create new forms of art and expression, all with acceptance of lust and sexual liberation at their core. "What a worrisome future our nation has!" Nose prophesized. "Lust turning red was more dangerous than ideologies turning to the left," and Japan was in the midst of both. "The earthquake," he concluded, was a "test given by the gods," and all Japanese had to "stand up resolutely and support a true restoration."[56]

Not surprisingly, few elites who voiced concern over lust, adulterous or otherwise, also expressed concern about the upper-class practice of having concubines or frequenting geisha. Moreover, male lust regulated within the confines of Japan's well-established, state-licensed prostitution system did not draw their wrath. Despite the fact that many hundreds of prostitutes burned to death or drowned within Yoshiwara because their "owners" would not let them flee, little opposition to either the system or its reestablishment was voiced. When individuals did speak against prostitution, their voices were ineffectual in restricting its return. Much to the dismay of people such as Abe Isoo, the government did not intervene with a heavy hand and outlaw prostitution following the destruction of Yoshiwara and the deaths of so many prostitutes.

In the Imperial Diet on December 22, 1923, member Matsuyama Tsunejirō introduced a representation against the reconstruction of the licensed prostitution quarters of Yoshiwara.[57] As a first step toward abolishing licensed prostitution altogether, Matsuyama urged the government to forbid the physical reconstruction of brothels and to disband those that had already reestablished their trade. Various newspapers reported that the brothels of Yoshiwara were in fact some of the first structures to be reconstructed in the capital.[58] Though Matsuyama received vocal support from Diet member Hoshijima Jirō, numerous other politicians challenged and ridiculed this representation. The lust that was seen as such a threat to Japan's moral order, as Sheldon Garon has suggested, was not "managed" male lust but rather unregulated female sexuality, which many believed urban, middle-class lifestyles in Tokyo fostered.[59]

Many individuals viewed sexual degeneracy, moral decay, and other lust-driven pursuits as symptomatic of a larger "corruption of the popular mind" that had, at its core, excessive emphasis on individualism and selfish gratification. Each affliction contributed to, yet resulted from, what select commentators saw as the excessive hedonism, debauchery, and atmosphere of decadence that permeated post–World War I cultural life. "The general trend of our society," Ichiki Kitokurō wrote, "was frivolous, decadent, materialistic, and luxurious, in which everyone pursued pleasure."[60] No place in Tokyo better illustrated Ichiki's point than Asakusa, the district where many critics believed immorality pervaded virtually every type of entertainment available and defined the everyday interactions of people. If the Ginza stood at the epicenter of upscale gratification where the new rich and those who aspired to material culture gathered to satisfy their habit of luxury and consumption, Asakusa stood at the epicenter of lowbrow hedonism, decadence, and frivolity. A number of commentators concluded that its destruction, along with that of the surrounding wards of Honjo, Fukagawa, Nihonbashi, and Kyōbashi, was a targeted sign from the heavens illustrating their contempt, anger, and displeasure at the social mores and actions of degenerate working-class Japanese.

Following the disaster, people spent as much time admonishing the purveyors and practitioners of debaucherous pursuits in Asakusa as they did the highbrow consumers who "devoured" luxuries in the Ginza. Okutani was one such individual. The "hedonistic lifestyle" epitomized by Asakusa, which, Okutani suggested, was nothing more than a "gathering place for the fulfillment of unhealthy desires and hedonistic pleasure," had led to a concentration of sin at the nation's core. The gods intervened, he argued, not only to admonish and destroy but to also help chart a new course, materially, physically, and socially, for Tokyo and Japan.[61] Economist Horie Ki'ichi was quick to point out that along with the expensive jewelry shops, department stores, and upscale restaurants that were destroyed in the Ginza, virtually all the lowbrow "cinemas, story-tellers' theaters," and other hedonistic entertainment centers had likewise "vanished" overnight. How "appropriate," he concluded in an article entitled "Tokyoshi no saigai to keizaiteki fukkōan" (The disaster of the city of Tokyo and the economic restoration plan) published in November 1923. Horie argued that Japan had reached the summit of international decadence, hedonism, and frivolity in the years following the First World War. Just look at the number of "dancers," "musicians," and "other entertainers" who could not get a showing in Europe because of "national exhaustion and frugality," but who came to Japan and "thrived." Though "thinking people" had warned government leaders of the dangerous and growing trend of hedonism, complaints had fallen on deaf ears. The downward spiral of degeneracy had only halted, he concluded, because the earthquake destroyed Tokyo's entertainment heartland. For the sake of Japan, he hoped it would never return.[62]

Well before 1923, Asakusa had developed a reputation as the hedonistic entertainment district of Tokyo. In 1915 travel writer Fujimoto Taizō devoted the first chapter of his book on the "nightside of Japan" to Asakusa. His first sentence summed up one popular perception of Asakusa shared by those who condemned and enjoyed it: "Asakusa is the centre of pleasure in Tokyo." Fujimoto continued, "People of every rank in the city crowd in the park day and night—old and young, high and low, male and female, rich and poor. It is also a haunt of ruffians, thieves, and pickpockets when the curtain of the dark comes down over the park."[63] Thereafter, Fujimoto spent the better part of thirty pages describing in rich color and detail the pleasure-oriented stimuli open to visitors of Asakusa, including movies, a circus, street performers, picture galleries, picture studios, alcohol, sex with prostitutes, tobacco, gambling, music, cakes, fruit, tea, and coffee. According to Fujimoto, there was no place better for a visitor, particularly of limited means, to waste money, forget the troubles of the day, and enjoy a hedonistic, dreamlike experience.

Asakusa emerged as a modern entertainment center in the mid-1880s with the concentration of new amusement venues in its fifth and sixth districts. With the addition of a small zoo, circus grounds, and a panorama hall where spectators viewed projections, Asakusa became a place known for fun. No building, however, better symbolized Asakusa's rise as a center for modern entertainment than the twelve-story tower Ryōunkaku (Cloud Surpassing Pavilion). Completed in 1890, this red brick, octagonal tower stood as one of the tallest buildings in the Eastern Hemisphere. It dominated Asakusa's skyline. It also possessed many firsts for a building in Japan, including an elevator, an observation deck, and telescopes

FIGURE 4.4 The twelve-story tower dominating the skyline of Asakusa

Source: Postcard, author's private collection.

for viewing the city. At night, along with light escaping from the tower's 176 windows, the building was illuminated by a series of large electric arc lamps. The icon of Asakusa also developed a noticeable tilt as a result of a large tremor that hit Tokyo in 1894, forcing its owner to erect reinforcing steel girders within the building. Inside, the tower housed small shops, restaurants, and art and picture galleries.

To many moralists, the twelve-story tower of Asakusa and its surroundings came to symbolize immorality, hedonism, decadence, and irresponsible behavior. In relaying one story set in the shadow of the tower, travel writer Fujimoto captured what many elite commentators felt was one of the chief curses of Asakusa: the damage that the pursuit of frivolity and hedonism—and the irresponsibility that it fostered—caused to many of Tokyo's poorest people and their families. The tale is set in the Kamiya Bar, a bar that became famous for serving "electric brandy" each night when the streetlights turned on. Amid rowdy drunkenness and consumption, "mostly [by] the lower middle-class people who had just dropped in here on their way home from their day-work," the writer placed a young girl. Fujimoto relates her story as follows.

A girl about ten years old is peeping into the hall [of the bar] from the entrance, and a bar boy having perceived her beckons her to enter. She comes in, and the boy asks her what she wants. The little girl, dressed in dirty cloths and with tousled hair, is shy by the brightness of light in the saloon, and hardly speaks to the kind boy. "I've come to find out my papa. Mamma told me he is in this bar." She is looking round the crowd, but could not find out her father. The boy brings a stool to one end of the hall and let the girl stand upon it easily to see faces of people in the room. Having found out her father at last, she jumps down the stool and, running up to a man sitting by the table and drinking whisky plucks him by the sleeve. Alarmed by the sudden attack the man of some forty years old, and in costume of fishmonger, looks back and, finding his daughter standing by him, he stands up and comes to a vacant space at a corner of the hall. Being somewhat intoxicated, he asks her: "What's the matter? Why have you come here?" "Mamma and I," whispers the daughter, with tears in her eyes, "have been waiting for you come back. We don't take supper yet. Mamma told me you must be in this bar again, and to find and take you home. Come home with me at once, papa." The father awakes from his dream and, after paying accounts, he drops out of the bar, accompanied by his obedient daughter.[64]

From a moralist's perspective, we have not only a lower- to middle-class laborer of limited means wasting money on unhealthy pursuits but also a father shrugging off family responsibilities. The girl's presence informs us of a mother who uses her child—placing her in an unhealthy environment—to summon her derelict husband. We also see a bar full of people, unfazed if not entirely accepting of a young child running wild among them, even standing on furniture.

Given Asakusa's persona and the symbolism attached to the twelve-story tower, it is not surprising that many popular depictions of the catastrophe focused on the destruction of this area and on the iconic tower in particular. Few visual images associated with the earthquake are more striking than lithograph number one of a series from the collection *Teito daishinsai gahō* (Pictorial account of the great imperial capital earthquake). Entitled *Asakusa kōen jūnikai oyobi hanayashiki fukin enshō no jōkyō* (The spread of fires around the twelve-story tower in Asakusa Park and Hanayashiki), this print depicts a scene to which virtually all Tokyoites who attempted to flee the fires could relate. It depicts the very definition of pandemonium. The prominent backdrop to the image, naturally, is the broken, smoldering tower. The frivolity and carnival-like atmosphere of the Hanayashiki district of Asakusa, as depicted by a large circus elephant, caged animals, and performers, has turned to complete chaos as crowds who once packed this area for fun and entertainment are trapped. The modern attractions that drew people to this place of fun, as symbolized by the goods stalls and shops, circus animals, performers, and buildings, restaurants and bars, power lines, poles, and bright street lamps, all suffered catastrophic damage. Where once people went almost exclusively to be stimulated by sights, sounds, and hedonistic experiences, customers were now fleeing in terror. Often considered an escape from everyday concerns and life, Hanayashiki was now a place to escape from by panicked people jolted back to reality.

Of all of the damaged buildings that personnel from the Imperial Japanese Army martial law headquarters demolished following the earthquake, the twelve-story tower assumed a place of notoriety. It was one of the first and most highly publicized demolitions conducted. In front of reporters and photographers, whom army officials had transported, demolition teams posed for photographs at the shell of the building. At 3.40 p.m. on September 23, as the *Tokyo nichi nichi shinbun* proclaimed, it was brought down forever.[65] Upon its final demise, reporters wrote that

FIGURE 4.5 The spread of fires around the twelve-story tower in Asakusa Park and Hanayashiki

Source: Lithograph print, author's private collection.

FIGURE 4.6 Hanayashiki after September 1, 1923

Source: Postcard, author's private collection.

crowds clapped and shouted with approval. This scene was vividly recreated in Kawabata Yasunari's novel *The Scarlet Gang of Asakusa*. In this novel the chief protagonist recounts, "We all cheered—hurray, hurray—and then burst out laughing. Remember [when the tower came down]?"[66] The symbolism of the tower prior to the disaster and the extent to which military personnel went to showcase its destruction suggest that this was more than just a simple demolition. Like the tower, its final fall was pregnant with meaning, perhaps symbolizing an end and certainly a new beginning for the site, and maybe for the nation.

DIVINE PUNISHMENT OF A DEGENERATE SOCIETY IN PERCEPTION AND REALITY

Given the large number of commentators who used the earthquake to suggest that Japan was sinking into an abyss of degeneracy, Nonomura Kaizō, vice principal of Waseda University High School No. 1, took it upon himself to investigate if these opinions reflected reality. He did so through an article entitled "The Destruction of Modern Civilization (*Kindai bunmei*

FIGURE 4.7 The twelve-story tower of Asakusa before and after the earthquake and fires

Source: Postcard, author's private collection.

no hakai)," published in the December 1923 issue of the popular education journal *Kyōiku jiron*. Nonomura argued that people in 1923 could debate the extent, causes, and symptoms of society's decline for years and reach no conclusion if they looked solely at contemporary events. Some people might see items considered luxuries as nothing more than modern conveniences, and pursuits classified as debaucherous as nothing more than lighthearted, harmless entertainment. Others, of course, saw evil in both. What people needed was historical perspective. Only with the help of history and in particular awareness of historical contexts, Nonomura added, could people see that Japan's "modern civilization was heading toward destruction."[67]

The history that Nonomura employed to help demonstrate his case of decline was that of Rome in the last years of Marcus Aurelius (A.D. 121–180). At "that time," Nonomura wrote, "Roman civilization had already begun to decline but Romans were unaware of it," just as the majority of the people in preearthquake Japan were likewise oblivious to the specter of decay lurking over society. In Nonomura's opinion, Rome exhibited many signs of decline, including a laxness in official discipline, increased governmental corruption, trends toward luxurious housing, clothing, and adornment, relaxed sexual morals, acceptance of divorce and marriage for the sake of love, a decline in religious faith, wealthy, educated people having fewer children, heightened rates of proletarian reproduction, increased incidents of civil strife associated with people agitating for greater freedoms and rights, and rapid population migration to cities as people increasingly believed urban centers to be the pinnacle of civilization. Nonomura asked whether Japan was not suffering from many of these same afflictions in 1923. "Yes," he replied. Unfortunately, many Japanese of 1923 as well as Romans saw such developments as nothing more than logical manifestations of an advancing, wealthy, increasingly participatory civilization. In fact, he argued, many in both time periods saw these manifestations as evolutionary, while only a small minority of educated elites saw them for what they were: symptoms of degeneration.

Japan, Nonomura concluded, was fortunate because it had been given a heavenly warning to alter course while Rome was given none. The 1923 catastrophe was an announcement that society was in the final stages of decline, and all Japanese, he suggested, should chart a new course. Okutani developed a similar interpretation. The Tenrikyō priest, however, went further than Nonomura and told audiences around Japan that they "must thank the gods for their infinite kindness" in unleashing this disaster upon Tokyo. The catastrophe had destroyed the most corrupting part of the nation and produced a clarion call for a Taishō restoration.[68]

Nonomura and the large number of elite critics who shared his view of societal decline based their conclusions on perception more than reality. Japan was experiencing many of the challenges associated with modernization. Moreover, many of the thoughts and actions of urban Japanese that drew the scorn of commentators were merely responses to the opportunities, dislocations, and alienation associated with Japan's emergence as an industrializing urban society. They were the by-products of a nation that was becoming increasingly urban, international, wealthy—however unevenly distributed this wealth was—and politically pluralistic.

Statistics show beyond doubt that Japanese society was indeed becoming more urban.[69] The percentage of Japanese who lived in towns or cities larger than 10,000 individuals increased from 18 percent to 32 percent of the total population between 1898 and 1920. The population of Japan's six largest cities (Tokyo, Osaka, Kyoto, Nagoya, Kobe, and Yokohama) more than doubled, from 3.04 million to 7.63 million, between 1897 and 1920.[70] Moreover, the number of cities with a population surpassing 100,000 doubled, from eight to sixteen, between 1898 and 1918. Part of this increase was due to Japan's overall population growth from roughly 42 million to 56 million over these years. Mostly, however, urban population growth was due to internal social migration. This trend accelerated throughout the 1920s. Between 1920 and 1925, 2.99 million individuals left towns and villages with a total population under 20,000. The vast majority of these internal migrants ended up in cities of over 40,000 inhabitants. During this five-year span, social migration of people accounted for 73.4 percent of the population increases in Japan's six largest cities with populations of over 400,000.[71]

Along with becoming more urban, Japan also industrialized markedly during the twenty-five years to 1923. Factories across the nation increased by 27 percent, from 32,124 in 1909 to 43,723 in 1919.[72] Factories also got larger and became more impersonal. The number of factories that employed over 1,000 workers nearly tripled, from 58 in 1909 to 160 in 1919, while the number of factories that employed 500 to 999 workers expanded from 82 to 202. As Mark Metzler has documented, manufacturing output increased by 54 percent in inflation-adjusted terms during the First World War alone, and exports increased by 47 percent in terms of volume and nearly 300 percent in terms of monetary value.[73] In short, the war amplified trends of urbanization and industrialization. Both factors helped Japan become an international creditor for the first time in its modern history.

The transformation of Japan from debtor to creditor was also reflected among the population. The proliferation of industry and increased profits created a groundswell of wealth in Japan that was heavily concentrated in the industrializing cities. Though wages for most industrial workers rose, as figures later in this chapter demonstrate, profits were concentrated and gave rise to the perception that a class of new rich, *narikin*, had almost exclusively benefited from the First World War. Popular cartoon images of *narikin* lighting cigars with Japanese banknotes, along with other published tales of excess, often drew derision from social critics, bureaucratic elites, and members of the press.

People who believed that the excessive consumption of luxuries was a sure sign of moral decline could easily have their perceptions sustained in more ordinary ways, such as a stroll through the Ginza. Tokyo became a center of consumer spending. It was not, however, just industrialists and new rich who spent money lavishly in wartime and post–World War I Japan. An upwardly mobile middle class with increasing amounts of wealth also emerged and spent. In 1920 businesses in Tokyo employed 21.4 percent of Japan's total white-collared salaried workers—now often referred to as salarymen—in Japan. This percentage had risen from just 5.6 percent in 1908. Tokyo also witnessed a virtual explosion in the number of civil servants and government employees, from 52,200 in 1907 to 308,200 in 1920.[74] To many concerned elites, these were the consumers, along with university students, professionals, industrialists, and merchants who "devoured luxuries."

Virtually everyone, however, was spending more money in the early 1920s compared to, say, 1912. Japanese urbanites spent more because they had more money as a result of higher wages. They likewise spent more because prices for all goods and services were substantially higher in the early 1920s than they had been before or during the war. To use a more technical description, Japan experienced a period of pronounced wealth creation and inflation from 1916 onward as a result of the European war. The great wave of inflation peaked in 1920, but prices and wages on the eve of the Great Kantō Earthquake were still significantly higher than they had been ten years earlier. If, for example, 1912 is used as a base of 100, the consumer price index for Japan stood at 221 in 1920 and 199 in 1922.[75] These figures represent a 121 percent and 99 percent increase, respectively, from the levels of 1912. The increase in expenditures on food was even steeper. If 1912 is again used as a base of 100, the amount spent on food stood at 276 for 1920 and 266 in 1922, representing sizable increases of 176 percent and 166 percent, respectively.[76]

Wages for a number of professions, however, rose higher than the increases witnessed in prices for a variety of essential goods and services over the same period. A *geta* maker, many of whom lived and worked in cottage or home industries in the ward of Asakusa, for instance, made on average 59 sen per day in 1912.[77] By 1920 the same *geta* maker was earning on average 193 sen. The figure for 1922 stood at 202 sen. These numbers represented increases of 227 percent and 242 percent, respectively. A typesetter, a profession not uncommon in the ward of Kanda, experienced a similar increase in salary from the 57 sen earned, on average, in 1912 to 215 in 1922, an increase of 277 percent. Even day laborers, a profession of necessity for many of the poorest inhabitants of Honjo and Fukagawa wards, experienced a significant rise in wages from 1912 to 1922. A day laborer who earned, on average, 58 sen per day in 1912 earned 201 sen per day in 1920 and 218 in 1922. These equated to increases of 247 percent and 276 percent.[78]

Did these individuals, Tokyo's new middle class, who experienced an increase in wages and profits "devour luxuries?" This is a loaded question, of course, because what some commentators derided as luxuries might not have been viewed as such by consumers then or historians now. The Japanese government, however, made the challenge of defining luxurious items—whether Japanese of 1924 or we today agree with such definitions—easier for historians. They did so through the passage of a tariff bill in July 1924 that explicitly singled out luxury goods. Tables 4.1–4.4 provide clues as to how much money Japanese spent on so-called luxuries during the wartime boom and postwar period.

Two items deemed luxuries by the 1924 tariff were phonographs and records. In 1912 Japanese consumers spent ¥2.99 million on these items. In 1920, however, consumers spent ¥21.56 million on the same products, equating to a significant rise of 719.4 percent. Spending on other luxuries increased by even steeper margins. Consumers spent ¥124,000 in 1912 on cameras, camera parts, and photographic film. In 1922 Japanese spent ¥6.83 million on these same goods, accounting for an astronomical increase of 5,405 percent.[79]

Another category of items classified as luxuries in 1924 was cosmetics. Japanese—one would presume upper- and middle-class women—spent ¥836,000 on cosmetics in 1912. In 1922 they purchased ¥28.25 million

TABLE 4.1 Spending on phonographs and records in Japan, 1912–1925

YEAR	AMOUNT SPENT (YEN)	INDEXED TO 1912	CONSUMER PRICE INDEX FOR ALL ITEMS (INDEXED TO 1912)
1912	2,997,000	100	100
1917	9,274,000	309	118
1920	21,561,000	719	221
1922	13,214,000	441	199
1925	11,713,000	391	201

Source: Ōkawa, Chōki keizai tōkei, 6:250–51.

TABLE 4.2 Spending on cosmetics in Japan, 1912–1925

YEAR	AMOUNT SPENT (YEN)	INDEXED TO 1912	CONSUMER PRICE INDEX FOR ALL ITEMS (INDEXED TO 1912)
1912	836,000	100	100
1917	7,237,000	866	118
1920	29,230,000	3,496	221
1922	28,249,000	3,379	199
1925	32,936,000	3,940	201

Source: Ōkawa, Chōki keizai tōkei, 6:249.

worth of cosmetics, representing a hefty 3,279 percent increase over the amount spent in 1912. While the total monetary volume of these luxury purchases was remarkably small in comparison to the total value of food purchased, which in 1922 stood at just over ¥7 billion, the increase in percentage terms far outpaced similar percentage increases in inflation, wages, or the total amount spent on other goods.

Two final items worth mentioning are sake and tobacco. Spending on both increased markedly and drew the scorn of many commentators. Unlike other luxuries that counted as only a small percentage of overall spending, however, spending on sake, which reached ¥1.35 billion in 1922, was no small amount. Total alcohol (sake, beer, and *shōchū*) and tobacco

TABLE 4.3 Spending on sake in Japan, 1912–1925

YEAR	AMOUNT SPENT (YEN)	INDEXED TO 1912	CONSUMER PRICE INDEX FOR ALL ITEMS (INDEXED TO 1912)
1912	378,772,000	100	100
1917	462,174,000	122	118
1920	1,159,699,000	306	221
1922	1,356,260,000	358	199
1925	1,268,053,000	335	201

Source: Ōkawa, *Chōki keizai tōkei*, 6:208–9.

TABLE 4.4 Spending on all tobacco products in Japan, 1912–1925

YEAR	AMOUNT SPENT (YEN)	INDEXED TO 1912	CONSUMER PRICE INDEX FOR ALL ITEMS (INDEXED TO 1912)
1912	91,994,000	100	100
1917	116,007,000	126	118
1920	248,848,000	271	221
1922	271,318,000	295	199
1925	275,801,000	300	201

Source: Ōkawa, *Chōki keizai tōkei*, 6:222–23.

(cigarettes, cigars, shredded tobacco, and cigarettes with tips) totaled ¥1.78 billion in 1922: this sum was larger than Japan's national budget that year.

Increased spending by such significant percentages helped foster the perception that people were spending far more on luxury items than ever before. This is because Japanese were spending more. But spending on luxuries, excluding alcohol and tobacco, still constituted a small percentage of overall spending. It was these dramatic increases in spending over previous years, however, that enabled commentators to argue that the evils of "luxury mindedness" had taken hold of Japanese: perceptions were important.

Securing accurate and reliable statistics on social behavior is more problematic than securing figures on spending, prices, and wages. As Sabine Früstück and David Ambaras have illustrated, most statistics on social behavior are fraught with inconsistencies, bias, and missing pieces.[80] We know, for instance, that the number of movie theaters in Tokyo increased from 44 in 1912 to 112 in 1922.[81] What is more difficult to measure, however, is how many people frequented cabarets, dance halls, or other venues that offered what many considered to be immoral entertainment. It is impossible to know how many individuals engaged in adultery or frequented houses of prostitution.

If increased levels of crime are an indication of a "decline of civilization," then statistics gathered by Takenobu Yoshitarō from the Home Ministry and published in the *Japan Yearbook* paint a picture that is at odds with popular perceptions about the decay of Japan. The number of arrests made by police for robbery, larceny, gambling, and rioting had decreased from 1912 to 1922.[82] Likewise, victims of robbery by force and larceny also declined, though those who suffered from fraud, blackmail, and pickpockets rose. Arrests made by police for arson, battery and assault, and obscenity likewise increased, but in most cases only marginally: cases of obscenity, for example increased from 2,138 in 1912 to 2,150 in 1922. This increase was hardly in keeping with population growth. The number of new convicts decreased from 1912 to 1922, though the number of suicides from love or jealousy—often pointed to by concerned commentators as an indication of the prevalence of freedom of love or lust in society—nearly tripled between 1912 and 1921, from 234 to 637. Statistics go only so far in measuring something as judgment-oriented or as nebulous as social "degeneration" or "decay."

Statistics aside, what we are really dealing with here are perceptions about the state of Japanese society during a transformative period in its history. The decline, degeneration, or "destruction of civilization," as Nonomura and others described the post–First World War era, was not an objective social, political, ideological, or economic fact. Rather, it was a perceived phenomenon constructed by concerned elites, journalists, reform-minded bureaucrats, and government officials. Images of a degenerate society were products primarily of their casual observations, though sometimes supported by more thorough investigations and analysis. All, however, were heavily influenced by prejudices and preconceived notions about people and society at large. As Barbara Sato has skillfully shown, the "modern girl" was in many ways an imaginary by-product of modernization; in much the same way, David Ambaras has illustrated the problem of "juvenile delinquency" as a constructed phenomenon influenced by perceptions. Individuals who had long believed and suggested that Japan was experiencing an era of decline were quick to grasp the earthquake and the devastation it inflicted—seemingly targeted—as heavenly validation for such assertions.

Even though hard facts might not demonstrate the impossible-to-measure decline of society, perceptions and the discourse and policies they eventually fostered were important. They give us clues as to how elites viewed society. To many Japanese, the construction of a declining, degenerate society was assisted greatly by the earthquake catastrophe. This catastrophe was artfully used to bolster and cosmologically legitimate many descriptions and interpretations of a society on the path to ruin. Takashima Heizaburō wrote that in the aftermath of disaster, he had "never [been] more worried about the future of our country."[83] In using the disaster in a selective, opportunistic, interpretative way, discourse makers imbued what was really the by-product of two moving tectonic plates with significant political, social, religious, cultural, economic, and moral meaning. Commentators, with the assistance of a burgeoning mass media, helped define the earthquake as a heaven-sent warning, a test for the people, and an act of divine punishment brought against a society that had strayed down a morally questionable path of immoderation. In doing so, they helped construct the impression of a divinely inspired and targeted tragedy.

This constructed catastrophe, however, also became described as a once-in-a-lifetime opportunity to arrest civilization's decline and ameliorate the afflictions of modern society. If plans for physical and spiritual reconstruction were used effectively to combat the accumulated social ills of Japan, philosopher Shimamoto Ainosuke suggested that the Great Kantō Earthquake would not be remembered as a calamity, however devastating it had been. Rather, people in the future would look back at the earthquake with a sense of nostalgia. Japan's earthquake calamity, he wrote, would be remembered "as the mother of all happiness."[84] Turning disaster into happiness, however, would require considerable effort. These labors began at the fountainhead of where many believed the great evils of modern society emanated: the city.

Five

OPTIMISM: DREAMS FOR A NEW METROPOLIS AMID A LANDSCAPE OF RUIN

We can also think that because Tokyo was the center of the disaster area, it is a blessing for the future of Japan. . . . Just like Tokyo having had an evil influence upon the whole country as the center of hedonism in the recent years, if it can have a good influence as a spiritual model after receiving its baptism by the earthquake, there is no doubt that the merit will be immeasurable.
—Hoashi Ri'ichirō, 1923

Our Tokyo, having gone through the epoch-making great earthquake and fires, is now in the position where we can construct a modern city to our heart's content. When on earth can we have another "Great Opportunity" if we miss this one?
—Anan Jō'ichi, 1923

After touring Tokyo's devastated landscape in autumn 1923, politician Wakatsuki Reijirō confessed that he was overcome by many emotions. Wakatsuki's feelings ranged from sheer disbelief at the enormity of destruction and displeasure over the chaos and disorder that engulfed Tokyo, to a profound sadness over the loss of life, property, and assets. The capital, he reflected, was left in an utterly miserable state. "Amid the ash and ruins," however, Wakatsuki also felt something stronger. What he felt most acutely was a sense of opportunity, if not full-fledged optimism. "Now," he wrote in the November 1923 edition of *Toshi kōron*, "we have the chance to make an almost brand new city." This was an opportunity that not even the leaders of the Meiji government possessed in 1868 when they transferred the capital from Kyoto to Tokyo and charted Japan on a revolutionary new trajectory. Though the earthquake killed and displaced humans on a scale not before seen in Japan's modern history, Wakatsuki emphasized the fact that it also burned down much of "old Tokyo." "Even though we were aware of the necessity of urban improvement," he articulated, "we could not make changes freely or even easily because the old city was there." The earthquake, he argued, had changed all of that in less than one week.[1]

Wakatsuki was not alone in viewing the destruction of Tokyo as an unparalleled opportunity. Economist Fukuda Tokuzō, social welfare advocate Abe Isoo, and bureaucrat extraordinaire Gotō Shinpei—to name just a few—shared the former finance minister's optimism. Though these three and the constellation of bureaucrats, planners, educators, journalists, and elites who associated with them were proud of Tokyo's development, they suggested that Tokyo needed thorough renovation to ameliorate the "social plagues" that were the scourge of the city.[2] Overcrowding, poor sanitation, poverty, a dearth of green space, and a lack of adequate public transport all conspired against the city and its inhabitants.[3] Abe, Fukuda, and Gotō, in fact, went so far as to suggest that rather than create opportunities for Tokyoites to improve their lives and livelihoods, Tokyo enmeshed lower-class workers and their families in a morass of economic and social disadvantage. Over the course of its modern emergence, they suggested, Tokyo had changed from being an enabler and a place of opportunity to an oppressor. More pointedly, Abe opined that Tokyo had come to represent "the dark side of civilization" and in doing so had become a "monster."[4]

In the minds of many bureaucrats, urban planners, and social welfare advocates, the Great Kantō Earthquake altered the landscape of Japan's capital literally and figuratively. Postdisaster Tokyo not only reflected destruction but also inspired dreams of what Tokyo could become. Many individuals viewed the earthquake as more than a human calamity and an act of divine punishment. These individuals saw the disaster as a destroyer of a city that Japan, the people of Tokyo, and the state had outgrown, but a city whose residents and leaders were unable and unwilling to alter if it meant personal loss or sacrifice. Gotō suggested that the earthquake created an "ideal opportunity to make a perfect new city" (*kanzen naru shinshiki toshi o tsukuru zekkō no kikai*) that would not only initiate Tokyo's renaissance but also serve as an icon that demonstrated Japan's modern emergence to the world.[5] The feeling of opportunity was perhaps best expressed by Abe, who confessed: "When we stand among the desolate ruins and imagine the birth of a new, ideal, Imperial Capital, we cannot help but feel inspired."[6]

An examination of a cross section of the postdisaster dreams about new Tokyo articulated by bureaucratic elites in 1923 reveals the optimism and opportunism that were unleashed with the destruction of Japan's capital. Often these dreams about what new Tokyo could and should become, and why, were published in leading academic and popular journals. If Tokyo was a cityscape on which concerned commentators projected fears and anxieties about Japanese modernity, an examination of reconstruction visions illustrates that devastated Tokyo also served as a powerful landscape from which dreams about a better capital and society were conjured. An analysis of these dreams suggests that the desire to renovate Tokyo in a substantial fashion did not spring just from the disaster itself. Rather, as the earthquake amplified fears about a decline of society, so too did the catastrophe intensify discussions about the necessity and importance of urban renewal. Most important, the catastrophe nurtured expectations that Tokyo could be redesigned and reconstructed in substantial ways. Where plans and visions had previously been stymied, many hoped that Tokyo's destruction would enable a radical transformation of the capital.

Analyzing these visions suggests that numerous reformers believed Japan possessed a golden opportunity (kōki) to plan and build a new Tokyo that might well guarantee that the breeding grounds for social ills that lurked in the recesses of Tokyo would not reappear. Moreover, many hoped that a new capital would reflect and reinforce values the government and its progressive and high-modernist allies sought to encourage among its subjects. Such a built environment would, they opined, assist the government in its longer-term attempts to inculcate individuals with values of sacrifice, frugality, diligence, loyalty, hygiene, fitness, temperance, and community. Fulfilling a "great reconstruction plan," in the words of progressive bureaucrat Nagai Tōru, would allow the state to "better respond to the needs of the new era materially and spiritually in order to renovate society."[7] To urban planners, social welfare advocates, and policy activists who saw the burned out remains of Tokyo as an empty landscape of destruction, the opportunities for creation—if only for an autumn—seemed endless, bound only by the limits of the imagination.

TOKYO'S VULNERABILITIES ON THE EVE OF APOCALYPSE

Why had so many individuals swelled with optimism at the destruction of Tokyo? Moreover, why did so many elites publish so many distinctive ideas about what new Tokyo could become rather than advocate a rebuilding of the capital along the lines as it stood before calamity? The answer lies in part with the fact that Tokyo of 1923 was a city with manifold problems, real and perceived. Many observers suggested that Tokyo's "afflictions" not only reflected but also contributed to the larger social, political, and economic challenges that confronted urban Japan. These problems had not developed overnight but had become more prominent and apparent during the first two decades of the twentieth century. For one, Tokyo was bigger and far more populous on the eve of disaster than it had been in 1900. The population of metropolitan Tokyo had virtually doubled, from 1.12 million in 1900 to 2.17 million by 1920. More remarkable still, the population of the area outside Tokyo that became part of greater Tokyo in 1932—an area of roughly eighty-two small towns and villages—expanded from 380,000 in 1900 to 1.18 million in 1920. This represented an increase of 369 percent.[8] Opportunity, in particular the prospect of securing industrial-related employment, drew people to Tokyo. The number of factories in Tokyo that employed five or more workers expanded from 768 in 1907 to 7,233 in 1919.[9] Total employee numbers likewise expanded from 55,944 to 188,786 over the same period. Tokyo of 1923 was not just Japan's political capital and center of consumption; it was also an industrial city of production.

These new factories and their workers often located in the wards that already possessed high concentrations of industry, cheap labor, and crowded housing. In 1919 Honjo ranked first as the ward with the largest number of factories in Tokyo, with 1,113; they employed nearly 28,000 workers. Following Honjo stood Fukagawa, with 822 factories; Kyōbashi, with 621; and Asakusa, with 546. These wards, as table 5.1 illustrates, also had some of the highest population densities in Tokyo. All four were packed, and even Asakusa, with the smallest number of factories, felt crowded, with 55,000 people per square mile. It seemed spacious, however, in comparison to Nihonbashi, which contained 170,000 individuals per square mile.

Apart from being industrial and densely populated, these wards also held a number of other distinctions. First, nearly all were located in what was considered "low Tokyo," along the banks of the Sumida River from Kyōbashi and Fukagawa in the south to Asakusa in the north and northwest over to Kanda. Second, from a social policy perspective, these wards were home to high concentrations of individuals considered people of scant means, *saimin*. Though industrial jobs paid better than most rural alternatives, the average factory wage was roughly ¥2.25 per day for men and less than half that for women.[10] Moreover, not everyone who flocked to Tokyo to secure factory employment was successful in obtaining or

TABLE 5.1 Number of factories, employees, population density, and saimin in Tokyo, 1919–1920

WARD	FACTORIES (1919)	EMPLOYEES (1919)	POPULATION DENSITY PER SQUARE MILE (1920)	SAIMIN (1920)
Honjo	1,113	27,892	112,118	11,704
Fukagawa	822	17,427	87,471	19,303
Kyōbashi	621	24,948	136,250	4,255
Asakusa	546	5,585	55,219	9,849
Shiba	427	11,645	163,969	n/a
Kanda	389	4,347	143,286	n/a
Shitaya	414	4,122	115,985	n/a
Koishikawa	223	4,958	72,234	7,719
Nihonbashi	209	n/a	169,970	n/a

Source: Yazaki, *Social Change and the City in Japan*, 447–50, 457–58.

keeping a job. Many who failed remained in Tokyo and became part of Tokyo's poor underclass.

Anxieties about the working conditions of industrial laborers and those who failed to secure employment during the industrial boom of the First World War became a focus of Tokyo metropolitan officials. Concerns grew to eventually encompass the day-to-day living conditions and the domestic lives of Tokyo's working poor and destitute.[11] In conjunction with the national census of 1920, Bureau of Social Affairs bureaucrats within Tokyo's government surveyed wards throughout the capital to gauge the prevalence of poverty and to document the numbers of working and unemployed poor. In the 1920 report, city officials classified nearly 75,000 people, from 18,000 families, as saimin. The total income of the family unit in these cases was between ¥50 and ¥60 per month. The study concluded that people of scant means were most heavily concentrated in the densely populated craft-manufacturing, semi-industrial, and industrial wards of Tokyo.[12] A majority of Tokyo's most down and out,

FIGURE 5.1 Tenement houses in Ryūsenji, Shitaya Ward

Source: Tokyo shi, *Tokyo shinai no saimin ni kansuru chōsa*, 1920.

just over 60 percent, had come to the city during the previous ten years in search of industrial or light industrial employment. Most troublesome, investigators concluded, apart from saimin with fixed addresses, a fluid substratum of roving poor resided in flophouses (11,140 individuals) and on river and harbor craft near Fukagawa (10,712). Finally, city officials cataloged an indeterminate number of vagabonds who changed locations on an almost nightly basis.[13] Just over 100,000 residents of Tokyo's overall 1920 population of 2,173,201—equating to nearly 5 percent—were classified as saimin or roving poor.

A description of the average living conditions of a saimin family reflected the degree of their scant means far better than a simple numeric classification of their meager monthly incomes. On average, a family of four shared a single room of roughly three meters by three meters (or 4.5 tatami mats). In most instances, fifteen to twenty families shared one outdoor toilet and a communal cooking area. Often fronting onto unpaved, narrow, bleak alleyways, saimin quarters also shared a general absence of natural light and suffered from extremely poor ventilation. Such living conditions, urban planners and social welfare advocates agreed, contributed to a vast array of health maladies and fueled social problems.[14]

Another way in which Tokyo of 1923 stood apart from Tokyo of 1900 was in terms of increased vulnerability: social as well as physical. The concentration of large numbers of working poor and unemployed in crowded industrial areas made Tokyo far more vulnerable to fires. As Tokyo became bigger, the number of adequate firebreaks in the form of open spaces and parks had not only failed to increase, but had actually declined. Government spending on social welfare and infrastructure likewise did not keep pace with population growth. During the early years of the Taishō era, therefore, numerous moralists, commentators, and bureaucratic elites saw sections of Tokyo as more than just dangerous places of temptations and hedonistic excesses. Many also described areas of the capital as overcrowded, unhygienic slums that served as breeding grounds for social ills, disease, and potential unrest. Fukuda suggested the slums of Tokyo not only comprised a "cursed inner life" in Japan's capital, but also served as "the center and origin for all cursed things in Japan."[15] In this light, commentators conflated many of the real and perceived social, political, and ideological problems of post-Russo-Japanese War society with the capital's urban space and built environment.[16]

ATTEMPTS TO COMBAT TOKYO'S AFFLICTIONS PRIOR TO 1923

While many Japanese lamented Tokyo's problems, few concrete policies had been initiated to combat slums or transform other parts of the capital into a more livable and healthy metropolis. Fewer still were successful. Failure was not due to invisibility, but rather economic neglect and political inaction. Labor reform advocate Kuwata Kunazō was one of the first individuals to urge government officials to undertake urban betterment.[17] As a member of the Japan Association for Social Policy (Nihon shakai seisaku gakkai), Kuwata called for leaders to create more parks, develop better water and sanitation facilities, and expand public transport options in Japan's major cities. Though such reforms might not eliminate slums or alleviate overcrowding, Kuwata believed that they would improve the livelihoods of all urban residents and might also mitigate against possible urban-based social and political unrest. A more proactive urban policy could, in Kuwata's opinion, not only treat the illnesses associated with modern, urban society, but also serve as a sound insurance policy.

Kuwata's clarion calls for urban reform were echoed by an ever-increasing number of state bureaucrats, metropolitan officials, and progressives during the First World War. One official who needed little rousing, however, was Gotō Shinpei. As home minister in Terauchi Masatake's cabinet (November 1916–April 1918), Gotō spearheaded efforts to combat growing social problems in Japan. Rather than view social problems in isolation from their environment, the home minister suggested that successful social reform would have to be carried out from the ground up. Gotō believed that successful social reform involved a cadre of middle- and upper-class reformers working together to build a better, healthful, stable, and prosperous society through top-down intervention and guided, bottom-up participation. Moreover, Gotō argued that if building or reconstructing society was the ultimate objective, urban reform and renewal of the built environment must accompany all social reform. As Gotō became a driving political force behind reconstruction planning in 1923, a discussion of his background and early experiences is warranted.

The interconnections among social reform, urban reform, welfare, health, and society that seemed so apparent to Gotō and many of his close associates in 1917 had percolated within this mercurial bureaucrat over the course of his eventful life. Born in northeastern Japan, Gotō's earliest passion was medicine.[18] Not content with curing illnesses of individual patients, Gotō directed his energies and spare time toward improving medical administration and delivery. He also worked to improve hygiene and sanitation facilities for public places, including inns, orphanages, restaurants, and schools. Gotō went so far as to propose that local officials throughout Japan establish medical and hygiene police forces to assess, manage, and maintain public health, sanitation, and personal hygiene.[19] Though neither the central government nor local governments warmed to Gotō's health force proposal, rejection rarely if ever curbed his enthusiasm or dissuaded Gotō from sharing his ideas—almost always in proposal or petition form—with those in positions of authority. After a brief meeting with Gotō in 1922, Poultney Bigelow, co-owner of the *New York Evening Post*, referred to him as "the busiest man in Japan."[20] Gotō's closest friends and admirers, however, eventually grew weary of his "habit of research" and what one called his "proposal mania." Katsura Tarō, his future mentor and confidant, asserted that Gotō rarely proposed anything "reasonable" but was worth listening to because "once in a hundred times he produced a gem of wisdom."[21]

A number of Gotō's "gems" were unearthed in Japan's colonies. Sent to Taiwan first in 1896 and then again in 1898, Gotō remained in overseas positions for the better part of ten years. It was an important period in Gotō's life that would influence his future ideas about urban planning, social reform, and governance. As chief of the civil government in Taiwan directly beneath the governor general, Gotō established the Committee on Municipal District Planning (1899) and promulgated numerous building and zoning laws governing colonial development.[22] He also drew up plans for medical clinics, sewage and water systems, and modern schools throughout the island. Gotō attributed his planning successes to the absence of political and landowner opposition, a talented pool of bureaucrats and staff, and noninterference from political elites in Japan. For Gotō, a social bureaucrat and urban planner with ambition, funding, and power, colonial Taiwan was a paradise.

Gotō's success in Taiwan cemented his legacy as an effective elite and aligned his career path to Japan's newest colonial acquisition: the South Manchurian Railway zone. Known as Mantetsu, the public-private partnership company was responsible for maintaining a key rail line that Japan secured from Russia following the Russo-Japanese War of 1904–1905. Gotō was appointed as the South Manchurian Railway Company's first president, and from this position he developed cities and infrastructure all along the main rail line from Port Arthur to Harbin.[23] To facilitate the creation of cities, Gotō used a planning technique by which Mantetsu purchased or appropriated large tracts of land along the rail line and developed them with state funds. After the projects were completed, Mantetsu recouped its fiscal outlays by either selling the improved land or leasing it to individuals and corporations.[24] Apart from constructing new facilities and cities, Gotō also applied this technique to existing cities in Manchuria to widen roads, construct sidewalks, and build schools (including a medical school), industry and port facilities, and hospitals.[25] Gotō's experience in Manchuria deepened his conviction that private interests and individual property rights must never be allowed to restrict the state's authority to construct or expand infrastructure for public betterment. Could such ideas be transplanted and successfully superimposed on an increasingly pluralistic and democratic Japan?

Though Gotō became minister for communications and held other posts in Japan after his return in 1908, his most important contribution to Tokyo came from positions he held between 1916 and 1923. First, as home minister in 1916, Gotō exhibited increasing concern about the health and welfare of Japan's citizens, and in particular Tokyo's inhabitants. In 1917 he created a relief section within the Local Affairs Bureau of the Home Ministry that dealt with employment matching, assistance to the poor, veteran's benefits, and child welfare.[26] This bureaucracy was later expanded and eventually transformed into the Social Affairs Bureau in 1920.[27] As Sheldon Garon, David Ambaras, and Sally Hastings have demonstrated, this bureau and the new middle-class bureaucrats who staffed it helped transform the whole concept of urban and social reform.[28] Social reform no longer focused on charity, assistance, and relief but rather encompassed an active social policy that encouraged citizen participation and the involvement of different agents in society to combat social problems. Such theories drew on Euro-American concepts of progressivism and embraced the notion that state and society held an obligation and responsibility to improve peoples' living and working conditions. Rather than merely relieve suffering, urban and social reformers looked to define and stamp out the causes of afflictions, be they economic, social, medical, or environmental.

Owing to his background and bureaucratic successes, Gotō championed urban betterment as an almost essential precondition to effective and lasting social reform. "The city," he claimed, was "an organic body" with the urban built environment serving as the skeleton.[29] With a weak skeleton, the effectiveness of all other reforms, including those of the head (education), the heart (social), the stomach (economic), and the central nervous system (bureaucratic and administrative), would fail. To emphasize the importance of urban reform, Gotō created the Urban Studies Association (Toshi kenkyūkai) within the Local Affairs Bureau in 1917. Here Gotō assembled urban planners, welfare advocates, social reformers, academics, bureaucrats, and politicians from Japan and abroad to study and define the totality of the "urban problem" Japan faced and to discuss potential solutions.[30] The association published a monthly journal, *Toshi kōron* (Urban review), that served as the organization's mouthpiece.

Soon after leaving the Home Ministry, Gotō had what he believed was a stage on which to put his ideas to work. From December 1920 to April 1923, he served as Tokyo's mayor and used his position to further the causes of urban renewal and social reform. Virtually every major initiative he directed for urban development, however, met with either political resistance or fiscal constraints: many experienced both. Gotō realized quickly that Tokyo was neither Taipei nor Mukden. None of his proposals attracted more criticism or suffered from penury more than what was to be the crowning achievement of his mayoral term: the Outline of the Administration of Tokyo (*Tokyo shisei yōkō*). Known figuratively and more commonly by the title "the ¥800 million plan" (*hachiokuen keikaku*), Gotō's outline was in fact a major urban betterment plan designed to build infrastructure and upgrade existing projects throughout the capital. First and foremost, Gotō's plan prioritized roads and other transport-related infrastructure projects. From the total proposed budget of ¥758 million, Gotō earmarked 71 percent of funds for roads, transportation, and infrastructure as follows: road improvement, which included widening, straightening, and the construction of sidewalks (¥144 million, or 19 percent of the total budget); road paving and sealing (¥64 million, or 8 percent); reinforcement, reconstruction, and rearrangement of all underground structures in the city, including gas lines, water pipes, and sewers (¥200 million, or 26 percent); complete renovation of the underground water supply system (¥84 million, or 11 percent); and improvements to or construction of harbor facilities for Tokyo, (¥50 million, or 7 percent).[31]

Gotō also proposed constructing facilities destined to improve the quality of life for Tokyo residents. The ¥800 million plan, for instance, allocated ¥91 million (12 percent) to improving sanitation in the capital, primarily through the construction of new sewers, sewage treatment plants, and garbage disposal and incineration facilities. Significantly, Gotō sought to make Tokyo a much greener city. Highlighting the fact that parks and other open green spaces constituted only 2 percent of Tokyo's overall environment (compared with 20 percent in Paris, 14 percent in Washington, D.C., and 9 percent in London), Gotō proposed spending ¥68 million (9 percent of the total plan) on the construction of public parks and green belts.[32] Such facilities, he suggested, would not only lead to better health and fitness of Tokyo's residents but also serve as firebreaks if a major conflagration started in the capital. Rounding off the proposal, Gotō earmarked ¥8 million for the establishment of crematoriums and public cemeteries, ¥6 million for market sites and slaughterhouses, ¥7 million for city office buildings, and ¥15 million for new primary schools.

Shortly after the announcement of his outline plan, critics challenged virtually every aspect of the proposal. Moreover, they also challenged the means by which Gotō hoped to implement it, financial and otherwise. First, national politicians and select journalists ridiculed the plan for its grandiosity. In an editorial published in the Tokyo-based newspaper *Yorozu chōhō*, the editors lampooned Gotō writing: "What grandiosity!

Can't you complete the city's projects without such an enormous amount of money?" The editors had a point. The ¥800 million was roughly half of the size of Japan's *national* budget in 1922 and significantly larger than Tokyo's current year municipal budget of ¥125 million. Though Gotō countered this broadside by arguing that ¥800 million over the proposed length of the project—fifteen years—would amount to roughly ¥55 million per year, the notion that Gotō's plan was an unrealistic ¥800 million dream stuck. Beyond finance, even some city councilors who worked directly under Gotō foresaw trouble from landowners who would be compelled to sell or relinquish part of their private property to clear space for parks, sidewalks and widened roads. They also suggested that shop and restaurant owners would condemn the disruption to business that major roadway renovation would cause.[33]

Gotō confronted both sets of criticisms. First, he suggested that the plan could be paid for, in part, through increased taxes. Gotō asked permission for the assessed value of land—and thus the tax intake for Tokyo—to be raised to 1921 valuation levels rather than remain at the purchase price level. Second, Gotō requested that the national government give him the power to implement new transport levies within the city's limits. To raise further revenue, Gotō also requested that the Finance Ministry allow the city of Tokyo to issue municipal bonds. Finally, Gotō requested that the national government lead by example and pay property tax on the land its offices and ministries occupied within the municipality. If these measures failed to rake in enough revenue, Gotō asked the national government to cover any shortfall through a direct subsidy. Apart from responding to cries of financial grandiosity, Gotō challenged potential city-based critics—who he believed might protest revenue disruption and financial loss, higher taxes, and property loss—to develop a public spirit and love for the city. Love came at a price. Gotō encouraged all citizens of Tokyo to accept a degree of sacrifice, whether it be higher taxes, smaller incomes, or decreased property holdings, to make Tokyo more livable, efficient, and modern. He also suggested that improvements to the city would increase the valuation of properties citywide and thus more than compensate for any short-term loss in revenue or property size.[34]

Many individuals disagreed. While few residents organized concerted campaigns against higher taxes, land readjustment, or land confiscation, few had to. Gotō's plan failed to secure political support and financial backing from the successive cabinets of Hara Takashi (1918–1921), Takahashi Korekiyo (1921–1922), and Katō Tomosaburō (1922–1923). Without such assistance, the ¥800 million plan died, punctuated by Gotō's resignation in April 1923. Gotō relinquished the reigns of municipal government, embittered by what he saw as shortsighted, interministerial factional politics and the general social and political ambivalence of Tokyoites. Moreover, he suggested that landholders and the elected politicians who represented them were too bound by convention and selfishness to look beyond what Tokyo was, to see Tokyo as it could be and, more precisely, as how Gotō envisioned it. The city was about to be overturned, and when it was, Gotō hoped that many of the conventions and mindsets that he believed trapped its residents in the past would topple with it. Not surprisingly, designs, visions, and ideas about what new Tokyo should be were circulated widely throughout the autumn of 1923. Often the only thing they lacked was a healthy degree of realism.

FIGURE 5.2 The earthquake as opportunity: a well-dressed catfish, symbolizing opportunity, shaking hands with Prime Minister Yamamoto Gonnohyōe

Source: Uchida Shigebumi, ed., Taishō daishin taika no kinen, 5.

VISIONS FOR A NEW CAPITAL: A HIGH MODERNIST CAPITAL FOR A NEW ERA

In the aftermath of the September 1 catastrophe, numerous urban planners, bureaucrats, and commentators believed that the disaster had created a unique opportunity to build the city of the future. Some believed that a new, well-planned city would enable the state to combat the accumulated urban afflictions of modern society and ensure that the breeding grounds for social ills did not reappear. Others argued that a new, purpose-built capital could actually assist the state in better managing its subjects on social, ideological, economic, and political levels. These individuals hoped that new Tokyo's urban space and state facilities would reflect and reinforce values that the government and its reform-minded allies sought to instill among its subjects. They included health, hygiene and physical fitness, frugality, sacrifice, diligence, temperance, orderliness, and community. Many bureaucratic elites in Japan had tapped into rich Euro-American discourses on urban life, city planning and design, and social management since the turn of the century.[35]

A significant number of urban planners and bureaucratic elites, moreover, were enchanted by an ideology that James Scott has since labeled "authoritarian high modernism." In Scott's appraisal, high modernism was a "sweeping vision of how the benefits of technical and scientific progress," might be applied through state management and intervention "to influence every field of human activity." Many high modernists, Scott suggests, shared a belief that a central purpose of the state was "the improvement of all members of society—their health, skills and education, longevity, productivity, morals, and family life." The scope of intervention, he argues, grew to include child rearing, posture, diet, personal hygiene, recreation and leisure, housing, and the genetic inheritance of the population. This expansion occurred as states became bigger, more bureaucratized, and more powerful, and as the technology to regulate people's lives advanced. Importantly, Scott documents that the urban built environment and layout of cities likewise fell under the gaze of many high modernists, epitomized by the writings of Swiss-French architect and urban planner Charles-Edouard Jeanneret-Gris (know by his pseudonym Le Corbusier). Scott illustrates how high modernist impulses influenced urban planning, architecture, and design throughout continental Europe and in Europe's colonies from the 1920s onward. Such high modernist impulses likewise influenced visions about what new Tokyo could become.[36]

One such bureaucrat and social reformer was Nagai Tōru. Writing in the November 1923 edition of *Shakai seisaku jihō*, Nagai suggested that the government had a unique postdisaster opportunity to make a lasting improvement to the health and social welfare of all Tokyoites. The expansion of state-run medical and welfare facilities stood as a centerpiece of Nagai's vision for a reconstructed capital. As a first step in the reconstruction program, Nagai argued that the government should build many more hospitals and emergency medical clinics that would be useful in the event of a future calamity. But more than just treating people for emergency situations, Nagai believed, the government should invest considerable effort into keeping people healthy from an early age. Nagai advocated that the state create free-access maternity hospitals, child-welfare clinics, old-age nursing and care homes, and community-based health cooperatives that could treat everyday sicknesses and seasonal ailments before they threatened the individual or spread to wider segments of the population. Keeping people safe from serious illness and providing free access to those who became sick, Nagai argued, was a relatively inexpensive insurance policy that would lead to a more healthy society and productive workforce.[37]

Keeping people healthy, in Nagai's worldview, was about far more than providing access to medical care, whether it was emergency, general, or preventive. Nagai also believed that the state should use its power and influence to foster physically healthy, strong, and hygienic-minded people. To keep people healthy, particularly lower-class workers who were paid little and thus, he suggested, often skimped on diet, Nagai urged the state to construct subsidized, public cafeterias and subsidized markets where fruit, vegetables, fish, meat, and other staple products could be made easily accessible and affordable. Nagai urged the government, apart from improving the diet of Tokyoites, to establish and maintain facilities that could strengthen the bodies and improve the physical fitness of both children and adults. To accomplish this, Nagai hoped, public sports facilities and small parks could be constructed throughout the city, but particularly in densely populated and heavily industrialized areas, including Honjo, Fukagawa, Kyōbashi, Kanda, and Nihonbashi. Along with strengthening the bodies of citizens and subjects, Nagai suggested, that it was also important for the state to build and operate public bathhouses to maintain the hygiene of its poorest residents.[38]

Nagai also championed the construction of all manner of infrastructure projects that would better enable the state to manage the lives of Tokyoites. To supplement the physical well-being of the city's residents, Nagai suggested that the state also focus on the minds of urban Japanese through the construction of adult education facilities, including public and mobile libraries and lecture halls in community centers. He also advocated creating an extensive network of public housing estates, to be operated by city and prefectural authorities, as a means to keep the privately owned, rental-based slums from returning. Finally, he suggested that the state build an integrated network of community-based pawnshops and credit unions across Tokyo that would provide financial services to the working poor and middle-class residents of the capital. These institutions were to be more than just banks for the poor. Rather, once completed, they would not only facilitate savings through the issuance of low denomination bonds but, more important, serve as financial education centers where officials could teach residents about the proper management of money and personal finance through classes, demonstrations, and workshops. If the lessons failed or dire situations of paucity arose, however, the credit unions would provide regulated, low-interest loans to residents in need, thus avoiding the "interest trap" that many urban poor experienced at the hands of private moneylenders.

Nagai's Tokyo emphasized the construction and use of state infrastructure to forge and maintain what he saw as a better Japan. In short, it was a high modernist's dreamscape. State facilities would be utilized to make all Tokyoites healthier, wealthier, and wiser in the hope that this would lead to a more productive, harmonious, and engaged citizenry. This Tokyo, Nagai concluded with no little modesty, represented not just a new city but the spatial and conceptual manifestation of a "new era" in planning defined by the state's willingness *and* ability to delve into, and regulate, the everyday lives of Tokyo's inhabitants. To emphasize the importance of the project as he envisaged it, Nagai warned all Japanese that "the future of our nation depends on how we use this opportunity for reconstruction."[39]

Other reformers also espoused the development of a city rich in state-built and -controlled institutions. Tomoeda Takahiko, social welfare advocate, ethicist, and professor at Tokyo University, also laid out his high modernist vision for Tokyo. Writing in the November 1923 issue of *Shakai seisaku jihō*, Tomoeda focused much of his attention on how the state should use its power to construct a vast array of public housing estates on a scale not before seen in Japan. Under the watchful eye and regulatory power of the state, Tomoeda, the future director of the Japan-German

Cultural Institute in Tokyo, suggested that well-intentioned bureaucrats and private welfare associations could end urban poverty and improve the health, stature, and mindset of Japan's urban-based working-class population within a generation. Tomoeda argued that the pre-earthquake slums that had developed in the most densely populated wards of eastern Tokyo had been a cancer for the city from which virtually all of society's ills sprang. Any walk through Fukagawa or Honjo in the years prior to the earthquake illustrated the fact that the vast majority of landlords lacked any civic pride, any sense of social or moral obligation, and, for all intents and purposes, human decency. In such a rapacious environment, Tomoeda argued, Japan's poorest workers and their families would never escape the poverty trap.[40]

The earthquake was, in Tomoeda's mind, a welcome occurrence for one reason: it burned away the slums. He suggested as much, writing that the "earthquake burned many slums and areas of low educational and socioeconomic standards." But, he cautioned, the fires had only accomplished the physical elimination of the slums. If Tokyo truly wanted to rid itself of slums for good, its leaders would have to tackle the problems that had created them in the first place. "Many more slums," he suggested, were already being "formed among the ruins," and eventually these "people would have to be moved elsewhere." Therefore Tomoeda recommended that the state purchase or appropriate large sections of private-held land and construct terrace housing or estate blocks modeled on English and American public housing projects. The estates, however, were to provide more than just a roof to sleep under. Tomoeda argued that the state must also build parks, day-care facilities, community centers, vocational training and employment-matching agencies, and nurseries within or adjacent to the housing projects. These facilities would enable the state to cater to all needs of an ever-increasing urban population. Improving the homes, living conditions, and local environments of at-risk neighborhoods in the capital was a core government responsibility and an important first step to "improve social peace and happiness" throughout society.[41]

To help fashion a more well-ordered and regulated society, urban planner Kikuchi Shinzō focused his plans for "new Tokyo" on transportation. In particular, Kikuchi spelled out a detailed plan for making new roads and a modern mass transportation system the basis for Tokyo's future development. Kikuchi argued that the most obvious place for reconstruction planners to focus was on the road network that served the capital and connected it to outlying suburban areas. New Tokyo needed to abolish the narrow, meandering streets of the Edo period and embrace a logical and ordered modern grid pattern for its roads. This, he suggested, would help alleviate future traffic congestion and facilitate the movement of people and goods. Kikuchi also argued that a grid pattern would allow the city to expand outward in a logical fashion and enable city and prefectural authorities to develop a much larger mass transit system.[42]

To Kikuchi, transportation planning was more than just imposing a grid pattern road network and mass transit system on the urban environment. It was also about moving, directing, and managing people on a daily basis. Kikuchi suggested that the urban environment people were exposed to in their own neighborhoods as well as on their daily commutes could influence behavior and attitudes long after a walk to the shops or a journey to and from work ended. Just as people exposed to high-end luxury items as they strolled along the main streets of the Ginza district might be tempted to purchase an item or adopt a luxury-minded lifestyle, so too, Kikuchi articulated, people might adopt good customs and practices if they were surrounded by positive stimuli. Such stimuli, he argued, were even more powerful if they were reinforced on a daily basis.

To drive this point home, Kikuchi wrote about schools. In the first instance, he argued that all new primary schools should be placed and nurtured in healthy settings across the city. One way to accomplish this was to establish small parks and other community-focused, state-run social infrastructure projects near the city's primary schools. Under the watchful eyes and caring hands of the state, such facilities could be used before, during, and after school by young children and their families, thus helping to forge a healthy local neighborhood identity. To keep older students from falling prey to the vices of the city, however, Kikuchi recommended that all inner-city middle and high schools be rebuilt and relocated in suburban districts. Apart from removing older children from potential danger points in the city, Kikuchi's vision also meant that they could be specifically reached by the state during their daily commutes on public transport that flowed against the normal movement of adults who ventured to work on city-bound trains in the morning and returned home to the suburbs during the nightly peak hours. A physically well-conceived and remodeled Tokyo would not only make transportation more convenient to all residents but give the state new opportunities to refashion citizens and subjects as they went about everyday activities.[43]

Abe Isoo shared Kikuchi's belief that Japan possessed a unique opportunity to construct a new capital that would go far beyond what he classified as makeshift improvements to the built environment. Writing essays for the October and November 1923 issues of *Kaizō*, Abe argued that what Tokyo needed following the disaster was a "fundamental project for the construction of a new city" that was influenced not solely by a "materialistic point of view but also from the viewpoint of social policy."[44] Japan was at the dawn of a new age that would be defined by the expansion of state-directed and -managed social welfare initiatives that could, if successful, lead to the perfection of individuals and society. In essence, Abe advocated the construction of a fundamentally new built environment that not only reflected new social values held by elites such as himself but also provided the state with the appropriate infrastructure to inculcate society with values and healthy practices every day.

Physical as well as spiritual health, hygiene, and fitness were three concepts that Abe believed state officials should nurture through a redesigned Tokyo. Rather than just focus on roads, parks, community centers, and clinics individually, he urged planners to develop an integrated approach to social and urban planning. Urban residents, Abe suggested, were always on the move, whether to and from work, school, markets and shops, or for leisure activities. Abe therefore urged planners to start with this premise and construct the new city with mobile people in mind. The numerous unpaved, narrow, sidewalkless streets that existed in pre-earthquake Tokyo discouraged walking and made journeys on foot increasingly dangerous as car, cart, and mass transit transportation increased and further congested a taxed road system. To make the city more hygienic, safe, and healthy, Abe therefore endorsed the construction of wider, paved streets with attached promenades connecting to train stations, markets, schools, parks and open spaces, government buildings, and even places of industrial employment. Such a system would encourage walking as a means of transportation and exercise. To Abe, these were more than just walkways. He envisaged that such promenades would be used each day—by men going to work, children going to school, and women going to market—and would thus become avenues of opportunity for the state to reach people. Instead of exposing citizens to consumer-oriented window displays, restaurants and bars, and places of unhealthy entertainment, promenades could move people along avenues fronted by community centers, lecture halls, libraries, public cafeterias, health clinics, sports grounds, and public markets. The possibilities for positive stimuli seemed endless.

If the measures above did not go far enough to satisfy government officials, Abe also articulated a number of other radical policies that he believed could lead to a fundamental transformation of Tokyo. Virtually every major city in the world, the consummate reformer suggested, had developed over an extended period of time. The mixed compositional character among residential, commercial, industrial, and administrative areas of most cities betrayed this phenomenon. While understandable, Abe decried, this fact was lamentable nevertheless. "But," he argued, postearthquake Tokyo had the potential "to be designed afresh" either by "creating different zones in one city" or "building separate, purpose--built cities" adjacent to one another joined by extensive mass, public transportation systems. At his boldest, Abe argued that the central area of new Tokyo exclusively comprise government and administrative buildings, banks and company headquarters, and higher- education facilities. This inner zone of the capital would fit into the areas surrounding the imperial palace. Its boundaries, in Abe's mind, would be demarcated by a circular system of expansive green belts, parks, and moats that could serve as firebreaks in case of future calamity. The heavy industries that had grown in the capital's eastern districts could be moved 2–3 *ri* (8–12 kilometers) north of Tokyo's new center. In between, vast commercial and residential areas could be established replete with government infrastructure and service facilities. On top of all this, Abe argued that once and for all order, austerity, and conformity could be imposed on all residential developments and hopefully, by extension, their inhabitants. This, he suggested, would provide cost-effective housing and give Tokyo a modern, beautiful, and harmonious feel that was entirely lacking in the old capital. To drive this point home, he asked his readers to consider the following:

> Take Berlin for example. The cost of house construction is relatively low, but those houses give everyone a clean and beautiful impression. Why? It is because the design of almost all houses—the model, height, shape, colors, and windows—is uniform. Suppose a soldier clothed in a khaki uniform is standing at a street corner—the look of him is not particularly beautiful. But if several hundred of them are standing in rows in perfect order, it exudes a kind of beauty. The harmonious feeling they display is the essence of beauty. The view of similar houses lined up on both sides of a street is unmistakably harmonious.[45]

To Abe, an urban modernity reflecting and imposing austerity, order, conformity, and harmony was waiting to be fashioned on the remains of old Tokyo. It would just require considerable political will and commensurate finances to implement. As for the residents who would inhabit new

Tokyo, Abe held a classic statist attitude. Residents, he believed, could be conditioned to accept their new environment whether they embraced it initially or not.

VISIONS FOR A NEW CAPITAL: A MORE LIVABLE AND CATASTROPHE-RESISTANT CITY

While highlighting the importance of order, harmony, and austerity, a number of planners and commentators also suggested that Tokyo be reconstructed with its future residents' comforts and sensibilities in mind. To these individuals, Tokyo could be refashioned as a more livable and disaster-resistant metropolis. Kobashi Ichita reiterated this point many times in an article published in the November issue of *Toshi kōron*. The former vice minister of home affairs (1918–1922) and prefectural bureaucrat of fifteen years service urged government officials to "make good use of nature" when planning Tokyo's reconstruction. If they failed to do so and rebuilt a city "that went excessively against nature," Kobashi articulated, "there would be no guarantee that Tokyo would not again suffer at the hands of an extraordinary disaster." Striking a proper balance between the needs of human society and the natural and built environments, Kobashi suggested, was "the first necessary condition of city planning."[46]

What precisely did Kobashi mean when he urged officials to take nature into consideration when rebuilding Tokyo? At the most basic level, Kobashi advocated making Tokyo a much greener city than it stood in 1922. To accomplish this goal, he urged government officials and citizens alike to plant and care for trees throughout new Tokyo. "Trees," he wrote, "have a natural effect of making people healthy" both physically and psychologically. Apart from providing fresh air and a calming influence on people, trees could also serve as useful firebreaks if planted on the boundaries of roads and properties and along riverbanks. As such, he suggested trees would "contribute to the fire-fighting ability" of Tokyo's residents and create a more resilient city for the future.[47]

The introduction of many trees in Tokyo's urban plan, however, was just the beginning of Kobashi's greener vision for the capital. He also called for the construction of numerous parks throughout the new city. Kobashi's parks would not only contain trees but also provide sports and recreation facilities and community gardens for Tokyo's residents.

Importantly, he also suggested such parks would help foster a closer community spirit because residents themselves would be required to look after and maintain such parks and gardens. "A healthy ideology," Kobashi wrote, "flourished from the spirit of local communities that exhibited a spirit of cooperation and mutual sacrifice," and few things fostered this better than sports, parks, and community gardens. In Kobashi's mind, an almost symbiotic relationship existed between the people and their environment, and to improve one would assist with the development of the other. Along with planting trees and constructing new parks and community gardens, Kobashi urged government officials to "base the new capital on an ideology of simplicity and fortitude, shunning frivolousness and ostentation" because the former were values he believed the people of Tokyo needed to embrace. The maintenance of a healthy city required that the people possess and embrace a healthy outlook, ideology, and worldview that could be fostered from the built environment upward. "A healthy city plan," he concluded, "would help a healthy ideology manifest itself" across the capital and eventually the nation.[48]

Mononobe Nagao, a member of Japan's Earthquake Prevention Investigation Committee, echoed many of Kobashi's recommendations in a ponderous essay entitled "Shinsai chihō fukkō ni taisuru kibō" (Hopes for the reconstruction of the earthquake disaster area), published in the January 1924 edition of *Chihō gyōsei*. Similar to Kobashi, Mononobe argued that the government must do more than simply devise a plan for the capital with short-term considerations in mind. Rather, he argued, the key to a successful reconstruction lay in building a city that was more resistant to future catastrophes. Ultimately, he suggested, this was how people in Japan and abroad would judge the success of the reconstruction project.

As a distinguished civil engineer and lecturer in hydrodynamics, Mononobe wrote with considerable authority. "Every single detail of the new capital," he penned, "must be designed to be sufficiently earthquake resistant from the ground up." To start with, the soil of the city must be tested, and all subsequent construction (including building style, height, and composition) should be dependent on the results of soil composition tests. As a general rule, he argued that reclaimed land and land adjacent to riverbanks was the most susceptible to earthquake damage, and thus tall buildings, schools, and railroad lines must be proscribed in such areas. These were ideal sites for parks and open spaces. In locations of firmer, more stable soil, Mononobe suggested that multistory buildings could be safely allowed, but he cautioned that no building in Tokyo should stand above five stories. If five-story buildings were constructed, he recommended that these buildings be constructed with reinforced concrete and steel frames. Moreover, he suggested that such buildings be limited to government offices and other administrative centers that would not house people around the clock. Mononobe further suggested that great care and planning should go into schools, shops, and residential housing units even if placed on the sturdiest land in Tokyo. None of these buildings should be built taller than three stories. Similar to government buildings, these structures must be fashioned with reinforced concrete, steel frames, and fireproof materials throughout. In giving his final recommendation on the composition and placement of buildings, Mononobe urged the government to eliminate the use of bricks and stones for any structure as these materials were simply not appropriate for an earthquake-prone area such as Tokyo. Using such materials, he suggested, was courting future disaster.[49]

Beyond the material composition of buildings, Mononobe also urged planners to develop a logical city plan with earthquake and fire resistance in mind. For one, he called for the construction of large, tree-lined streets for Tokyo. More specifically, he suggested that at a bare minimum, all streets must be twice as wide as the height of buildings on both sides of the street. Finally, Mononobe suggested that earthquake- and fire-resistant disaster shelters be constructed throughout the city so as to avoid the total breakdown of order and pandemonium that erupted on September 1 when people attempted to flee the city by every means possible. While Mononobe confessed that planning and building this Tokyo from the basis of the soil up with modern architectural designs and materials and stringent planning regulations would be expensive and contentious, it was the only way to make Tokyo more resistant and its citizens better prepared for the next calamity that was certain to strike. Mononobe was prescient in both regards.

Planning and rebuilding Tokyo with the next catastrophe in mind was also close to the hearts and minds of others who advocated constructing a new, more resilient capital. Tokyo's mayor, Nagata Hidejirō, for one, opined that the creation of an extensive subway system would have manifold benefits for Tokyo's present and future inhabitants. Writing for the November 1923 issue of *Toshi kōron*, the municipal leader suggested that a mass-transit system should help move people within the capital and to its outlying suburbs quickly and efficiently in times of peace *or* disaster. He implied that reinforced concrete subway stations could serve as shelters if Tokyo were targeted by an enemy air attack. Writing that Tokyo could find itself targeted in a future war as "aviation technology has developed considerably in recent years," Nagata argued that subways would become an important civil defense arterial system for years to come if developed with fire and bomb resistant features.[50] Such construction, he concluded, was a cheap insurance policy for the residents of Tokyo to adopt.

Shitennō Nobutaka was even more animated and expressive in championing the idea of reconstructing Tokyo with earthquake, fire, and bomb resistance as the central, underlying premise of the new city. Shitennō's advocacy was not surprising given his background as an engineer, as an Imperial Japanese Army colonel who had toured European cities and battlefields following the First World War, and as a burgeoning air-power advocate. Shitennō's focus on the importance of planning and building for future generations also makes his opinions worth exploring, particularly since he became a spokesman for developing an air and civil defense network for Tokyo.

The decorated army colonel criticized the fact that Japanese people, in his appraisal, focused only on the present and shunned most long-term considerations because of either convenience, convention, or economic necessity. To drive this point home, Shitennō relayed a conversation with a colleague in France who had asked him whether it was true that Japan's frequent cycle of fire, reconstruction, and fire pointed to an "absence of the concept of eternity" among Japanese. The question, or so Shitennō relayed, caused only a brief moment of reflection, which was followed in a straightforward manner by, "Yes, indeed it is true." He continued, "We build houses like matchboxes only to let them burn; then, we build the same things again only to let them burn again."[51] Such a mentality, he argued, was already manifesting itself in Tokyo as demonstrated by the proliferation of temporary, makeshift homes of wood and tin that had sprung up like "morning glories" all around the burned out areas of the city.

While Shitennō lauded the reconstruction spirit of the people for constructing temporary houses, he suggested it was a "sin of lax people such as we Japanese who think only of the present" to reconstruct the city as it stood prior to the earthquake disaster. If it was a sin for the people, then it was an unfathomable act of cowardice for politicians to allow such makeshift rebuilding to take place. Shitennō believed that the world was at the dawn of a new age in which knowledge and technology existed for a more permanent, modern, and disaster-resistant city to be forged "with the country's long-term future in perspective." He therefore urged city, prefectural, and national officials to devise and implement a long-term plan that started with the construction and logical layout of fire- and earthquake-resistant buildings as a foundation. To emphasize what Tokyo could become, he recounted another story from Europe, this time from a subordinate of his from the army who had studied in Petrograd during the war. The student told him how different the Russian and Japanese mindsets were in relation to fire: Russian students simply refused to panic or even evacuate a dormitory when a fire broke out in the building. The excited Japanese student, however, was about to flee when the building supervisor entered his room and informed him to stay inside and close the window. When the Japanese student replied that he had to gather his belongings and leave as the building was on fire, she replied: "The fire is three storeys below and will not reach here because the floors and walls are made of reinforced concrete. Just calm down and return to your studies." The student, as Shitennō informed his readers, did as told and the fire was extinguished with little difficulty, no loss of life, and little panic.[52]

Why was this the case? Shitennō suggested that the modern, reinforced concrete building—a structure built with authority and permanence in mind—bolstered people's confidence and allowed them to overcome the fear of fire and conquer its real manifestations with ease. How different the situation was in Japan, Shitennō wrote, with its buildings made from timber, bamboo, and paper. While a small fire in Petrograd was of little concern to Russians, a similar event in Tokyo, he argued, would lead to widespread panic and, quite possibly, a major conflagration. "How," Shitennō asked, "could we resign ourselves to the thought that we cannot do what the Imperial Russians could do?" If Japan's leaders took decisive steps and built a more resilient Tokyo, they would no longer have to accept such an inferior position. In no uncertain terms, Shitennō proclaimed that "by improving the material aspects of people's homes [and by extension their cities]," not only could the government make Tokyo less susceptible to fire and calamity in a physical sense but it could also "help stabilize peoples' spiritual and psychological mindsets" as well. This could go a long way toward preparing society for future catastrophes, whether from a violent earthquake or from enemy air attack. Shitennō ended his essay with a challenge and a warning to all Japanese leaders who endorsed a limited reconstruction project that aimed only at restoring Tokyo as it stood in 1922. He wrote: "If the imperial capital fails to be equipped with fire, earthquake, and bomb resistance and is constructed with timber buildings that are the same as or inferior to the pre-earthquake capital, then in the future historians will judge that there was no one who had a discerning eye, or a politician worthy of name in Japan during the Taishō earthquake period." Accepting a scaled-back reconstruction, Shitennō concluded, would be a waste of a great and unprecedented opportunity to forge Tokyo as a world leader of cities.[53]

VISIONS FOR A NEW CAPITAL: URBAN RECONSTRUCTION ON AN INTERNATIONAL STAGE

Apart from Shitennō and Kobashi, other commentators likewise stressed the importance of reconstructing Tokyo with a long-term perspective in mind. More than a few of these individuals emphasized the concept of "constructing for eternity" because of Tokyo's place within Japan—as the center of Japan's empire—as well as its potential place within a new, modern world. Many hoped that reconstructed Tokyo would reflect Japan's emerging power and prestige within East Asia and the world. More than this, new Tokyo could project Japan's confidence and power to all who visited or saw the imperial capital. Just as war and empire had served as a measure of Japan's development in the past, construction of a new, modern city would allow the world to measure Japan's progress as a modern, civilized nation. In this role, new Tokyo would be perceived of as an international benchmark of progress.

Gotō Shinpei was one individual who believed that reconstructed Tokyo could serve as an important physical manifestation of Japan's values and reflect the nation's self-perception. New Tokyo, Gotō suggested, must therefore embody the position that Tokyo had secured within Japan since the Meiji Restoration as the capital of a modern and successful nation-state. Moreover, the new capital should likewise provide an impressive visual representation of Tokyo's emergence as the center of a regional empire governed by Japan since 1895. Finally, Gotō hoped that the capital might also symbolize what he believed Japan could become over the course of the next generation: a world power. It was time for Tokyo to become something far greater than Edo, dressed-up around the edges, since modern Japan had become so much more than a reconstructed or refashioned Tokugawa bakufu. Japan's progress, he argued, had surpassed even what the most ambitious and forward thinking Meiji Restoration leaders had dreamed possible. As a consequence, Gotō reiterated that the forging of new Tokyo was a project of unparalleled national significance that would shape the nature and direction of Japan's continued transformation—and foreigners' perceptions of it—over the next phase of development.[54]

To emphasize the importance of this project to the Japanese people, Gotō used the emperor's September 12 Imperial Edict to legitimate his ideas and plans. Gotō employed the text of the edict—an edict he had helped write—to support calls for the construction of a grand, imperial capital vastly unlike how Tokyo stood on August 31, 1923. First, Gotō drew attention to how the edict referred to Tokyo not as a city, but the "capital of the empire" and the "fount of national culture." Gotō further utilized the emperor's statement that "remedial measures should not be taken simply for the recovery of the old state—instead, the city must undergo a complete transformation for the sake of future development." He did this to buttress calls for the adoption of a transformative plan for the capital *and* the nation.[55] As a man who seldom needed encouragement to undertake big, high-profile projects, Gotō took the imperial edict as a manifesto to devise big dreams for new Tokyo.

Tokyo, in Gotō's mind, was and would forever be Japan's "center of politics, culture, and economy." The reconstruction plans, he argued, must take these factors into consideration. From a political standpoint, the former mayor of Tokyo suggested that new government buildings be fashioned as visually impressive structures that exuded authority and confidence. To facilitate this visual and spatial impression, Gotō recommended that key government buildings including parliament, cabinet and ministerial offices, and municipal structures be constructed in a modern, impressive architectural style. Each major government building, he suggested, should rest at the center of a large, concentric circle connected to other government structures through wide imperial avenues. Such a capital could, Gotō believed, reinforce and reflect political and administrative power relationships through the design and layout of government buildings. Moreover, Gotō envisaged that Tokyo, like Washington, D.C., could also house impressive imperial monuments to honor past national accomplishments and national museums, performance halls, and universities to emphasize Tokyo's position as the cultural and intellectual capital of Japan's empire. Within the economic sphere, the home minister argued for the creation of grand harbor and transport facilities and the establishment of an extensive mass-transportation network that not only served the capital but connected it more effectively to all corners of Japan. Above all else, Gotō urged that buildings, facilities, and infrastructure be built with future generations in mind. Tokyo must not be built "clinging to old values" or "constructed with temporary measures." If old Tokyo was reconstructed, Gotō predicted that current inhabitants and future generations of Tokyoites would eventually regret such a shortsighted, narrow-minded refashioning.[56]

Aside from the wording used in the emperor's September 12 edict, Gotō employed one other argument on behalf of his ambitious reconstruction dreams: international considerations. Projecting national power and modernity through a magnificent modern capital was central to Gotō's vision. Writing in the December 1923 issue of *Toshi kōron*, he asked his fellow countrymen, "if we only restore the pre-earthquake state of the city, how could we maintain our face in front of the nations of the world." Gotō claimed that the large amount of sympathy and direct aid that the world had donated to Japan following the earthquake was an "indication that the world appreciated the international value of our imperial capital." Anything other than a "flawless construction" of a new, resplendent capital that illustrated Japan's resolve, embodied its modern values, and projected the nation's stature would "fall short of the world's expectations" and be a total waste of a golden opportunity.[57] If Japan wanted to be viewed as a true equal among powers, the reconstruction of a world city provided a unique challenge and unprecedented opportunity to demonstrate its postwar position.

Anan Jōichi, urban planner and Tokyo metropolitan politician, was even more explicit in linking a successful reconstruction to heightened international recognition of Japan as a civilized nation. In the November 1923 issue of *Toshi kōron*, Anan argued with passion that Japan was known in the world only because of its previous successes in war. He suggested that the Sino-Japanese War of 1894–1895 had "made Japan's presence felt by the world," while the Russo-Japanese War of 1904–1905 had allowed "Japan to join the great powers." While previous benchmarks based on conquest were understandable, in the new age of internationalism status would be judged using different standards. Previous leaders may have taken pride "in conquering the world on horseback and being acknowledged for military power," but Japan now must show that it could be a steward of civilization, not merely a destroyer of armies and navies. No better opportunity existed than with the reconstruction of Tokyo. By building a new, modern, and rationally planned capital, he argued, Japan possessed the opportunity to illustrate that it could be a creator of civilization. Moreover, it could show the world that the Japanese race included "perfectly competent builders of great cities."[58]

The notion that the world would follow the reconstruction project closely and judge Japan's progress as a civilized imperial power was also embraced by other commentators who advocated a thorough reconstruction of Tokyo. Keio University professor Horie Kiichi wrote in no uncertain terms that many European countries and the United States were "watching how Japan would overcome its current difficulties." Specifically, he suggested that the "reconstruction of Tokyo would be carried out under the close observation of these nations." If the reconstruction was too slow, failed to transform Tokyo into a modern metropolis, or fell prey to bureaucratic infighting, Japan would lose the political and financial capital it had secured during the prosperous war years. In short, Japan's international position had a chance to go forward or recede in the eyes of the world as a result of the reconstruction process. "Whether we gain honor or shame," Horie concluded, "depends on how well or how badly we carry out the reconstruction." Horie hoped that a successful reconstruction of Tokyo, devoid of political contestation and economic mismanagement, would foster worldwide confidence in Japan's economic and political position.[59]

Two others, Ugaki Kazushige and Takashima Beihō, endorsed Horie's arguments. Takashima claimed that many foreigners would draw conclusions about the spirit and character of Japanese people based on how Tokyo was reconstructed. In an article published in the October 1923 edition of *Chūō kōron*, Takashima informed his readers that history was full of great natural disasters, and the manner in which people responded to such catastrophes provided unmistakable clues as to the power and resilience of the human spirit. In some instances, society succumbed to depression and defeatism in the face of calamities and allowed nature to overcome human civilization. Other times, people merely persevered and rebuilt civilization only to have it annihilated again at a future time. Only a few instances existed where people actively overcame nature and reconstructed a city that illustrated civilization's triumph, resolve, and mastery over nature. If Japan wanted to overcome this great challenge of civilization, Takashima suggested, it would come down to spirit and resolve.[60]

The reconstruction of San Francisco after the April 18, 1906, earthquake and fire was an example Takashima employed to prove his point. The world, he claimed, watched how the people of San Francisco rebuilt America's gateway city to the Pacific. While the United States was assisted by wealth and an abundance of natural resources, it was the "sturdy American spirit" that Takashima believed had led to the transformative reconstruction of San Francisco. Takashima suggested that the successful reconstruction project illustrated to the world American power, resilience, ingenuity, and spirit. Just as Japanese people had drawn conclusions about the United States following 1906, so too, he argued, people would characterize all Japan by the reconstruction of Tokyo. While the Meiji Restoration, the Sino-Japanese War, and the Russo-Japanese War had changed international opinions about Japan, people would now focus on Tokyo's reconstruction to determine if Japan was truly a world power or a nation whose people could not deal with the "great modern test" of the Taishō era. Because Japan did not possess the wealth or resources of the United States—a fact everyone in the world knew—Takashima suggested that Tokyo's reconstruction would depend on the Japanese spirit and its people's willingness to sacrifice. If Japan failed, people around the world would know that Japan was nothing more than "a Buddhist statue without a soul, or a painted dragon without eyes."[61] In short, a reconstruction of old Tokyo would illustrate Japan's weakness to the international community and the superficial hollowness of its development as a modern nation state.

As a military man concerned with foreign perceptions of Japan, Ugaki Kazushige was also worried how a "failed reconstruction" would look to the outside world. Put simply, "the world watches us," Ugaki wrote in his diary of September 18, 1923. Ugaki concluded that the world would see Japan as spiritually weak, economically poor, and politically irresponsible if it failed to forge a new, modern capital on the ruins of Tokyo. More problematic and worrying, he believed, was that political contestation at the moment of the 1923 crisis would expose the fractured state of Japan's polity to a world audience. In such an environment, Japan would be seen as a country devoid of decisive leaders and a nation that would crumble if faced with the challenge of a modern war on its own soil. For the sake of disaster sufferers today and the inhabitants of Japan tomorrow, Ugaki pleaded, "let the reconstruction be successful." Ugaki believed that the earthquake disaster and subsequent reconstruction project had placed Japan "at a watershed moment between prosperity or decline of the imperial state," and that the world would judge Japan based on what course of action it followed.[62]

The notion that the world was carefully watching and judging Japan's response to the catastrophe and subsequent reconstruction process was strengthened by the well-publicized arrival of Charles and Mary Beard in devastated Tokyo. Gotō, in one of his first acts as home minister, invited

Beard to visit Tokyo to examine the immense problems facing the city and to provide recommendations as to its reconstruction. Beard was no stranger to Japan. At the previous invitation of Gotō, Beard spent the winter of 1922–1923 in Tokyo conducting research into the politics and administration of Japan's capital. Upon returning to New York, Beard wrote a book-length study on Tokyo's governance and submitted it to Gotō on June 12, 1923. Published in Japanese and English, Beard's study was used by municipal officials to recommend changes to, and the modernization of, Tokyo's finances, zoning and building laws, administrative structure, and urban layout.[63]

Though familiar with Tokyo, the city that Charles and Mary Beard arrived in on October 6, 1923, was, as Charles Beard wrote, "completely changed" from his last visit.[64] Throughout October Beard toured various parts of the devastated city and met with government officials to discuss reconstruction planning.[65] He echoed virtually all major suggestions put forth by Gotō and others, who called for an infrastructure-rich, grandiose-looking city to be built on the ruins of the capital. Beard called for the construction of wide, tree-lined streets, modern public-housing estates, and parks and gave recommendations on the design and construction of buildings, mass-transit facilities, and, importantly, the ways in which the state should finance such a vast reconstruction project.[66] Beard's vision of Tokyo comprised equal parts progressivism, modern architectural design, and high modernism.

Members of Japan's media followed the Beards, unlike other commentators, like celebrities. Charles, in particular, became such a well-known individual of the postdisaster landscape that his image is a prominent feature in the immense oil-on-canvas mural of the disaster produced by Arishima Ikua, entitled *Daishinsai no inshō* (Impression of the great earthquake disaster). The painting, which today hangs in the Earthquake Memorial Hall, placed Beard in his customary cream-colored suit next to Prime Minister Yamamoto, dressed in his admiral's uniform, and Gotō, wearing a Boy Scout uniform.[67]

On more than one occasion, whether in public or private, Beard informed anyone who would listen that Japan was at an important crossroad in its history. Moreover, he reminded many people that the world was watching the reconstruction of Tokyo. Japan would, he argued, be judged on how well or how poorly it reconstructed Tokyo. More specifically, he suggested that Gotō as home minister and head of reconstruction planning would either garner praise or receive criticism in relation to Tokyo's redevelopment. "The world's eyes are upon you, Gotō," Beard informed the home minister in a November 3, 1923, letter. "If you make a great plan and carry it out, the Japanese people will remember you with gratitude for your vision and courage." "But," Beard continued "if you aren't able to make the above plan and instead accept a compromise . . . you will eventually be forgotten by your people or will be remembered as a person who was unable and did not want to fight." Beard ended his letter with a firm recommendation: "My dear friend, do not just look at today; construct for eternity."[68] Building for future generations, let alone eternity, would prove more difficult than almost anyone in the autumn of 1923 imagined.

INSPIRED BY DREAMS, BLINDED BY DESOLATION

In examining the ideas, visions, and dreams for reconstructed Tokyo espoused by urban planners, commentators, and bureaucratic elites, a number of points become apparent. On the most basic level, these plans reveal that apart from destroying, Japan's earthquake calamity also inspired. The destruction of "old Tokyo" fostered a belief among numerous individuals that a resplendent, modern city could finally be forged. Numerous elites believed that new Tokyo could become a capital that not only reflected new values and a new urban modernity but also would provide the state with the infrastructure necessary to introduce and reinforce these values among the population on a daily basis. Visionary planners believed that a new Tokyo, rich in state facilities, could be used to combat the baneful side effects of urbanism and modern industrial capitalism. Many suggested that a rationally planned and constructed city could keep the disreputable entertainment quarters, disenfranchising slums, and luxury-minded consumer districts from reappearing to entrap and entice Japan's populace. Virtually all the plans thus embraced the tantalizingly modern but ultimately elusive prospect of being able to design, develop, and use the urban built environment to further the aims and objectives of the state.

On another level, plans exposed the top-down, elite level nature of urban planning in Japan. Commentators, bureaucrats, and politicians rarely, if ever, consulted citizens—at least in the planning stage—to inquire as to what they wanted in or from the city of the future.[69] Rather, influenced by doses of progressivism, high modernism, and moralism, commentators and bureaucrats drafted plans to create an idealized city that would counter what they saw as the real and perceived problems of modern Japanese urbanism. Likewise, elites conceptualized many visions of new Tokyo to reflect what they believed the people needed. What people actually wanted was rarely, if ever, considered. As the following chapters will illustrate, such plans or ideas of what the citizens of Tokyo needed did not always rest comfortably with what many residents actually desired or were willing to endorse and accept. While many residents embraced the idea of more parks, wider streets, paved sidewalks, and better state facilities in theory, they did not always support such plans in practice. This was especially true if it meant losing even a small amount of property. In espousing designs for a new Tokyo that were radical departures from the existing city, elite planners and commentators not only looked beyond the existing property lines that remained but also failed to grasp the everyday wishes and concerns of the citizenry. A multitude of planners and elites were, in fact, blinded by the desolation of postdisaster Tokyo. While many elements of virtually every plan may have seemed reasonable, if not inspired, overall plans for new Tokyo more often than not resembled castles in the air and idealized projections rather than achievable blueprints for a sustainable urban landscape inhabited by people with diverse interests and aspirations. The disjuncture between dreams of what many believed Tokyo could become and what was realistically possible were rarely so apparent as they were in the autumn of 1923.

Many commentators, planners, and bureaucratic elites who perceived the opportunity to construct the city of the future looked beyond what was feasible given the scale of destruction, the finances of the nation, and the fact that Japan was a burgeoning pluralistic democracy rather than a one-party authoritarian state. Once actual planning—as opposed to dreaming—began, planners soon found little uniformity among elites in power about what Tokyo should become when issues of funding, legislation, and the proper reach of government in peoples' lives emerged. The opportunity to radically construct a new capital proved to be far more illusionary than real. Nor did Japanese elites have the economic, political, or legal means to reconstruct the capital or the nation in such a radical fashion as many wished following the calamity.

However desolate old Tokyo looked to observers in September 1923, it was not a blank slate awaiting creation. The capital of tomorrow could not be conjured into existence through innovative plans alone. Myriad property owners, shopkeepers, and families returned to where their lives were once centered and sought a return to normalcy, not the beginnings of a brave new metropolis. Even in an unparalleled emergency situation created by Japan's most destructive disaster, the state did not have the political power or will to brush aside property rights, override a constitutional structure, or ignore the concerns of people outside the disaster zone and their elected representatives who worried about the regional and national consequences that might result from a large, state-funded reconstruction project. Moreover, many politicians underestimated the bureaucratic rivalries and political conflicts that debate over reconstruction would trigger. However powerful and persuasive calls for the forging of a resplendent new Tokyo may have seemed, reconstruction was a hotbed of contestation waiting to ignite. Over the autumn of 1923, reconstruction debates exacerbated and found themselves held hostage to larger ideological, political, economic, geographic, and social tensions that defined the sociopolitical world of late Taishō and early Shōwa Japan.

Six

CONTESTATION: THE FRACTIOUS POLITICS OF RECONSTRUCTION PLANNING

Since before the earthquake, the general public has been tired of political strife. . . . I presume that at this time of crisis, the political parties will agree that in order to be loyal to the nation, they should suspend their hostilities and collaborate to carry out the reconstruction project.
—Tawara Magoichi, 1923

All members of the Diet saw with their own eyes how the proud city of Tokyo now lies in ruins, and how numerous sufferers are reduced to distressing conditions. In such circumstances, one would have expected them to discuss all the measures . . . with much greater earnestness and sincerity, but what was the actual state of things? Party sentiment ran even higher than in ordinary sessions, all parties shaping their course with the promotion of party interests primarily in view.
—*Tokyo nichi nichi shinbun*, December 1923

Within hours of becoming Japan's fortieth home minister on September 2, 1923, Gotō Shinpei shared his aspirations with his predecessor, Mizuno Rentarō. Though much of Tokyo was a smoldering wreck, Gotō informed Mizuno that he looked beyond the destruction, death, and suffering and beheld an opportunity to transform Japan's capital into an awe-inspiring modern metropolis. "Now," he told Mizuno, "was our best chance to completely remodel and reconstruct Tokyo."[1] Gotō, in fact, predicted that his tenure as home minister would be defined by lasting accomplishments. Under his leadership, Gotō claimed that fellow political elites and the people of Japan would finally embrace a bold reconstruction plan for Tokyo. Mizuno saw things differently. He cautioned Gotō: "you may well say this now, but such a reconstruction is going to require substantial funds." Gotō dismissed Mizuno's caveat. The new home minister boasted in no uncertain terms that "we will secure the funds somehow; I will get as much as I need."

Three months later, Gotō's optimism was nonexistent. It had vanished over the course of a politically bruising autumn. The politics surrounding Tokyo's reconstruction planning and budgeting had been defined by contestation, not by the optimism that Gotō had predicted. The starry-eyed bureaucrat who referred to the September 1 calamity as Japan's unprecedented opportunity to reconstruct Tokyo left office as little more than a broken, dejected shell of a once ambitious man. Over three frantic months of planning, intense negotiations, and outright political warfare, Gotō witnessed his wildly expensive plans, and those of his equally opportunistic colleagues, get whittled back from nearly ¥4 billion to a meager ¥468 million. Until his death in 1929, Gotō told all who would listen—journalists, politicians, bureaucrats, foreign diplomats, and even the national assembly of Boy Scouts—that Japan would eventually lament and forever regret failing to grasp the "golden opportunity" that presented itself following Japan's greatest natural disaster.[2]

Rather than unite the political establishment, debates surrounding reconstruction planning and budgeting exposed and widened many of the political and ideological fissures that defined the increasingly contoured landscape of interwar Japan. Reconstruction planning also exacerbated many long-standing urban-rural tensions and amplified manifold bureaucratic rivalries, bringing both to the center of ministerial and parliamentary politics. In doing so, the contestation of reconstruction revealed many aspects of Japan's interwar political culture. First, the disaster exposed the financial weaknesses—real and perceived—of the Japanese state following the boom and bust associated with the war. Had Japan possessed a large surplus of money or been imbued with elites willing to deficit spend in a Keynesian-inspired fashion, far fewer political clashes might have occurred. The earthquake thus revealed Japan's economic fragility, or it at least illuminated the perceived financial limitations that many of Japan's most orthodox economic elites possessed about their country.

Volatile reconstruction debates, ministerial conflicts, and parliamentary challenges also exposed the inability or unwillingness of many different agencies to work together for the large and complex task of reconstruction. This facet of the disaster highlighted significant structural weakness of Japan's highly bureaucratized, increasingly pluralistic, yet still semioligarchic state. Japan of 1923 was not democratic enough to compel or even encourage appointed bureaucratic elites to work together with other elites, parliamentarians elected from a limited franchise electorate, or the citizens of eastern Japan to forge a popular mandate or groundswell of support for the reconstruction project. Gotō ignored and later disparaged the majority party in parliament, the Seiyūkai, as an "unnatural majority party" that only sought power and singularly championed the interests of its wealthy, tax-paying supporters. Alternatively, Japan's political structure was not authoritarian enough for elites with ambitious or grandiose reconstruction dreams to coerce other actors or to run roughshod over the Meiji Constitution of 1889. Even during this time of crisis and catastrophe, political power remained diffuse and spread formally among many bureaucratic actors, government ministers, and parliamentarians, and informally among influential elder statesmen who elected officials suggested had little claim to political legitimacy.

While elites and commentators expended considerable effort constructing the disaster and subsequent reconstruction as national events for citizens across Japan, they spent far less effort lobbying other actors who often held competing and contradictory visions for the future. This, and the general political contestation that the reconstruction debates fostered, demonstrated that the oft-employed war analogy was not nearly as effective in attaining the degree of elite-level cooperation and unity that were hallmarks of Japan's wars against China in 1894–1895 or Russia in 1904–1905.

UNSTABLE GROUND: GRANDIOSE EXPECTATIONS AMID THE RUINS

In the days immediately after he became home minister, Gotō let his imagination and ambitions run wild. Though most cabinet ministers focused their efforts devising policies to reestablish order and provide relief, Gotō spent a significant amount of energy contemplating Tokyo's resurrection. Just three days after the quake, Gotō outlined the principles that he believed should serve as the foundation for reconstruction.[3] First, Gotō encouraged the prime minister to issue a statement that the capital would not be moved. Second, he urged his fellow cabinet ministers to embrace the adoption of the most advanced Euro-American urban planning and construction methods in designing and rebuilding the capital. Finally, he articulated the need for the government to prepare itself to deal firmly with landowners who, Gotō predicted, would resist radical reconstruction plans. When the home minister finished his presentation,

Finance Minister Inoue questioned him on one thing: how much money was needed to reconstruct Tokyo. Gotō suggested at least ¥3 billion. Awestruck by this figure, Inoue questioned the home minister no further.

Gotō followed this presentation by introducing more specific, yet even bolder, reconstruction recommendations at a September 6 cabinet meeting. There, the home minister suggested that the national government should purchase all of the burned out land of Tokyo, roughly 11 million tsubo (33 million square meters).[4] If virtually no private land plots remained in the destroyed sections of Tokyo, Gotō argued that large thoroughfares, wide roads, sidewalks, greenbelts, parks, bridges, schools, hospitals, and other buildings that housed social welfare infrastructure could be constructed in a rational, planned pattern without resistance. Moreover, he suggested that because of the emergency situation that existed in postdisaster Tokyo, the government could implement measures to secure such a plan through extraordinary means. Specifically, he intimated that the government could bypass the constitution and compel recalcitrant landholders to sell or exchange their land through the threat of uncompensated confiscation if they failed to heed government directives. While admitting—if not boasting—that the adoption of such a policy was unprecedented in modern Japan's history, he concluded that similar policies had been overwhelmingly successful along the South Manchurian Railway. When again asked by Inoue how much such a plan would cost, Gotō confessed it would be even more expensive than his first estimate: his dream budget now stood at roughly ¥4.5 billion.

Inoue challenged this idea immediately and was joined by every cabinet colleague. He first confronted Gotō on the issue of finance. How, asked Inoue, did Gotō suggest that the government pay for such an expensive undertaking? Inoue then reminded Gotō that his plan would cost more than what Japan had spent with its three previous annual budgets combined. In the first instance, Gotō suggested that the government could issue public bonds, just as it had done earlier during its wars against China and Russia. If citizens did not purchase enough bonds to cover anticipated costs, Gotō suggested, the government could recoup initial outlays and even make a profit once the reconstruction was complete by selling or leasing parcels after reconstruction. Though no other minister saw the merits of Gotō's proposal, discussion moved from financial objections to matters of jurisprudence. Ministers queried the constitutionality of Gotō's scorched-earth purchase and reconstruction plan and suggested that it was nothing more than a despotic, Tokugawa era (1600–1868) land grab. Sensing that a political maelstrom was about to ignite within the cabinet, Prime Minister Yamamoto recommended that no large-scale decisions be taken concerning Tokyo's future until the government created a formal institution to oversee reconstruction.[5]

The rebuff that cabinet colleagues gave Gotō on September 6 did little to dampen his enthusiasm, restrict his grandiose ambitions, or lessen his desire to use the earthquake in an opportunistic fashion. At 9 a.m. on the following day, Gotō hosted a meeting of officials from the City Planning Bureau of the Home Ministry at his home. After an introductory oration in which Gotō again emphasized the golden opportunity that Japan possessed, he led urban planner Ikeda Hiroshi, architect Sano Toshikata, and engineer Yamada Hiroyoshi (Hakuai) on an extensive tour of devastated Tokyo. The enormity of destruction witnessed in eastern Tokyo convinced all three men that a once-in-a-lifetime opportunity had indeed materialized to radically redesign and reconstruct a new Tokyo.

Though each individual was entranced by the prospect of devising a comprehensive master plan for a new Tokyo, seeds of doubt concerning the reality of Gotō's ambitions began to germinate in the minds of one of his closest colleagues, Yamada. "After grasping the true impact of the earthquake on our capital" during his first formal inspection tour, Yamada admitted that he labored for nearly forty-eight straight hours "conceiving and drafting an ideal plan." As head of the First Engineering Section of the City Planning Division within the Home Ministry—a section that dealt with roads, traffic facilities, health-related infrastructure, and sewer and water systems—Yamada established a solid outline that other planners added to on an almost hourly basis. Throughout all the discussions and amendments, however, Yamada kept one thing in the back of his mind: money. On the evening of September 9 when he tallied the rough budgetary estimates for this initial plan, his demeanor turned sullen. He estimated that the ideal plan his team had developed would cost approximately ¥4.1 billion. This was just too expensive. Although Yamada confessed that he "was just a layperson in budgetary politics," he told his colleague Ikeda that the 4.1 billion figure was "impractical."[6] Regardless of what Gotō kept repeating, Yamada predicted that politicians would never support such an expensive reconstruction.

The difference of opinion that existed concerning the scope and scale of the proposed reconstruction project was brought home to Yamada over the course of two tense meetings held in mid-September. After scaling back his initial plan to a more realistic 1.7–2.1 billion yen proposal,

Yamada introduced his sketches and an outline budget to members of the construction department of the Railways Ministry on September 17. Unexpectedly, Yamada found himself taunted by these bureaucrats for embracing "a small and limited" reconstruction plan. "Is this," one bureaucrat interjected, "as big of a plan" that all of the planners in the Home Ministry "working together" could devise?[7] In the face of such derision, an exhausted Yamada grew incensed and excused himself from the meeting. He faced stronger critiques, which he could not ignore, however, when he presented his plans to Gotō and the minister of education, Okano Keijirō, on September 19. Rather than ridicule him, Gotō admonished Yamada, informing him that his plan was just too small to capitalize on the catastrophe and turn Tokyo into a true imperial capital (*teito*). Gotō directed Yamada and his planners to redouble their efforts and to revisit their plans with a "broader perspective" in mind.

Yamada refused to let zeal overcome his acute pragmatism. The following day he directed his subordinates to draw up four additional reconstruction proposals, each with differing price tags.[8] Along with the ¥4.1 billion initial plan favored by Gotō, Yamada's subordinates devised plans with the following budgets: ¥3 billion, ¥2 billion, ¥1.5 billion, and ¥1 billion. Each plan earmarked different amounts for roads, canals, bridges, parks, land readjustment, and other facilities. The most marked difference between the plans was the amount of land that paved roads would comprise in the new capital. In the ¥3 billion plan, 31 percent of Tokyo's urban space would be taken up by roads. In the ¥2 billion plan, roads would comprise 24 percent of Tokyo's built environment; in the ¥1.5 billion plan, 22 percent; and in the ¥1 billion plan, 20 percent. These figures compared well to the Tokyo of 1922, in which just 11.7 percent of the capital's urban space was paved roadway. Under only the most expensive plans, however, would Tokyo secure paved roadway space that matched or superseded the cities of London (23 percent) or Berlin (26 percent) as they stood in 1923.[9] When they were completed, Yamada discussed each plan in detail with Horikiri Zenjirō, chief of the Home Ministry Accounting Department, and agreed that the ¥1.5 billion plan was the most realistic one to develop.

Aware that a divergence of opinions, not to mention the nation's financial limitations, might limit the scope and scale of his grand reconstruction dreams, Gotō contemplated ways to secure the support he believed was required to secure a large reconstruction program. Gotō requested that cabinet colleagues endorse a novel plan to create an Imperial Capital

Reconstruction Ministry (Teito fukkōshō) that would possess responsibility for all aspects of the reconstruction project, including budgeting, planning, and implementation.[10] Gotō's motives were obvious. From his earlier discussions with Yamada and his fellow Home Ministry bureaucrats, Gotō realized that Tokyo's reconstruction would be fraught with political, ideological, and budgetary disputes. The creation of a reconstruction ministry that concentrated all power and authority and was answerable only to the emperor and the prime minister, Gotō believed, would be the only way to avoid political contestation and insulate his grandiose dreams.

Rather than introduce his proposal for a reconstruction ministry to cabinet colleagues as the naked, ambitious power play that it was, however, Gotō used a number of other approaches to justify its creation. In a proposal he submitted to the cabinet, Gotō wrote that for the sake of "speed, integrity, and efficiency," all aspects of reconstruction planning and implementation must fall under one minister.[11] If one agency was responsible for planning while another possessed power over implementation, no guarantee could be given that reconstruction plans would be followed as proposed. Moreover, he suggested that the amount of time spent coordinating activities between planners and different implementing agencies would be considerable and ultimately wasteful. Finally, Gotō alluded to the symbolic power of creating a new ministry following Japan's unprecedented calamity. The project of reconstruction, he argued, was a national undertaking similar to the vast programs of reconstruction implemented by European powers following the First World War. Gotō claimed that by creating a new ministry, the government would demonstrate to people in Japan and abroad that its leaders understood the importance of reconstruction and its significance to the nation.

Cabinet ministers refused to endorse Gotō's proposal. On numerous occasions in late September, heated exchanges took place between Gotō and his fellow ministers. Ministers saw this as a bureaucratic gambit and refused to relinquish control over reconstruction projects that were likely to fall under their respective jurisdictions. Bureaucrats from the Railways Ministry, for example, repeatedly stated their desire to oversee railroad station and track reconstruction. More important, Finance Minister Inoue countered that he would never endorse a proposal that would take appropriations authority away from his ministry. In Inoue's mind, such a policy was inconceivable.[12] Gotō had lost his first major battle. Japan's polity would not become as authoritarian as Gotō wished. On September 27 the cabinet endorsed an alternative proposal and created an Imperial Capital

Reconstruction Institute (Teito fukkōin) that possessed little of the authority that Gotō desired.

Over the course of October and November 1923, the Reconstruction Institute became the most important planning organization for the reconstruction project. Gotō served as president of this body and was given complete control over the selection of its staff. He appointed two vice presidents, each of whom held administrative control over three distinct bureaus within the organization: Vice President Miyao Shunji oversaw the planning section, the land readjustment section, and the building section, while Vice President Matsuki Kanichirō administered the civil engineering section, the supply section, and the accounting section.[13] Under these three individuals, the government created a sizable executive bureaucracy, which included a chief engineer, Naoki Rintarō; 7 directors; 15 secretaries; 30 administrative officers; 105 engineers; 150 junior administrative officers; and 350 assistant engineers. Together members of this institute were instructed to further develop and refine plans for Tokyo's reconstruction. Though all drafting activities rested under the auspices of the institute, it did not possess power in determining appropriations for the overall project. This power remained in the hands of Japan's finance minister and would prove to be a major obstacle to Gotō and his grand ambitions. Moreover, despite the handpicked assemblage of an impressive cast of pro-Gotō people within the institute, little consensus existed or emerged among reconstruction planners.

AGREEING TO DISAGREE: DEVISING SIMULTANEOUS RECONSTRUCTION PLANS

The creation of the Reconstruction Institute with subordinate advisory agencies did little to build consensus. Rather, disputes became more pronounced as bureaucratic positions became formalized and planning took on a real and urgent dimension. The vice president of the Reconstruction Institute, Miyao Shunji, witnessed virtually every debate. Since Miyao's portfolio included planning, land readjustment, and building, he was responsible for drafting and overseeing the most contentious parts of the overall project and submitting a finished plan to Gotō. This was no small task. Miyao found balancing idealism with pragmatism one of the most difficult parts of his administrative job. The personnel beneath him, he reflected in a 1930 article published in journal *Toshi mondai* (Urban problems), were almost equally divided between those who argued that "we should promote an ideal program irrespective of expenses" and those "who championed a modest, attainable plan made in accordance with financial realities."[14] Little common ground could be found to bridge this challenging divide.

During the first three weeks of October, planners and bureaucrats within the Reconstruction Institute fought with one another over virtually every aspect of each Tokyo reconstruction plan devised. As Home Ministry bureaucrat Yamada Hiroyoshi described it, "every day we met to discuss planning, and every day we had all sorts of arguments. We never reached consensus."[15] Some individuals, for instance, advocated plans for new Tokyo to be connected by a vast thoroughfare network with wide streets, paved sidewalks, and mass transit linking virtually every residential neighborhood. Others, however, ridiculed such ideas and claimed that drafting unrealistic plans was wasteful and counterproductive. The government, these planners argued, could not even afford to purchase the land necessary for such grandiose plans let alone pay for the associated construction costs. To Yamada, the creation of a Reconstruction Institute had done nothing to forge consensus.

Internal disputes within the Reconstruction Institute were amplified by the uninvited visits of businessmen, industrialists, journalists, and commentators with no formal link to the organization. Upon arrival, these visitors gave unsolicited and often unwelcome advice. On more than a few occasions, moreover, jealous bureaucrats not associated with the institute arrived and demanded that their opinions be considered. In 1930 Vice President Miyao recounted one memorable exchange that highlighted the fractious politics associated with reconstruction planning. When news leaked out that Reconstruction Institute planners had started discussing school reconstruction plans, Miyao soon found himself face to face with an angry mob of Education Ministry bureaucrats who demanded to know why they had not been consulted. After a heated debate, the bureaucrats warned Miyao, "Don't take the rebuilding of schools into [your] consideration; it is not your job."[16]

Such visits often inflamed an already tense situation within the institute. Planners who debated the size of the proposed reconstruction plan used the opinions expressed by visitors to bolster arguments for or against a large reconstruction plan. Yamada described the chaos and contestation that defined the planning process: "It was like a war. Even though we made a plan in the morning, it was changed in the afternoon and then it was changed again in the evening. We often joked 'what is the morning plan, the afternoon plan, the evening plan' and so forth. Every time the plans changed so too the proposed budgets changed accordingly . . . we never stopped working."[17]

They never stopped fighting either. Miyao attributed much of the contestation he witnessed to the simple fact that the Reconstruction Institute "needed a concrete reconstruction budget" to work with. When he asked Gotō for even an outline budget, Miyao claimed that Gotō never gave a specific figure. Why? Miyao concluded it was because the home minister had no idea how much money would actually become available. This simple fact, Miyao told Gotō, "put us in an extremely difficult position" as everything came down to how much money would be budgeted for reconstruction.[18] To drive this point home, Miyao informed Gotō that his planners could not even reach agreement on how wide to make one main trunk road that was to run across Tokyo from Kitasenji in the north to Shinagawa in the south via Ginza because no formal budget existed.[19] "A difference of only one *ken* (1.8 meters) in this street's width," he informed Gotō, would have a considerable impact on the overall reconstruction budget.[20] The ambiguity of finances led institute members to fight over the smallest parts of reconstruction plans, as planners knew that money was not finite and that, for example, a wider road in location A meant that there might not be enough money for a park in location B. Until the budget was sorted out, planning was worse than a futile exercise; it was poisonous.

Though Gotō responded to Miyao's complaints by reiterating his dream-large-and-plan-big mantra, Miyao took it upon himself to get more accurate budgetary information. The Reconstruction Institute official met secretly with Finance Minister Inoue on a number of occasions in mid-October. On each occasion, Miyao asked Inoue for an estimate of the reconstruction budget. Inoue was blunt to the point of being rude. He suggested repeatedly that ¥800 million would likely be the maximum amount of funding made available.[21] This devastated Miyao. When he countered that very little could be accomplished if the budget was less than a bare minimum of ¥1.2 billion, Miyao recounted how Inoue "scolded me" for being selfish and putting the interests of Tokyo over those of national finance.[22] Armed with this budgetary information, Miyao enlisted Yamada to help him wear down many of the idealists who continued to follow Gotō's advice. Even then, it was not until October 24 that Reconstruction Institute planners reached a general agreement on the scope and scale of a basic reconstruction plan.

In drafting the Reconstruction Institute's formal plan, Miyao, Yamada, and Matsuki found themselves in a difficult position. On the one hand, they knew that Gotō wanted a big, all-encompassing plan, but on the other, they understood that Inoue was poised to deflate Gotō's ambitions with a single stroke of his pen. The plan they produced was a far cry from Gotō's dream, but it was still larger than what Inoue had intimated he would fund.[23] The five-year plan stood at roughly ¥1.3 billion, with ¥200 million of this figure earmarked for reconstructing Yokohama. Planners directed the lion's share of Tokyo's budget, nearly ¥700 million out of ¥1.1 billion, toward roads and bridges. Another ¥110 million was to be spent on the reconstruction of intercity canals, the construction of a major canal between Tokyo and Yokohama, and harbor and estuary rebuilding and construction projects. Under the reconstruction plan, Tokyo would get a new sewage system at a cost of ¥110 million. Planners directed ¥5 million worth of reconstruction funds to upgrade water mains throughout the city. Rounding out the planned appropriations were ¥10 million for rubbish disposal and incineration facilities, ¥30 million for parks, ¥30 million for administrative costs associated with the Reconstruction Institute, ¥40 million for land readjustment, and ¥60 million for direct construction subsidies.

Along with providing a budgetary outline, the planners articulated a caveat related to funding and discussed a number of fundamental principles that they argued underpinned their reconstruction plan. First, Miyao suggested that the outline budget was smaller than what was needed to reconstruct Tokyo in accordance with the emperor's wishes as articulated in the September 12 Imperial Rescript. More funds were necessary to fulfill this instruction. The plan they created, he claimed, reflected what institute planners believed the national government should fund and oversee in relation to reconstruction. Miyao concluded that other critically important reconstruction projects, such as primary schools, small parks, hospitals, and social welfare facilities—projects that municipal and prefectural governments had traditionally funded—would have to be covered by local and prefectural governments. In Miyao's mind, however, the national government would have to provide loans to back this funding.

Otherwise the debt burden for municipal and prefectural government would hinder any future development plans.

Beyond this important point, the bureaucrats who drafted the first plan suggested that a well-conceptualized road system should be the foundation of Tokyo's future growth and prosperity. Planners created a Tokyo with two large thoroughfares of 24 *ken* (43 meters) in width bisecting the city from north to south and east to west. Hundreds of "smaller arterial roads" between 6 and 20 *ken* in width were planned to connect these trunk lines. Numerous modern bridges would also be built, and they, along with canals, trams, and underground subways, would supplement the skeleton of new Tokyo. Next, planners argued that buildings and houses would have to be constructed to conform to the rational road network. Roads in old Tokyo had often been constructed around buildings rather than through them, giving Tokyo a meandering, serpentine-like transport system. This practice, they argued, had to end. Such rebuilding with roads at the center would require significant land readjustment and the possible whole-scale transfer or exchange of land. In particular, they suggested that a number of middle and high schools, government offices, markets, and slaughterhouses be moved to Tokyo's suburbs to open up central city land for roads, greenbelts, and parks. Finally, they appealed for the adoption of a strict building code that would require new buildings to adhere to new fire and earthquake standards.

Would a compromise plan that delighted no one be broad enough to satisfy everyone? Reconstruction Institute planners soon found out the answer to this question. When Gotō received the plan, he was underwhelmed. Though he assumed that Inoue would refuse to fund a large-scale plan, Gotō was disappointed that planners did not present an ideal, all-encompassing master plan. He agreed to introduce the plan to the cabinet but informed Reconstruction Institute planners to continue to develop a master plan with a price tag of roughly ¥3 billion. Gotō still believed he could convince cabinet colleagues to support such a plan.

When Gotō submitted the ¥1.3 billion plan to his cabinet colleagues, it drew widespread support. Why? It was a smaller-scale plan than many in the cabinet had expected Gotō to champion. One important component, however, remained in doubt: the budget. Inoue punctured the consensus by suggesting that the draft reconstruction plan was impossible to accept because it would be "over budget." He therefore reiterated his opinion that no decision on reconstruction could be finalized until a clearer picture emerged as to how much money would be available to plan and rebuild the capital.

At this juncture, tensions between Gotō and Inoue reached the boiling point. In front of all his cabinet colleagues, Gotō demanded to know why a special session of parliament could not be called immediately and a total budget for the reconstruction endorsed, as had been done at the beginning of both the Sino-Japanese and Russo-Japanese Wars. Was Japan not facing a similar challenge that necessitated the quick passage of an overall budget, the specifics of which could be determined later? Why was Inoue dragging his feet? Gotō reminded colleagues that nearly eight weeks had passed since the catastrophe and yet no decision on the important matter of reconstruction or its budget had been decided. Gotō declared that the delay was an embarrassment and asked whether Inoue was up to the important task of national finance.[24]

A normally calm, collected Inoue fought back. A special session of the Imperial Diet could not be called to endorse an overall reconstruction budget until the ordinary budget for 1924 was decided. As a result, ministries would have to accept budget cuts that could be steep to pay for reconstruction. Inoue reminded his cabinet colleagues that some ministers, including Home Minister Gotō, had refused to reduce their ministry's budget for 1924, despite requests to do so. This type of bureaucratic intransigence, Inoue claimed, slowed the process of reconstruction budget planning. Gotō hit back and stated that Inoue's slavish devotion to reducing expenditures for fiscal year 1924 was not only irrational but also absurd, given the state of emergency and uncertainty that existed in Japan. Gotō argued that Inoue's sluggish and overly conservative response jeopardized national stability. Gotō concluded that Inoue's chief responsibility to the emperor and nation was the budget, and he asked how much longer it would take for him to complete his job.

Inoue did not let this broadside go unanswered. The finance minister argued that Gotō had refused to accept any cuts in the Home Ministry's budget and that until he did, neither the budget for 1924 nor the reconstruction budget could be formalized. Inoue then gave Gotō and the cabinet two options. One was to have the government wait until he received ministerial support for voluntary budget reductions for 1924 and beyond. The second option was to call a special Diet session immediately with the condition that the reconstruction budget be no larger than ¥300 million. This figure, Inoue claimed, was all that the nation could afford without a commitment by ministers to reduce their yearly spending from 1924 to 1928. "If this option is fine," Inoue concluded, "I will submit a budget today." Inoue's ultimatum "cast a chill over the meeting."[25]

Upon hearing these words, the color of Gotō's face drained from red to pale in a matter of seconds. Gotō claimed that it was clear that "cabinet ministers had lost confidence in him." Moreover, he confessed that he "no longer trusted his cabinet colleagues and was therefore not willing to take responsibility for the administration of the reconstruction project." In simple terms, he tendered his resignation. In dead silence, Gotō then stood up, bowed to Prime Minister Yamamoto, and proceeded to walk out of the room. Numerous sources suggest that only the quick and spirited intervention by Den Kenjirō thwarted Gotō's physical exit and his resignation from becoming permanent. Den, the agriculture and commerce minister, immediately chased after Gotō and grabbed hold of the embittered minister's frock coat. He told Gotō not to "act silly," but rather "to calm down, take a seat, and discuss things" in a mature fashion. When Gotō resisted, Den tightened his grip on the home minister and informed him that resignation was the ultimate act of selfishness. Refusing to release Gotō from his grasp until the home minister returned to his seat and retracted his resignation, Den mustered all his physical might and pushed Gotō back into his seat. He then worked to calm Gotō, stating, "There, there, there, just calm down and return to business." Gotō struggled at first but ultimately obliged. Inoue then agreed to produce a reconstruction budget as soon as the 1924 budget was finalized. Over the course of this eventful encounter, Yamamoto and other cabinet ministers stared in silent disbelief.

Soon after this heated exchange, Gotō hoped that before Inoue agreed on a budget, he could create a groundswell of elite-level support for a larger reconstruction plan than even the Reconstruction Institute had recommended. His hopes rested in the first instance with one of two advisory bodies created under the Reconstruction Institute. The first was known formally as the Imperial Capital Reconstruction Councilors Committee (Teito fukkō sanyokai). Gotō had created this organization on October 18, but it had been called to meet only after his open political brawl with Inoue.[26] Chaired by Baron Sakatani Yoshio, the committee comprised thirty-two senior councilors, including businessmen (Matsumoto Jōji and Kirishima Shōichi), military men (Ugaki Kazushige and Okada Keisuke), municipal politicians (Nagata Hidejirō), urban planners (Ikeda Hiroshi), and academics (Honda Seiroku), along with sixty-three junior councilors.

Their primary responsibility was to provide "honest, frank, and sound advice" to members of the Reconstruction Institute in all matters that were brought before them by Gotō.[27]

In general terms, councilors supported Gotō's claims that the plan they received from the Reconstruction Institute was inadequate for the capital's future needs. At their November 1 meeting, councilors suggested that more attention was required to expand and enlarge the road and canal network. Specifically, members recommended that virtually all the planned smaller roads be widened and that the canal network of eastern Tokyo be expanded to allow greater transport capacity.[28] They also encouraged planners to emphasize fire resistance and disaster resilience more uniformly throughout the capital's reconstruction plan. At Army Minister Ugaki Kazushige's urging, they thus supported the creation of more open spaces, greenbelts, and underground public shelters that could protect people from future calamities. While their initial findings pleased Gotō, the members ultimately refused to enter into the contentious war of words and ideas over the reconstruction budget. Concrete budgetary matters, they concluded, were outside their jurisdiction. All elites who read their recommendations, however, understood that if roads were to be expanded further than the original proposal suggested and more open spaces created throughout the capital, more money would be needed. Securing these funds became Gotō's ultimate challenge.

THE FINAL BATTLE OVER THE RECONSTRUCTION BUDGET

If Gotō remained the "big opportunistic dreamer" during the autumn of 1923, he was countered at almost every turn and eventually trumped by the "big deflationary pragmatist," Inoue Junnosuke. Inoue guarded his ministry's control over all financial and budgetary matters, particularly those associated with Tokyo's reconstruction, for one simple reason: he believed that the nation could not afford anything near what Gotō and many of his starry-eyed associates wanted for this project. Inoue was a model of pre-Keynesian, classical economic restraint.[29] Prior to his appointment as finance minister in 1923, Inoue served as governor of the Bank of Japan from 1919 to 1923, where he championed measures to rein in inflation through induced deflation. He had done so in part by restricting the supply of yen in circulation. Beyond this measure, Inoue also supported initiatives to trim government spending and encouraged politicians to work in unison with families across Japan to reduce expenditure, embrace thrift and frugality, and support a moderate financial path. Inoue hoped that such measures would foster greater economic stability and eventually enable Japan to return to the international gold standard at some point in the 1920s. Reducing debt and balancing budgets was not only a fiscal necessity, it was also a moral imperative for both the government and individual households across Japan. In short, Inoue embraced what he saw as a "contractionary tide" (*kinshuku no fūchō*), and the earthquake calamity did nothing but reinforce his belief in thrift and moderation.

Rather than champion a large reconstruction plan that would stimulate Japan's economy, Inoue emphasized the importance of austerity in the face of calamity. His reasoning was simple, and it betrayed his retrenchment-minded, conservative fiscal outlook better than almost any other policy he had supported over the course of his entire career. He understood that in normal circumstances Tokyo provided roughly 30 percent of Japan's tax revenue. As a result of the earthquake disaster, Inoue predicted that the government would collect approximately ¥130 million less in tax revenue for fiscal year 1923 than originally anticipated. He therefore directed all ministers to retrench their operating budgets to make up for this unexpected shortfall. Reducing expenditures by ¥130 million, just under 10 percent of Japan's overall national budget, was no easy task, as ministries had already earmarked funds that Inoue now wanted back. Inoue refused to even consider developing a reconstruction budget or a national budget for 1924 until the budget for 1923, a budget originally decided upon in November 1922, had been readjusted to reflect the unexpected revenue shortfall brought about by the disaster.[30] Predictably, every minister except Navy Minister Takarabe Takeshi—whose ministerial budget was already set to fall precipitously after the Washington Naval Arms Limitation Treaty took effect—groaned, resisted, but ultimately complied with Inoue's directive. This exercise, however, took a precious two weeks to accomplish. Moreover, it jaundiced cabinet ministers against a large-scale reconstruction program.

While waiting for ministers to scale back their individual budgets took time and frustrated planners who wanted a firm reconstruction budget, Inoue's next decision caused greater difficulties. In the course of revising the 1923 fiscal year budget downward, the finance minister laid the groundwork for the reconstruction budget. Rather than fund a dream plan comprising what planners desired, Inoue prepared his budget by determining what he believed the nation could afford. These were two vastly different, irreconcilable approaches that caused resentment and fostered contestation. Inoue concluded that planners would have to make do with the funds he provided rather than plan the ideal reconstructed city and then ask for the appropriations to make it possible.

Why was this the case? For one, Inoue remained adamant that the government would not introduce new taxes or raise existing taxes to fund reconstruction.[31] The finance minister argued that the only prudent and morally responsible way to raise reconstruction funds would be to issue bonds or public loans. Inoue, however, attached an important condition to this public funding plan. He announced that the yearly interest payable on these reconstruction bonds would have to be paid for out of existing revenue. Inoue therefore told each minister that he would calculate the total appropriations that he believed each ministry could trim from yearly budgetary appropriations between fiscal years 1924 and 1928. The surplus amount saved from future yearly budgets, Inoue declared, would then be used to cover the interest due on reconstruction bonds calculated at 5 percent per year.[32] Under his scheme, Inoue boasted that reconstruction could be carried out without adding to Japan's debt, increasing taxes, or risking inflationary pressures on the economy. Once calculations were determined, Inoue declared he would ask for unqualified support from each ministry in question.

These interrelated decisions had enormous political and budgetary consequences for the reconstruction of Tokyo. First, they slowed the reconstruction process further as planners in the Reconstruction Institute were not given an official budgetary framework with which to effectively plan until the overall budget for 1924 could be decided in early November. More damaging, Inoue's funding formula encouraged nearly every cabinet minister to question and eventually challenge calls for a large-scale reconstruction program for the simple reason that a larger reconstruction budget would translate into decreased funding for individual ministries after 1923. Army Minister Tanaka Giichi was one of the first, and perhaps most concerned, cabinet ministers who approached Inoue after it became clear that budgetary retrenchment would be required to cover interest payments on the reconstruction bonds.[33] Inoue reflected on his encounter with Tanaka as follows:

> Once, Army Minister Tanaka visited me and said, "I would like to talk to you." When I enquired as to what it was about, he replied, "You know Gotō is talking about ¥3.5 billion or ¥1.5 billion, but you as finance

minister, do you support such a plan?" I responded, "Whether I agreed with his plan or not, I cannot support it because we do not have enough money." "That's right," interjected Tanaka, "we also need a lot of money for the restoration of the army." Tanaka was worried that the entire national budget would be taken up with the reconstruction of Tokyo, but I reassured him that we wouldn't do such a foolish thing and would keep money for other necessities. We parted, with Tanaka telling me that he "felt relieved."[34]

Other ministers with similar concerns followed in the wake of Tanaka.

During the second week of November, the finance minister finished the national budget for 1924 and drew up an extraordinary reconstruction budget for Tokyo and Yokohama. On November 16 he introduced the government's reconstruction budget. Inoue informed his colleagues that the government could budget no more than ¥703 million for the reconstruction project. This amount, he concluded, was not negotiable. Announcing his budget, Inoue reiterated that his decision to reduce the budget from a hoped for ¥1.3 billion to ¥703 million had not been based on abstract debate or a personal whim. Rather, this figure reflected what Inoue believed was the maximum amount available to fund reconstruction given Japan's limited finances and already high debt levels.[35] His proclamation infuriated Gotō and angered virtually every planner within the Reconstruction Institute. Not only was the ¥703 million a far cry from the ¥1.3 billion requested in the final plan submitted by the Reconstruction Institute, it was nearly ¥100 million less than what Inoue had intimated would most likely be available for reconstruction during secret discussions with Miyao in October. When faced with this unexpected turn of events, Reconstruction Institute planners had to scale back their reconstruction plan. Once again, they were asked to do so with utmost speed because the final plan had to pass two further bureaucratic and political hurdles: endorsement by a committee of elder statesmen and acceptance by the Imperial Diet.

AFTERSHOCKS AT THE EPICENTER: THE ANGRY ELDER STATESMEN OF THE SHINGIKAI

Though neither Gotō's individual efforts nor the opinions expressed by members of the Sanyokai convinced Inoue to reevaluate his budgetary decision, the home minister persisted in believing that he possessed two opportunities to persuade the finance minister to expand the reconstruction budget. Gotō refused to give up until all his options were extinguished. His first hope rested with the Imperial Capital Reconstruction Deliberative Council (Teito fukkō shingikai). Yamamoto appointed this council of elder statesmen, a throwback to the *genrō* councils of the Meiji and early Taishō eras, on September 19 at the request of the emperor.[36] The council's mandate was straightforward: it was to serve as the highest body of advisers in relation to the overall reconstruction project. Gotō assumed that if the council recommended expansion of the reconstruction plan, Inoue would be compelled to revise the budget upward.

Membership on this twenty-person council resembled a "Who's Who" of elite level movers and shakers in 1923 Japan. Eleven cabinet ministers held a formal position on the council, including Yamamoto Gonnohyōe, Gotō Shinpei, Inoue Junnosuke, Okano Keijirō, Den Kenjirō, Ijūin Hikokichi, Takarabe Takeshi, Tanaka Giichi, Hiranuma Kiichirō, Inukai Tsuyoshi, and Yamanouchi Kazutsugu. Membership also extended to include the leaders of both major political parties in the lower house of parliament, Takahashi Korekiyo (Seiyūkai) and Katō Takaaki (Kenseikai); and two influential members of the upper house, Egi Kazuyuki and Aoki Nobumitsu. Privy Councilor Itō Miyoji, businessmen Shibusawa Eiichi and Wada Hirofumi, and bureaucrats Ōishi Masami and Ichiki Otohiko were also members. Each noncabinet member was a leader in his field or profession. Moreover, each possessed a wealth of experience in politics, business, or finance.

Yamamoto viewed the Deliberative Council as an elite-level advisory body and hoped that it could accomplish two important political goals. First, he predicted that the reconstruction project would be contentious, despite the rhetoric of the national unity that was produced following calamity. The prime minister believed that the carefully handpicked council could imbue the government's final reconstruction plan with an air of political unity.[37] Second, Yamamoto anticipated political difficulties with parliamentarians. Gaining endorsement by the leaders of both political parties at the Shingikai, he believed would improve chances for final passage of the reconstruction bill in the lower house of the Diet.

In a theoretical and practical sense, Yamamoto's reasoning was sound. Cabinet-Diet conflict in Japan was particularly intense given the structure of Japan's polity. Unlike other parliamentary democracies that existed in 1923, the majority party in Japan's lower house had no constitutional right to form a cabinet or to determine the makeup of Japan's government; this was a right that the emperor possessed through the Meiji Constitution of 1889. Though Japan was moving in the direction of "party cabinets," namely, cabinets that reflected the composition of parliament, it was not a guarantee in 1923, as Yamamoto's cabinet illustrated. The historical legacy of this system was often acrimonious political disputes and governmental instability. Japanese cabinets found themselves repeatedly at odds with the lower house of the Diet. This was particularly problematic for cabinets because the Imperial Diet possessed the final power over the budget. In the minds of many parliamentarians, Yamamoto's earthquake cabinet had no political legitimacy, and Diet members were willing to use the budget issue to embarrass and weaken the government. The prime minister was cognizant of this fact and hoped that the Deliberative Council could keep contentious politics from hijacking the reconstruction project.

When the prime minister convened the Deliberative Council in late November, Gotō believed he could convince a majority of its members to advise the cabinet to reopen discussions related to the scope and scale of the reconstruction project, particularly its budget.[38] At its first meeting on November 24, Miyao Shunji read out the entire draft reconstruction bill and presented each member with a detailed budgetary outline of the proposal. Gotō then opened discussions with a plea to expand the scope and scale of the project in order to build a true imperial capital, as requested by the emperor in his September 12 rescript.[39]

Egi Kazuyuki was the first Deliberative Council member to respond to the draft reconstruction proposal. He did so in a spirited fashion through a speech that lasted for nearly one hour. Unfortunately for Gotō, he did not critique the plan for being too small or haphazard, but rather savaged it for being too grand.[40] In the first instance, Egi suggested that simple restoration of the city was needed, not radical construction of an ideal, grandiose capital. He advocated this position for two reasons: first, he believed that the city needed to be restored quickly for the sake of the sufferers; and second, he suggested that state finances could not afford a large-scale reconstruction as envisaged by Gotō. Egi thereafter focused at length on two projects that could be altered or excised from the proposal. He claimed that it was extravagant to propose widening major roads to a width of 30 *ken* (54.6 meters). This would require too much money to cover construction materials, labor, as well as funds to purchase the land needed to enable expansion. Egi also believed it was unwarranted and unethical to incorporate plans for the construction of extensive Tokyo port facilities and a major canal between Yokohama and Tokyo. The canal and port facilities, in Egi's opinion, were not reconstruction projects but rather long-sought public works projects that Gotō had incorporated into the emergency reconstruction bill. Not only was this unwarranted, it was also unethical. Egi therefore concluded that both should be removed from the budget and raised as separate construction projects during the next fiscal year.

After Egi fired his initial salvo at the reconstruction plan, Itō Miyoji launched an equally spirited frontal assault against the government's proposal.[41] Itō stated that he opposed virtually every aspect of the proposal without reservation. This forthright volley set the tone for his extended oration. First, Itō suggested that Gotō's plan—even this scaled-back ¥703 million version—was too idealistic. Planning and building a grandiose capital, he argued, would take too long to complete. What provisions existed for people who wanted to return quickly, rebuild, and resume their lives as they stood before September 1? How long could they wait for the new city to be built around them? Had anyone contemplated the economic costs of a slow-speed reconstruction? For the economic sake of sufferers and the nation, Itō opined that Tokyo needed to be restored as soon as possible. He also raised a second practical issue: the illogic of building an ideal capital in such an earthquake prone region. If Gotō believed in the importance of Japan possessing a capital that would project modern imperial power well into the future, Itō questioned why Gotō did not support moving the capital to a less earthquake-prone region of Japan.[42] Did not history suggest that another calamitous earthquake would destroy Tokyo in the future?

The elder statesman's most vitriolic attack on the reconstruction plan, however, centered on the question of private property, landownership, and the Reconstruction Institute's plans for uncompensated land appropriation. In the first instance, Itō cautioned against the precedent the government would set with the reconstruction project. It would be far too easy for future politicians to seize land in the name of "public interest" without compensation. Second, in relation to planned government land purchases, he railed against paying a flat rate of ¥197 per tsubo rather than compensating landholders with what would be construed a fair market price. Itō believed that the government should pay a price that reflected land valuations immediately prior to the earthquake. Both government proposals, Itō concluded, violated not only the spirit of the constitution but specifically article 27, which safeguarded the right of individuals to own property. He concluded this challenge by suggesting that while the previous Tokugawa government took land from feudal lords on a whim without compensation, the Meiji Constitution allowed for no such despotic practices.[43] He asked Gotō if this was the type of world to which Japan should return. Soon after Itō concluded his assault, a somewhat dazed and bewildered Yamamoto asked for a recess, hoping that a break would allow Gotō and the other government officials a chance to regroup and rebut the challenges articulated by Itō and Egi.[44]

When the Shingikai reconvened in the afternoon of November 24, Gotō had his first opportunity to counter the charges raised by Egi and Itō. By all accounts he did a poor job, with even his official biographer and son-in-law, Tsurumi Yūsuke, claiming that Gotō's rebuttal "sounded like a theological conundrum."[45] Tsurumi suggested that Gotō failed to convince anyone but himself. Inoue's interpretation of the meeting supports Tsurumi's assertion. Inoue reflected in 1930 that "only Mr. Gotō understood [Mr. Gotō], nobody else." He lamented that Gotō's performance was simply "laughable."[46]

Lack of style aside, was there any substance to Gotō's response? In the first instance, he argued that the reconstruction budget was anything but extravagant. He did a miserable job proving this point, however. Gotō claimed that the nation had rallied to help Kagoshima reconstruct after the destructive volcanic eruption of Sakurajima in 1914. Why, he pondered, could Japan not do this again, yet in a much bigger fashion, given the importance of Tokyo?[47] The task Japan now faced, he argued, was even more pressing because Tokyo was the "center of the nation's politics, economics, and culture."[48] Next, he claimed that constructing a canal between Tokyo and Yokohama and the development of Tokyo port facilities were central to the success of the overall reconstruction project. Both would help facilitate the flow of key reconstruction materials to the disaster zone.[49]

Gotō thereafter launched a spirited and at times personal attack on Itō. He claimed that while there were many large private landholders such as Itō who desired to cling to their individual property rights even in the face of such an unprecedented natural disaster, this group represented a minority. Other large landholders, Gotō claimed, such as the publishing magnate Ōhashi Shintarō, supported land rezoning, readjustment, and the construction of new roads, sidewalks, and social welfare facilities for the city's residents. Gotō argued that these landholders, unlike Itō, understood that urban betterment programs would not only improve Tokyo as a livable city but also increase land values in the future.[50] Gotō's oration, and in particular his closing remarks aimed at Itō, did nothing but further politicize the discussions.

In a less political manner, Inoue Junnosuke followed Gotō and spoke on behalf of the government's position. The finance minister anticipated that committee members would pressure him to expand the overall budget, and he was not prepared to counter charges of extravagance. Inoue therefore did not add direct support to any of Gotō's claims but rather kept his presentation tied closely to what he knew best, financial matters.[51] Inoue argued that the proposed budget of ¥703 million fell within the confines of sound financial policy. With no little modesty, Inoue claimed that the budget was balanced to perfection and that it could be implemented without any harm to the nation's financial position either in the short or long term. The budget, he concluded, could not be challenged on economic grounds.

Though Inoue spoke with authority, discussions again became heated when Aoki, Takahashi, and Katō argued against the government's plan. Katō, in fact, recommended that the Deliberative Council disband and that all issues related to reconstruction be discussed and finalized before the full parliament.[52] At this moment it seemed possible that the advisory body established to help imbue the government's reconstruction plan with an air of transcendental political unity would, at best, reject the government's plan or, at worst, disband altogether. Both Yamamoto and Gotō knew this outcome would put monumental pressure on the cabinet to resign for their failure to gain Shingikai approval for the government's proposition.

What motivated some of the most prominent Shingikai members to challenge the government's reconstruction proposal in such an aggressive manner? First, despite Inoue's assurances, finances played an important role in members' deliberations. Itō pointed out that the total government debt of Japan in 1923 stood at ¥4.3 billion. Japan was still paying off loans secured from Britain in 1899 and 1905 as well as a sizable amount of domestic and international loans issued between 1905 and 1911.[53] Itō wondered if Japan could safely add more debt. He also suggested that the government was, in effect, dreaming if it believed this would be the only funding request made in relation to the reconstruction project. Cost overruns would develop, and the final price tag of reconstruction would not be known until after the reconstruction was complete. Only the absolutely essential items should be planned and budgeted at the beginning. More could be added to the program as it developed in accordance with state finance. Egi agreed wholeheartedly with this argument and interjected, "When the time is right, we can add other facilities."[54] Second, both men also worried about the ethical and constitutional issues of a reconstruction plan in which land was taken without compensation or at rates below what was fair market value.

Personal antagonisms, bureaucratic rivalries, and a feeling of underappreciation may also have influenced the behavior of many committee members. Despite creating the Deliberative Council in mid-September, neither the Home Ministry planning department nor the Reconstruction Institute consulted, briefed, or even invited Shingikai members to take part in any reconstruction plan discussion. This was a major political oversight. While Yamamoto and Gotō saw the Deliberative Council as an elite-level, unity-imbuing rubber stamp, many members saw things differently. The emperor's instructions to Yamamoto on the formation of this body stated that it would be "the body of highest decision-making authority" in reference to the reconstruction project. The only decision-making power that some members felt they were given was a yes or no on the final proposal, and this, they believed, was not enough consultative input.[55]

Did age play a role? Some commentators in 1923 and a few historians in Japan since have suggested that age influenced outlook and the ultimate decision-making process within the Shingikai. Commentator Annan Jōichi wrote in 1923 that the group of "stubborn, old, rude rulers" who comprised the Deliberative Council cared only about the present. Knowing that they would "visit heaven" within ten or twenty years, they failed to grasp the opportunity that had presented itself to transform Tokyo into a true imperial capital for both the present and future generations.[56] Historian Fukuoka Shunji quoted a November 28, 1923, *Tokyo asahi shinbun* editorial that supported this assertion: "the old group of men who have no understanding of city planning" failed the government.[57] Historian Nakamura Akira also raised the age factor and calculated that the average age of the Shingikai members was 62.7.[58]

Using age, however, has its limitations. Many other contemporary political, economic, and ideological factors accounted for the angry reception the government received at the first meeting of the council. In fact, its oldest member, eighty-three-year-old Shibusawa Eiichi, kept the Shingikai from disbanding after its morning meeting.[59] Shibusawa, a well-respected business tycoon, countered Katō's suggestion to disband and claimed that it was the duty of the Deliberative Council to advise and amend if need be, but ultimately to endorse a reconstruction proposal. Anything less, he argued, would be shirking the responsibility the emperor had given them. He thereafter recommended that the Shingikai create a special committee (*tokubetsu iinkai*) to hammer out a concrete proposal that could be submitted to the Diet. Seconded by Itō, Yamamoto agreed and appointed an eleven-person special committee with Itō as chairman. The ten other members included Takahashi, Katō, Egi, Aoki, and Shibusawa from outside the cabinet, and Gotō, Inoue, Okano, Den, and Inukai from within the cabinet.

When the special committee of the Deliberative Council opened on November 25, the tensions that had fueled the flames of contestation the day before burned with even greater intensity. Over the course of two fractious days, the members of the Shingikai continued to fight bitterly. Consensus was reached only because of the political machinations of Chairman Itō. Itō argued that votes on individual items would only foster further contestation, and that the goal of the Deliberative Council was to reach an agreement. The chairman then recommended a daring proposal: he asked all cabinet ministers who sat on the council to recuse themselves from further discussions.[60] In Itō's mind, the logic behind his request was simple. Cabinet ministers had already reached agreement on the government's reconstruction plan, and if any meaningful alterations were to be made, they would have to come about from the noncabinet ministers present. Members accepted Itō's proposal begrudgingly. Itō, Egi, Shibusawa, Aoki, Takahashi, and Katō then went through the reconstruction proposal and deliberated over each item. By narrowing the actors down to six, five of whom shared the desire to scale back the government's plan, the process became far less contentious. By the end of the second day, the six members reached agreement on the reconstruction plan.

Itō announced the revised reconstruction plan on November 27. Presented in the form of an amendment to the government's original bill, the new proposal reduced the reconstruction budget from ¥703 million to ¥598 million.[61] The special committee rejected plans for Tokyo harbor construction facilities and the proposed canal between Tokyo and Yokohama. This reduction trimmed nearly ¥40 million from the overall budget. Next, the special committee recommended that the width of most roads be reduced throughout the capital and, importantly, decreed that all "existing road lines were to be retained" wherever possible in the reconstructed city. This, they argued, would expedite reconstruction and save nearly ¥70 million.[62]

When Itō presented these recommendations, the prime minister and the cabinet faced a moment of decision. One option Yamamoto possessed was to encourage the cabinet to accept the proposal and hope that parliamentarians, as the final arbiters in the budgetary process, might increase the reconstruction budget to fund undertakings rejected by the Shingikai. The second option was to advise the cabinet to reject these recommendations outright. The latter course, Yamamoto predicted, would result in the cabinet's premature demise.

Every day that the Shingikai met, Gotō communicated with Yamamoto in private and encouraged the prime minister to refuse any compromise with the Deliberative Council. On November 25 Gotō submitted a long, and at times rambling, opinion piece to Yamamoto.[63] He encouraged Yamamoto not to cling to power for the sake of political survival or as a means to resurrect his political career, but rather to uphold the principle of imperial reconstruction. He argued this even if it "meant the cabinet would collapse in three days." "Even if we are blamed by the public today," Gotō continued, "future generations [who inhabit the grand imperial city] will thank us for our resolve."

Gotō used a number of specific and well-reasoned arguments in his attempt to persuade Yamamoto. First, he suggested that the Shingikai's proposal cut so severely into the already scaled-back reconstruction plan that the city could not be reconstructed in a significant way. Second, Gotō claimed that adopting a narrow approach to reconstruction would not only throw away the unique opportunity to build a glorious capital but also weaken the prospects for national renewal. The home minister's most assertive written appeal to Yamamoto resembled a letter that Charles Beard submitted to Gotō on November 3. In it, Beard warned Gotō to ready himself for the fiercest political battle of his life. Beard predicted that the forces of compromise and restriction would ambush the Yamamoto cabinet. If the cabinet capitulated to such pressure, Beard cautioned Gotō, "your failure would result in your nation's failure." Beard encouraged Gotō to look beyond the present and to plan for future generations. "The world's eyes are upon you," Beard informed Gotō, "and history shows that the world remembers men who had vision and courage to leave later generations a city such as Paris."[64]

While no doubt moved by Gotō's idealism and persistence, Yamamoto did not budge from his pragmatic determination to secure a reconstruction compromise. Given the state of Tokyo, the prime minister suggested that the immediate needs of rebuilding the capital took precedence over holding out for an ideal reconstruction plan. Yamamoto informed Gotō that refusal of the Shingikai's proposal and resignation of the cabinet would throw Japan into political chaos and halt the reconstruction planning process until after a new cabinet was formed. Yamamoto also informed Gotō that such an act might not lead to any new proposal being accepted. There was a high probability that the same debates would shape the next government's reconstruction deliberations. The end result, he informed Gotō, did not hold out enough possibility for change, and therefore resignation of the cabinet was a risk not worth taking. Yamamoto then advised Gotō that he would encourage the cabinet to endorse the Shingikai's proposal no matter how unsatisfactory it might seem. Gotō accepted the prime minister's decision but remained unmoved by his justification. As Finance Minister Inoue reflected in 1930, the Shingikai threw "cold water" over the government's well-balanced and rational, reconstruction plan, but Yamamoto accepted it nevertheless.[65] Unbeknownst to many observers in late November 1923, the Imperial Diet was only days away from not only throwing more cold water on the plan but actually submerging it within an ocean of political contestation.

THE CLARION CALL OF RURAL PARLIAMENTARIANS: THE COUNTRYSIDE AS THE FOUNDATION OF THE NATION

On the evening of December 12, Inoue assembled his closest Finance Ministry subordinates for a simple dinner.[66] In an austere room adjacent to the finance minister's temporary office in the Mitsubishi Building, Inoue informed his guests that he required their unqualified support in the political ordeal that was about to unfold. Tomorrow, Inoue told his colleagues, he would introduce the government's reconstruction bill to parliament. Over the following three weeks, Inoue suggested that all Finance Ministry officials would have to be prepared for questions, challenges, and attacks on the government's reconstruction bill and its budget. Everyone present understood the gravity of the situation they faced. The serious nature of the task that stood before them led one subordinate to remark that he felt as if he was attending the gathering of the forty-seven rōnin (masterless samurai) on the eve of the military raid of revenge against Kira Yoshinaka in 1703.⁶⁷ Everyone laughed at this admission until another young bureaucrat pointed out that it was the eve of Japan's forty-seventh Diet session. At that moment, the jocularity vanished.

A somber seriousness likewise defined the opening orations that Prime Minister Yamamoto and Finance Minister Inoue delivered before a packed Diet the following day. Yamamoto's speech was short and precise. He thanked the emperor and all those who had given aid in support of Japan's worst disaster in history.⁶⁸ He also emphasized the fact that the government had approached reconstruction in a cautious manner, not wishing to jeopardize the financial standing of Japan. The government, in Yamamoto's words, had prioritized "practical" and pragmatic reconstruction initiatives, eschewing extravagant undertakings that would add only to the "appearance" of the capital. Everything in the existing plan was well justified. Where Yamamoto was brief, Inoue, on the other hand, went to great lengths to explain and legitimate the government's reconstruction budget. He stressed that the government plan emphasized a reconstruction project that fit within the confines of national fiscal responsibility. Under his plan, he boasted, no new taxes would be raised. Moreover, he argued that the savings generated from spending cuts, which would be used to cover interest payments on public bonds, would not negatively affect Japan's overall financial health and well-being.⁶⁹ He ended his speech, like Yamamoto before him, asking parliamentarians for support and a timely passage of the government's program for the sake of the sufferers and for the benefit of the nation.

The "benefit of the nation" argument was turned almost immediately against Inoue and the government. Yoshiue Shōichirō (Seiyūkai MP from Chiba Prefecture) asked the prime minister during members' question time to comment on the potential impact the reconstruction bill would have on rural areas across Japan.⁷⁰ While admitting that the earthquake disaster was a remarkable incident, Yoshiue claimed that there were "more important national issues" that confronted Japan's politicians. When asked to explain, the parliamentarian declared that the reconstruction program threatened the health and vitality of rural Japan. Yoshiue stated that the abolition of tariffs on reconstruction materials, a measure adopted by Inoue in the days after the calamity, would not only destroy the rural economy but also threaten the long-term prosperity of Japan. The influx of cheap overseas materials, particularly lumber, cement, bricks, nails, glass, and screws, Yoshiue implied, would result not in a capital constructed from the countryside with benefit for all, but rather a capital built on the back of rural Japan. Rural areas, he suggested, would face an agonizing decline. While they would not be turned upside down overnight, as the capital had been, their death would be slower and more painful. Rural Japan would suffer like a patient stricken by "tuberculosis."

Yoshiue's assertion brought cheers and cries of support throughout the Diet chamber and embarrassed the government from the outset. Gotō replied that reconstruction could be seen from both a narrow and a wide perspective and that Yoshiue saw the issue from only a narrow rural perspective.[71] The home minister's response was ineffectual at best, antagonizing at worst. Diet members issued further calls for the government to explain its position. Inoue obliged. He stated that between September 11 and 17 the government had issued imperial ordinances 407, 408, 411, and 417, which removed the import duty on rice, eggs, and other essential foodstuffs such as barley, wheat, beans, condensed milk, infant food, and mineral oils, as well as on items deemed necessary for relief, including woolen blankets, cotton and wool clothing, boots, shoes, hydrogen peroxide, chloroform, and cotton tissues used in operations. He also stated that the government had abolished import tariffs on key items necessary for reconstruction, including cement, bricks, plate glass, iron bars and rods, nails, rivets, screws, bolts, and, importantly, all lumber products.[72] While the government suggested that its policy was needed to ensure rapid reconstruction and to assist sufferers, it served as a political lightening rod for those opposed to the government's reconstruction program. Even before parliament had opened, the majority Seiyūkai Party had declared that abolition of the tariff ordinance was one key policy it wished to overturn.[73]

Over the following seven days, whether in open question time on the floor of the Diet or in closed-door Budget Committee negotiations, the health of rural Japan was never far from the minds or tongues of many Diet members. Tsuhara Takeshi (Kenseikai, from Kyoto) asked government representatives if they had planned for any concrete development initiatives for rural areas in Japan under the slogan of national reconstruction.[74] When Inoue replied that he had not yet budgeted any spending increases for areas outside the disaster zone, Tsuhara and other Diet members cried foul and suggested that national reconstruction was nothing but a convenient slogan to secure nationwide support for what was ostensibly a narrow, capital-focused reconstruction project. Moreover, members who rallied behind Tsuhara claimed that reconstructing Tokyo would break the back of rural Japan. Inoue suggested that despite the abolition of tariffs, rural areas would still benefit from the purchase of reconstruction material as overseas supplies could not possibly satisfy the demand that would soon present itself. This was met with a chorus of taunts to which Inoue replied that the tariff abolition was only a temporary—albeit necessary—dressing to alleviate suffering in Japan's capital.

Diet members again turned Inoue's words against the government. More than a mere emergency dressing was needed to staunch the flow of blood that was draining the life from rural Japan claimed Miwa Ichitarō (Seiyūkai, from Aichi Prefecture). In an extended speech on the floor of parliament under the guise of six open-ended questions, Miwa declared, with no exaggerated gestures: "Sufferers of the earthquake look terrible and we feel sorry for them for sure, [but we act] as if we are shocked to see the blood of the injured. They will soon recover. I regret, however, to see that the government ignores the issues of rural areas, which may jeopardize the foundation of the nation."[75]

Miwa asked the government officials present if any of them cared about farmers who were always "fatigued" and almost always on the precipice of "extreme poverty." Miwa suggested they did not. He claimed that of the 464 Diet members, only fifty-six were true "farmers" who could be relied on to advance the concerns of agriculturalists consistently. When an unnamed member shouted, "What is your profession," Miwa replied proudly, "Well, I am a farmer." This was met with raucous cheers and boisterous calls, "Hiya Hiya" (Hear, Hear). Miwa ended his grandstanding "questions" by stating that Tokyo would, no doubt, be rebuilt as the government wished with wide avenues, paved sidewalks, and all the modern conveniences that made city life a world removed from the hardships faced by those in rural Japan. Would farmers who asked for paved roads or wider carriageways on which to ship the fruits of their labor have their requests granted? No, he lamented they would not. The present government, and all successive governments, he predicted, would reject such requests on the grounds of fiscal austerity.

Tagi Kumejirō (Seiyūkai, from Hyōgo Prefecture) likewise challenged government plans for a "grand" reconstruction of Tokyo on the basis of what it would cost rural Japan. "I wish the government could create a city like Berlin, Paris, or New York," he declared.[76] "But," he asked, "do we have the necessary resources or national power to achieve such a task?" Interrupted by a fellow MP who shouted, "Chairman, warn him, he may be drunk," Tagi shot back, "Don't worry whether I am drunk or not . . . I will speak clearly so please listen." Laughter erupted at Tagi's retort. Tagi then returned to his speech in a spirited fashion and claimed that the abolition of tariffs would not only further weaken the countryside but destroy the foundation of agriculture and resources for future generations. Consumers in urban areas would become hooked on cheap imports and prioritize them over Japanese-made goods and supplies. This would accomplish nothing, he suggested, but to further hollow out the economy of rural areas and accelerate the flow of Japanese money overseas. It was a lose-lose proposition for Japan. What did the "abolition of cotton tariffs" in 1922 accomplish, and could this not provide a preview of what would happen to other parts of Japan's rural economy if tariff abolition was not repealed? Some 80,000 *chō* (nearly 200,000 acres) of cotton-producing fields had gone fallow in 1923, and this had done nothing but fray the fabric of rural society. The government's reconstruction plan, he concluded, would "threaten the nation."

It was not just rural concerns that caused parliamentarians to attack the government's reconstruction bill. Numerous Diet members took issue with the government's desire not just to oversee but to direct land readjustment and land exchanges throughout the disaster zone. Hata Toyosuke (Seiyūkai, from Saitama Prefecture) argued that land readjustment was not the job of the central government under any circumstance, even following Japan's greatest natural disaster.[77] Local landowners' unions or cooperatives had carried out land readjustment in rural areas previously and he saw no reason for the central government to meddle in this undertaking now. "Land readjustment," he argued, "should not be the [central] government's job."[78]

Hata was joined by other Diet members who challenged the government's plan to involve itself in what they considered to be the "local task" of land readjustment. Both Mito Tadazo (Seiyūkai, from Kagawa Prefecture) and Shimada Toshio (Seiyūkai, from Shimane Prefecture) argued that not even local government had a right to force landowners into land readjustment schemes. In many regards they were echoing the stated policy of the Seiyūkai Party. Even before parliament began, this party agreed to challenge any attempt at what they perceived as government intervention into land readjustment. They argued through their official publication, *Seiyū*, that

> landholders need to organize cooperatives and operate them independently [from the state]. This type of program [land readjustment] must respect the rights of individual ownership. It is not right for the state or

local government to forcefully undertake land readjustment. We need to rely upon the autonomy of citizens to deal with land readjustment in burnt out areas. Only when the landowners' cooperatives cannot manage this project or landholders themselves do not intend to do so may the state or local government initiate the project within the legislation of city planning.[79]

Two issues motivated Seiyūkai parliamentarians to raise the land readjustment issue in such a vociferous manner. First, landholders were their chief constituents. Many landholders feared direct government intervention into the practice of land readjustment in 1923 and worried that it could set a threatening precedent for future government involvement. Second, MPs assumed correctly that government directed land readjustment would be a costly undertaking—upward of ¥45 million by the government's own account—and sought ways to reduce the overall cost of the reconstruction program.[80] In such an environment, land readjustment was an easy target of opportunity for the Seiyūkai.

One last component of the government's reconstruction plan that parliamentarians challenged repeatedly was the very existence of the organization assigned to oversee the project: the Reconstruction Institute. On the second day of members' question time, Inoue Takaya (Seiyūkai, from Gifu Prefecture) asked Prime Minister Yamamoto why the City Planning Bureau of the Home Ministry could not carry out planning and oversight of the reconstruction project. Were they not professionals experienced in city planning? Suzuki Jōzō (Seiyūkai, from Ibaraki Prefecture) followed up this question and asked the government if creating a whole new bureaucratic entity to direct city planning and reconstruction was really the most cost effective means of using money. Other Diet members, almost all Seiyūkai members, likewise challenged the rationale in forging a new government organization when money spent on administration did nothing but take funds away from the actual reconstruction project itself. This led various members, including Hata and Shimada, to describe the Reconstruction Institute as "extravagant," "unwarranted," and "unnecessary."[81]

Beneath the rhetoric of frugality, however, existed a burning desire on the part of many Seiyūkai members to humiliate Gotō Shinpei. Gotō and the Seiyūkai despised one another. On numerous occasions Gotō directed considerable scorn and antipathy toward the Seiyūkai, claiming that it always pursued power over principle, behaving, as statesman Ōkuma Shigenobu had earlier described it, like a "parasite."[82] Gotō believed that the adoption of universal male suffrage would destroy the Seiyūkai's super majority stranglehold on parliamentary politics and usher in a new era of true representative democracy in Japan. Gotō, in fact, encouraged Yamamoto to introduce a universal male suffrage bill to parliament prior to the earthquake bill because he believed this would greatly weaken the Seiyūkai's legitimacy if its members were to vote such a measure down immediately following the earthquake.[83] Yamamoto refused, however, claiming that only earthquake-related items could be discussed and debated by the special Diet session. For their part, most Seiyūkai members refused to recognize the legitimacy of the "transcendental cabinet" of Yamamoto and directed their anger at its most public face and their greatest nemesis, Gotō.

Contestation over the reconstruction plan and budget climaxed in the Budget Committee of the lower house. On December 19 committee member Shimada proposed an amended reconstruction budget that included sizable spending cuts. The reductions he suggested equated to nearly 22 percent of the government's original request.[84] Shimada eliminated the entire budget for the Reconstruction Institute, namely, ¥22.9 million. He likewise recommended a reduction of just over ¥106 million earmarked for reconstruction projects in Tokyo and Yokohama. Nearly ¥36 million of this figure came from scaling back government-directed land readjustment initiatives. The national budget for street reconstruction was also slashed by just under ¥70 million. To accomplish these savings, Shimada proposed that the city government take responsibility for the reconstruction of all roads less than 12 *ken* (21.82 meters) in width.

Seiyūkai Diet members rallied to Shimada's proposal. When Budget Committee chairman Yamamoto Teijirō (Seiyūkai, from Niigata Prefecture) brought the resolution to a vote on the afternoon of December 19, it passed with no Seiyūkai member opposed. It next went to the full Diet and met strong opposition from only a few members. Tagawa Daikichirō (Independent, from Nagasaki Prefecture) challenged the Seiyūkai-backed bill on almost every point. Tagawa, who had published a detailed pamphlet entitled *Tsukuraru beki Tokyo* (The Tokyo that should be built), claimed that Japan was missing a golden opportunity by not embracing a large, well-funded reconstruction project. His words failed to move any Seiyūkai parliamentarian. Diet members endorsed the revised budget by a vote of 241 to 106.[85] Parliamentarians, exercising the one sacrosanct legislative power they possessed—that of final arbiter of the budget—reduced the size of the government's reconstruction project from ¥598 million to

¥468 million.[86] Reflecting on the Diet's final decision, Miyao Shunji wrote, "Many politicians held the mistaken belief that the Yamamoto cabinet considered Tokyo as the whole nation and would sacrifice rural areas for its benefit."[87]

When Diet members reduced the budget, Yamamoto's cabinet possessed two realistic alternatives. On the one hand, the cabinet could swallow its pride, accept the revised budget, and attempt to expand the reconstruction program through the issuance of new reconstruction legislation in the next Diet session scheduled to open on December 27, 1923. On the other hand, the government could reject the revised budget, dissolve parliament, and call a general election. Each alternative held out potential promise as well as peril. Cabinet ministers debated both options on December 18 when it appeared likely that the Seiyūkai-controlled parliament would scale back the government's bill.[88] Hardline members of the cabinet such as Communications Minister Inukai Tsuyoshi and Justice Minister Hiranuma Kiichirō wanted to use the reconstruction issue to confront the Seiyūkai and hopefully break its stranglehold on power. Such an action would enable the government to save face and perhaps give Gotō the opportunity to shape the subsequent election results from his position as home minister. Would people across Japan really support a party that sought to restrict reconstruction funds for what had been packaged as Japan's greatest challenge since the Russo-Japanese War? Hiranuma and Inukai professed faith in the Japanese public to punish the Seiyūkai for their party-oriented political antics at the next election.

This course of action, however, held two political pitfalls. First, a rejection of the earthquake bill would prolong the commencement of official reconstruction yet further until the Imperial Diet passed a bill in 1924. Second, the dissolution of the Diet and new elections provided no guarantee that the Seiyūkai would be forced from its majority position. A return of the Seiyūkai as a majority party, Navy Minister Takarabe Takeshi (the prime minister's son-in-law) suggested, might well embolden its members to attack the government's plan with greater tenacity in the forty-eighth Diet session. Could the government risk this outcome? He and another conciliation-minded, pro-Satsuma cabinet member, Railways Minister Yamanouchi Kazutsugu, therefore encouraged acceptance of the revised budget. Yamamoto agreed but in doing so issued the following statement on December 19:

> It is deplorable that the reconstruction budget that the government submitted has been revised at today's meeting of the House of Representa-

tives. We are afraid that through this revision, the reconstruction project will be greatly hindered. If the [revised] budget is rejected, however, it will lead to difficulties in implementing the reconstruction plan. Therefore, considering the people's suffering, we have decided to accept the revision so that the budget will be approved.[89]

Gotō rationalized this defeat as the result of selfish behavior on the part of the Seiyūkai. He informed Yamamoto to accept the amended budget and to ready the cabinet for a much larger showdown in the forty-eighth Diet session. The home minister told Yamamoto that the reconstruction bill could be expanded in the new parliament and if the Seiyūkai refused to accept the government's proposed amendments, the cabinet would be justified in dissolving the Diet and calling an election. Gotō went so far as to devise a dissolution notice to be used at a moment's notice.[90]

The ever-ambitious Gotō, however, would never get another chance to confront the Seiyūkai as a cabinet minister or to expand upon the scaled-back reconstruction program. At 10:42 a.m. on December 27, Nanba Daisuke fired bullets at the crown prince as he was driven to the opening session of the Diet. Though no physical harm came to the future emperor, the Yamamoto cabinet took responsibility for failing to protect the crown prince and resigned en masse at 4 p.m. on the same day. Gotō's elite-level political career ended and his dreams for a new Tokyo forged from the cauldron of calamity went unfulfilled. Gotō's sudden and complete political demise was best symbolized by two visitors who called at his home in Azabu Ward, Tokyo, on the evening of December 28.[91] The men in question approached the front gate of his residence with a coffin and told one of his servants that they wished to deliver Gotō's final resting place as a New Year's present to usher in 1924: his political life, they implied, was finished. Police who guarded the home minister's residence were not amused by this political prank rich in meaning. They arrested both men on the spot and sought, unsuccessfully, to remove the coffin before the disturbance drew Gotō's attention.

Gotō did not take his final political demise well. In an agitated and confrontational state of mind, he recorded his thoughts on his time as home minister in an extended essay entitled *Fukkō jigyō no seihai ni kanshite Tokyo shimin ni kokuhaku su* (Confessions to the citizens of Tokyo concerning the success and failure of the reconstruction project).[92] In attributing blame for the failure, Gotō cast aspersions at virtually every political actor in Japan. While he understandably directed considerable vitriol at the Seiyūkai and members of the Shingikai for what he claimed were their selfish, short-sided, and politically motivated attacks against his government's plan, he also savaged the "media and academics" who "failed to educate the public" on the merits of a grand reconstruction. In Gotō's mind, citizens were also to blame as they "did not thoroughly understand the meaning of reconstruction" or its importance for the nation. Only a few were willing to "serve society" and accept sacrifice for the greater good of reconstruction. Too many citizens had become agitated over superficial political scare tactics about land confiscation and readjustment. Gotō, in a moment of hindsight and historical amnesia, also blamed the Yamamoto government for capitulating to the Seiyūkai even though he had supported this policy at the time. Had the government dissolved the Diet and called an election, he concluded, the assassination attempt would not have occurred as the crown prince would not have had an opening ceremony to attend on December 27. In essence, Gotō blamed everyone but himself.

DISASTERS, POLITICS, AND THE ILLUSION OF TRANSFORMATION

When the fires that ravaged Tokyo extinguished themselves and a semblance of political order returned to eastern Japan in mid-September, Gotō was not alone in believing that a unique, once-in-a-lifetime opportunity existed to transform the ruins of Tokyo into a grandiose metropolis. From the safe distance of Kobe, editors for the *Japan Weekly Chronicle* asserted that the difficult project of reconstruction "was a great opportunity for Viscount Gotō . . . to make a great name for himself as the rebuilder of a greater Tokyo and a finer Yokohama."[93] Others, such as bureaucrat Tawara Magoichi, hoped that all political disputes and elite-level personality clashes would be placed in abeyance or trumped outright by the demanding task of reconstruction. Each was mistaken and misguided by the illusory belief in the transformative power of disasters to alter political landscapes as thoroughly as they damage the physical environment and destroy the structures built by humans.

Rather than unite Japan's polity in a "postdisaster utopia," as some scholars claim has emerged following other natural disasters, in Tokyo, reconstruction plans, their budgets, and the government's final bill exacerbated political tensions.[94] Planners themselves could not agree on the ideal scale and scope of the reconstruction project and fought on an almost daily basis about the efficacy of pursuing a grandiose plan or one that was more realistic in a financial and political sense. Many cabinet ministers cooled to the idea of constructing a grandiose imperial capital once they realized that a large reconstruction bill would result in less funding for individual ministries. The urban-focused reconstruction program that the government finally submitted to the Diet enabled members from rural areas to highlight a plight—the poverty of rural Japan—that they believed was equal to, if not more acute than, what victims in the capital suffered. When the government implemented a temporary abolition of the tariff ordinance to assist reconstruction efforts and to help feed and cloth those in need, parliamentarians from rural areas challenged such measures as a threat to the social and economic well-being of the nation. Few actors could find common ground on anything related to the reconstruction project.

Japan's democratic-bureaucratic-oligarchic political system, with all its accumulated idiosyncratic structures, likewise remained immune from change immediately following the unprecedented disaster. It proved to be a poor form of government to respond effectively and efficiently to a catastrophic natural disaster or the subsequent reconstruction program. However pluralistic Japan's polity had become by 1923, it was not democratic enough to compel bureaucratic elites such as Gotō or Inoue and planners such as Yamada and Ikeda to work with parliamentary elites such as Takahashi or Katō to forge a comprehensive reconstruction bill that would appeal to all actors who would ultimately be asked to endorse the project. Party leaders and the elder statesmen of the Shingikai were not even invited to discuss or involve themselves in the reconstruction- or budget-planning process. Important leaders in the cabinet and within the Imperial Diet held nothing but contempt for one another. Often each side claimed that the other possessed no political legitimacy. Finally, in this quasi-democratic system in which only 3,069,000 people out of a total population of 55,961,140 possessed the right to vote (5.4 per cent of the population), few if any elites in 1923 saw any need to inform the public about the scope and scale of the proposed reconstruction, let alone to ask them to participate or even voice an opinion about how the city they would inhabit should be reconstructed. Throughout the autumn of 1923, elites looked beyond what the people might have wanted and planned a city that they believed the people needed.

For those elites who desired sweeping change, however, Japan's polity was not authoritarian enough to enable them to coerce other actors with competing and often conflicting views. The diffuse nature of Japan's polity was a continual source of friction to bureaucratic elites with high modernist ambitions. They could not implement far-reaching policies that ran counter to the spirit of the 1889 Meiji Constitution. Japan of 1923 was a pluralistic parliamentary polity, not an absolutist state comprised of authoritarian rulers in which one individual or one group of power-brokers could compel or run roughshod over another to secure political objectives. Political power was spread formally among many bureaucratic actors, ministries, and parliamentarians, and informally among influential elder statesmen. These groups guarded the limited powers they each respectively possessed with fierce, often recalcitrant tenacity. Attempts to bring representatives from these various foci of political influence together under the guise of committees, subcommittees, or institutes—when they occurred at all—often resulted in greater political contestation.

Charles Beard, perhaps better than any other observer of Japan's government and the Great Kantō Earthquake, understood the desultory role that Japan's peculiar political system played in hindering the reconstruction effort. Like Gotō, Beard was dismayed but not surprised by passage of a scaled-back reconstruction bill in Japan's Diet. Writing in *Our World* magazine in 1924, Beard informed his English-reading audience that implementing a major reconstruction of Tokyo as envisaged by Gotō was a task of appalling magnitude that required considerable coordination of social and economic forces. Most of all, Beard argued, Japan's cabinet—and in fact Japan's whole government—lacked the power that he believed was the essential element required to give effect to large city plans. He suggested that urban landowners and property-owning rural elites and their representatives in the Seiyūkai sabotaged a once-in-a-lifetime opportunity to transform Tokyo into a model city. "We seem to be between two worlds," Beard wrote, "the day when a dictator could by the wave of a wand order a new city to spring up in a wilderness or upon the ruins of an old metropolis is past: hydra-headed democracy or near-democracy is not ready for the task." Highlighting his high authoritarian impulses rather than the progressivism for which he is better known, Beard concluded that the failed reconstruction plans for Tokyo caused him to question "whether in an age when the people have a voice in affairs there can be effected a concert of powers sufficiently potent to carry out a comprehensive scheme of city planning in the face of organized, short-sighted private interests and political ineptitude."[95]

Others in Japan shared Beard's conclusions even if they did not articulate them in such a provocative manner. What commentators described as the "failed reconstruction plan for Tokyo" led many elites to suggest that Japan's polity and its people needed to change in fundamental ways to ensure that the necessary program of national moral regeneration and spiritual renewal did not meet a similar fate as the grandiose dreams of Japan's starry-eyed city planners. While no degree of unanimity emerged as to how Japan's polity should change, elites of all political persuasions and from diverse backgrounds agreed that citizens across the nation would need to embrace ideas of sacrifice, diligence, frugality, and contribution now more than ever if the misfortune of the disaster was to be turned into a blessing and an opportunity for rebirth.

Seven

REGENERATION: FORGING A NEW JAPAN THROUGH SPIRITUAL RENEWAL AND FISCAL RETRENCHMENT

The recent earthquake disaster was a most powerful stimulus. It would be wonderful if at every home parents and elders take the initiative in quitting an old bad habit or starting something new and good.
—Takashima Heizaburō, 1923

It is difficult for people to discontinue the use of goods to which they have become accustomed, those who have sufficient means to buy must continue to buy even though their price may be double or even treble.
—Kurachi Seifu, 1924

Munakata Itsurō was a lifelong advocate of discipline, training, simplicity, and fortitude. Born in 1866, Munakata devoted his career and much of his spare time to two pursuits that he believed were interconnected on an intimate level: education and the martial art of judō. From his first formal teaching position as headmaster of Omura Junior High School in Nagasaki (1893) to his appointment at the Tokyo Advanced Teacher Training School (1907) to his final post as headmaster at the Sendai First Junior High School (1916), Munakata championed the philosophy that mental development, physical prowess, and character could be cultivated best through exercise, spiritual discipline, and formal education. Though he retired from education in 1920 to spread the practice of judō throughout Japan, Munakata was delighted to learn that the education journal *Kyōiku* (Education) selected his entry as the anthem for national reconstruction in their November 1923 issue.[1]

To the retired teacher, the editors of this journal, and many commentators across Japan, Munakata's song embodied what they hoped would be the spirit of a new Japan. Entitled "Saigo kokumin no kakugo sanken no uta" (The resolution of the nation after the disaster—A song of three ken), the three *ken* alluded to in the title were the heart and soul of Munakata's song. They were also the three practices or virtues that he believed the people of Japan must embrace to forge a new national spirit. The three ken were *shinken* (earnestness), *kinken* (diligence and thrift), and *kōken* (contribution). "Everyone," as one line from verse 11 of this fifteen-verse song suggested, "must turn over a new leaf" and internalize the values of the three ken. Though Tokyo had become a "hell on earth" as a result of "divine punishment," Munakata's lyrics suggested that by making the three ken "the foundation of everything in the world," Japan would become the "brightest of bright lights" and a beacon of "humanity, justice," and "international peace." The greatest disaster of history provided Japan with an unparalleled opportunity to "accomplish its mission."

Hyperbole aside, Munakata was not alone in believing that the people of Japan needed to welcome a new era defined by moral regeneration, spiritual reconstruction (*seishin fukkō*), and social and economic moderation. Since the end of the Russo-Japanese War, but particularly from 1918 onward, numerous political elites advocated prescriptive measures with the aim, as David Ambaras described it, "to create a Japan free from the dislocations, alienation, and moral confusion" that many believed accompanied the emergence of a "modern, industrial society."[2] These individuals, often from diverse social and ideological backgrounds, made proclamations and debated policies that they hoped would arrest the perceived degeneration of Japan. The earthquake calamity, which many people constructed as an act of divine punishment against the people of Japan for their moral shortcomings, amplified these concerns about the spiritual decline of society and the economic vulnerability of this nation. Japan's greatest natural disaster unleashed a cacophony of voices that advocated the urgent need to reinvigorate the "popular mind" (*minshin*) to counter perceived laxity, extravagance, hedonism, flippancy, and selfishness and retrench economic excess. Takashima Beihō best summarized this mindset. Writing in the October issue of the popular journal *Chūō kōron*, Takashima argued that both the shell (Tokyo) and the values of the 1868 Meiji Restoration had vanished by the autumn of 1923. It was therefore necessary for Japan's current leaders and citizens to implement a glorious Taishō Restoration that would put Japan on a new ideological and economic trajectory. Japan was at a watershed moment or crossroad that would determine the future prosperity or decline of the nation. To put "true life" into both the new city and the nation, the task of restoration was every bit as important as reconstructing the city on a physical level.

If the former failed, Takashima concluded, Tokyo would be nothing more than a statue of a Buddha with no soul; an object "not worth appreciating let alone worshipping."[3]

In examining the proclamations made and policies implemented immediately following Japan's earthquake calamity to counter the perceived crises—moral as well as economic—confronting Japanese society, it is clear that the Great Kantō Earthquake compelled intense introspection across Japan. It also led concerned people from all classes, regions, and ideological backgrounds to champion, as Privy Councilor Ichiki Kitokurō articulated, a reinvigoration of the popular mind and a return to a more moderate, disciplined financial path.[4] While virtually every commentator who spoke on this issue agreed on the efficacy of promoting moral regeneration, economic retrenchment, and spiritual renewal, prescriptions for obtaining this end were often markedly divergent. Differences foreshadowed the larger political and ideological debates that erupted with increased regularity in late Taishō and early Shōwa Japan between liberal progressives and those with more authoritarian impulses who sought not only to better manage but often to perfect society. Some saw the postearthquake period of emergency as the perfect time for the heavy hand of the state to flex its muscles and eliminate so-called luxuries, vices, and restrict questionable behaviors and dangerous thoughts that tempted individuals and weakened the nation. Others resisted proscriptive measures and advocated more progressive and inclusive, yet still interventionist, policies aimed at cultivating a spirit of guided citizen autonomy and participation. However urgent elites believed the need for regeneration was, and despite embracing a shared common objective, advocates for spiritual renewal realized quickly that their dreams were often as elusive as those who sought the construction of a grandiose, modern new capital.

THE IDEOLOGICAL FOUNDATIONS OF A NEW ERA: FROM THE TOP DOWN

Though many political elites believed that the postearthquake period heralded the possibility of a new era defined by transformative change, they endorsed a tried and tested means to announce it in the first instance: an imperial rescript. Drafted with considerable input from Education Minister Okano, Home Minister Gotō, and Chief Cabinet Secretary Kabayama

Sukehide, the Imperial Rescript Regarding the Invigoration of the National Spirit (Kokumin seishin sakkō ni kansuru shōsho) was introduced by government officials as one of the three most important rescripts issued by the imperial household in its history. Its authors suggested that its stature was equal to the Imperial Rescript on Education (1890) and the Imperial Rescript Concerning Sincerity, Thrift, and Diligence of 1908, often referred to as the Boshin Rescript.[5] Along with elevating its historical significance, governing elites also described the 1923 rescript as the foundation document that articulated how and why Japanese society must be reconstructed spiritually in the turbulent period following the September 1 disaster.

Government officials who wrote this rescript devoted roughly the first third of its text to demonstrating its historical significance. Making direct reference to both the 1890 and 1908 rescripts, its authors suggested that the 1923 document was the third in an imperial continuum that sought to correct deficiencies and strengthen the spiritual, economic, and social foundation of society. While great progress had been made after each earlier proclamation, the authors of the 1923 document argued that once again Japanese people had fallen prey to materialism and adopted "extravagant habits and extreme tendencies." The authors argued that if these afflictions were not arrested, they would "threaten the future of the country." "The accomplishments of our predecessors," the rescript continued, "could be jeopardized unless the evils of our time" were rectified.[6]

Not surprisingly, the following two-thirds of this document highlighted the desultory habits and tendencies that its authors argued needed jettisoning. The people of Japan, proclaimed the rescript, must reject frivolousness, self-indulgence, extravagance, and ostentation and aim for a life of simplicity, sincerity, and fortitude. Above all else, this document encouraged people to value diligence, thrift, moderation, loyalty, filial piety, discipline, and resolve.[7] Along with invigorating the popular mind by embracing these values, it suggested that people must also render services for the sake of public interest, deepen the virtue of philanthropy, and at all times observe proper public morals. Only by doing so could a sturdy national spirit emerge as the foundation of national prosperity.

Released late in the afternoon on November 10, 1923, the official proclamation of this rescript was timed to maximize its political influence. Its promulgation occurred two days before Prime Minister Yamamoto convened the national conference of governors. This annual event brought elite and midlevel bureaucrats from every prefecture in Japan together to discuss politics and state administration. The rescript and the instructions on its implementation set the tone of virtually every speech made by high-ranking governmental officials during this conference. Yamamoto opened the conference stating that he had already informed the people of Japan—through his published instructions on receiving the rescript—precisely what he expected from them; namely, the purging of frivolousness and economic extravagance, the elimination of extremism in thought and action, and the pursuit of cooperation, service, and loyalty.[8] All bureaucrats, from governors down to teachers, must assist the government in securing these objectives. Yamamoto claimed that despite the urgent tasks of reconstruction the nation faced, an "atmosphere of laxity" still hung over society. The situation demanded, in Yamamoto's opinion, that new public morals be defined and enhanced and that the popular mind be strengthened. To facilitate this objective, he requested that government officials set examples of exemplary behavior in everyday pursuits by observing discipline, exercising moderation, and, most important, endorsing austerity.

Other cabinet-level speakers echoed Yamamoto's ideas but calibrated them to their specific ministerial agendas. Speaking on November 13, Gotō obliged local elites not only to rally the people behind spiritual reconstruction but also to forge a groundswell of popular opinion behind the physical reconstruction of the capital. Tokyo's reconstruction, Gotō argued, was the first phase of an overall project of national development, and he encouraged local bureaucrats to eliminate "jealous thoughts" toward the reconstruction effort, especially while embracing thrift over local administration. In his conclusion, he pleaded with governors not to let "trivial matters" or "minor political disagreements" weaken national solidarity. Finance Minister Inoue asked all bureaucrats to become models of fiscal austerity in administration and in daily life and to lead by example in their respective communities. Education Minister Okano asked all bureaucrats present to assist future education initiatives that sought to "embed the fine customs of simplicity, fortitude, and diligence" in the minds of all students across Japan. Okano ended his oration with an announcement that he would shortly send a personal message to every principal across Japan directing them to "devote themselves day and night in order to comply with the emperor's wishes."[9] This closing remark triggered instantaneous applause from cabinet officials and the invited gubernatorial delegates.

Newspaper editorials published in the days following November 10 provide some clues as to how others saw the rescript and its calls for Japanese people to embrace moderation in thought and action, to cultivate a sturdy national spirit, and to practice thrift, diligence, and public mindedness.[10] The editors of the *Miyako shinbun* gave a ringing endorsement of the rescript. The paper claimed that frivolity, indulgence, and extremism had, unfortunately, become commonplace in Japan since the end of the European War. Those who desired but could not afford an unconstrained or extravagant lifestyle, the paper claimed, were often driven to radical political and economic ideologies out of simple envy and discontent. Numerous commentators and elites had made such critiques about the degeneration of Japan in the past, the paper suggested, but no sense of urgency had overcome political leaders to address these pressing issues until the earthquake tragedy unfolded. In this regard the disaster was a welcome blessing. The editors suggested that the rescript was an important first step in dealing with the challenges the nation faced.

Other newspapers were less laudatory to the government for its use of an imperial rescript. The Seiyūkai-backed paper, *Chūō*, condemned the government for dragging its feet in relation to addressing the problems that contributed to what the paper called the spiritual decline of Japan. That the current cabinet had to respond with an imperial rescript illustrated just how deficient its leaders had been in guiding the nation through the turbulent postdisaster period and how little political legitimacy it held among the public. They asked if the cabinet would continue to shield itself and its failings behind the imperial household. Tokutomi Sohō's newspaper, the *Kokumin shinbun*, stressed that acceptance of the rescript by the people depended on how government officials themselves fulfilled the obligations of the rescript and lived up to its high ideals. Tokutomi suggested that neither the current government nor the leaders of Japan's most important political parties could become exemplars of virtue and proper behavior as prescribed by the rescript.

Of all newspapers in Japan, the editors of the *Tokyo asahi shinbun* provided the most penetrating analysis of the rescript and the contemporary political and ideological situation that gave birth to its issuance. Its editors suggested that while extravagance and laxity had become more common since 1918, no generalization could adequately describe Japanese society following the European war. Striking a populist tone, the editors asked if farmers or factory workers who lived on the precipice of poverty every day really had the ability to further economize their lifestyles and embrace frugality as the government suggested. When discussing the decline of society in more specific terms, they argued that considerable blame could be directed toward successive governments since 1918 for their failures to deal effectively with the manifold social, economic, and political problems that afflicted Japan. Why had leaders waited until November 1923 to devise proclamations aimed at ameliorating society's ills? Was this, they asked, because the leaders of business and industry who virtually controlled the economy and the political parties had profited in the wartime and postwar years? Were the new plutocrats aware, the editors continued, of how their extravagant lifestyles endangered Japan on a moral, economic, and political level? Given these trends, the editors suggested that not only prescriptions, but also actual changes in behavior, must originate from the top down.

Apart from newspaper editors, numerous commentators published journal articles and essays on the rescript over the course of 1923 and 1924. Virtually every commentator embraced the aims of the rescript but argued that words, however lofty, would have to be followed by concrete policies to ensure its successful implementation. Arima Sukemasa, a longtime advocate for enhanced moral education and ethical finance, wrote that "fact was more eloquent than argument" in a special edition of the journal *Tōa no hikari* (Light on the east).[11] Writing to coincide with the one-year anniversary of the rescript in 1924, Arima declared that more needed to be done to guarantee the spiritual reconstruction of society. He suggested that the successive governments of Yamamoto Gonnohyōe (September–December 1923), Kiyoura Keigō (January–June 1924), and Katō Takaaki (1924) had spoken much about spiritual renewal but had implemented few policies to facilitate this end.

Did any political figure exist in today's Japan, Arima asked, who embodied the spirit of the rescript or could be elevated as a symbol of new, postearthquake Japan? "No," he replied, with no little bravado. Only if Japan looked to its past could the people find one individual who possessed all the correct traits, virtues, and pureness of spirit needed to serve as a symbol of Taishō renovation. The individual Arima championed was General Nogi Maresuke.[12] In Arima's opinion, Nogi was the "embodiment of true, fortitudinous national spirit," a man who possessed valor and one who endorsed frugality, simplicity, and resoluteness. Most important, however, Arima suggested that Nogi personified the greatest value that all Japanese must emulate: sacrifice. Arima took great pains to remind his readers that Nogi, as a father who lost two sons in the Russo-Japanese War and as a loyal soldier who committed suicide upon the death of the Meiji emperor in 1912, was the quintessence of national spirit. A cult of Nogi, Arima implied, should be fostered through schools and in public life to facilitate spiritual promotion and renewal. Unsurprisingly, no mention was made of how poor a military tactician Nogi was as he sent tens of thousands of Japanese soldiers to their deaths during the five-month siege of Port Arthur during the Russo-Japanese War.

Fukasaku Yasubumi agreed with Arima that Japan lacked an individual of "high caliber who was universally acknowledged to have both intelligence and virtue" who could serve as a symbol of the new era. He disagreed that simply lionizing General Nogi was enough to fulfill this objective of encouraging spiritual reconstruction. Writing in the same issue of *Tōa no hikari*, the future chair of the ethics department at Tokyo Imperial University argued that Japan's current and future leaders must redouble their efforts at cultivating subjects and citizens of sound mind and exceptional character. While "engraining the teaching of the rescript forever as the national bible" in education was a solid first step, Fukasaku suggested that much more work was required to reach and change Japanese attitudes in the 1920s than had been the case in 1890 or 1908. Fukasaku was aware that Taishō society was more complex than at any time before in Japan's history and that the insalubrious temptations associated with modern urban life were ever-present in 1924. People's moral underpinning and character were no longer influenced solely by what they learned in school or from their parents, the military, or the government. Rather they were manipulated by movies, books and magazines, and products associated with modern consumer society. While this might seem like an insurmountable obstacle, he argued that it also provided an opportunity for the government if it proved wise enough to seize it.[13]

PROPOSING SPIRITUAL RENEWAL THROUGH PROGRESSIVE INTERVENTION AND SUASION

Fukasaku expanded on his idea of using new means and technologies to reach people in a book published in 1924 entitled *Shakai sōsaku e no michi* (The path toward creating society). Throughout chapter 17, a chapter Fukasaku devoted entirely to the postearthquake period, he suggested that a golden opportunity existed for the state to use all the creative and entertaining mediums open in the modern age to foster an atmosphere conducive to renewing the popular mind.[14] Fukasaku argued that motion pictures (*katsudō shashin*) were perhaps the best way to reach people across society and inculcate them with new values, ideas, and mindsets. Film was a cheap, novel medium of communication that could replicate a single message across the nation, and it could also be emotive. In short, it offered exciting new possibilities to reach and move people like never before.

To Fukasaku, the moral and ideological character of people had slackened since the end of the European War, and the degenerate type of entertainment common in Japan both reflected and accelerated the decline of the popular mind. Fukasaku therefore argued that if people were exposed to "clean and pure entertainment," there was a chance that the people who received it "would become clean and pure in thought." Fukasaku urged the government to create movies that were as "enlightening as possible to guide audiences across Japan" toward a more moderate, serious, reflective, and frugal path. A perfect opportunity existed to influence the poorest inhabitants of Tokyo, many of whom still resided in barrack housing. "People living a simplified life in the shelters" would be most primed to internalize values presented by government movies. This was particularly the case, he argued, if officials could make movies that advanced positive morals by incorporating themes and images associated with the emotive topic of the earthquake disaster, an event that was all too familiar to the homeless sufferers.[15]

Apart from movies, Fukasaku listed a number of other creative ways to facilitate a state-directed spiritual rebirth of people across the nation. Uplifting and positive music could not only create a "pleasant atmosphere" in the barracks, schools, and at community-sponsored events, but it could also unite people in their daily tasks. Music also possessed a calming quality that, if used effectively, could make people more open to other forms of suasion. Fukasaku also advocated the production and free distribution of simple pamphlets and other reading materials with positive moral messages. If successful, he argued, this could counter the proliferation of frivolous, sensational, and pedagogically hollow reading materials that had emerged prior to the earthquake disaster. Children should be targeted specifically as they had become avid consumers of pointless, if not morally suspect, magazines and other reading materials in the years prior to the disaster.[16]

Well-choreographed public lecture tours and museums were two final mediums that Fukasaku advocated to facilitate the project of spiritual reconstruction. Tours that incorporated emotive visual images, motion pictures, and memorabilia—much like Kurushima Takehiko's postearthquake tour—could sell concepts of reconstruction throughout Japan. Visits to local regional and rural areas could be packaged by local elites as "community events" to maximize their impact. Fukasaku implied that the earthquake and the various reconstruction projects associated with it were ideal topics for tours as they would allow bureaucrats to mix, in artful ways, "expert speakers and intellectuals" with visual displays, movies, and items from the disaster zone. Once traveling displays were completed, artifacts, visual images, and historical records could be housed in museums, which Fukasaku described as "containers of knowledge and treasure houses of the truth."[17] If used correctly, he concluded, museums could inspire people to learn about the earthquake disaster, the reconstruction efforts, and the importance of stemming the decline of Japan's society.

Self-proclaimed child education expert Takashima Heizaburō felt that spiritual reconstruction must begin with children and their parents in the confines of the home. Focusing on a middle- and upper-class audience, Takashima emphasized the simple steps that could, if implemented throughout the nation, help reorient society away from extravagance family by family. Like many postearthquake commentators, Takashima believed that the calamity provided the nation's psyche with a strong, if not unprecedented, stimulus. As such, it served as an ideal opportunity to encourage all people to make a resolution, jettison a bad habit, and start afresh. Parents should set positive examples and break a bad or wasteful habit, such as smoking or drinking in front of their children. Moreover, the entire family should initiate new patterns of behavior, all with the earthquake in mind. Takashima criticized what he saw as a practice that helped instill a mindset of luxury and privilege among children, namely, the "the competition among parents to buy children expensive clothes, school supplies, and toys."[18] Such actions simply had to be changed, and the earthquake, he believed, could be the initiator of that needed transformation.

Takashima also believed it was important to remind children of the hardship and loss following the earthquake. As time passed, people would lose awareness of this catastrophe, proving correct the expression "danger past, the gods forgotten" (*nodo moto sugireba atsusa o wasureru*). He therefore suggested that simple practices associated with the disaster be established in every home across Japan. To start with, Takashima encouraged his readers "to eat [only] brown rice on the first of each month to commemorate the earthquake."[19] This could serve as a gentle reminder of those who went for days without food after September 1 only to receive brown rice distributed by state officials at relief centers. Once such practices were ingrained in family life, further reforms could be carried out with less difficulty. Everyone could aspire to and eventually follow the well-publicized example of thrift adhered to by the imperial household. On every second day, he claimed, they consumed a plain diet of only one bowl of soup and one side dish. Once frugality was internalized through the most prevalent form of consumption, diet, it could be expanded to cover dress and all work and leisure activities. If a "revolution in the popular mind" was "carried out in every home," Takashima argued, "an atmosphere of diligence and thrift" would eventually encompass all aspects of society and lay the foundations for a much-needed national renovation.[20]

The schoolchildren and teachers of Kobayashi Junior High School in Miyazaki Prefecture suggested that school was the ideal place for the government to promote the reinvigoration of the national spirit. The plans and policies they agreed on in early 1924 were articulated so clearly that the national education journal Kyōiku (Education) published their ideas in the August 1924 issue. Four key principles underpinned their program: (1) to create a warm, relaxed, family-like environment for teaching and learning; (2) to emphasize the cultivation and nurturing of moral character in all pursuits; (3) to value and respect educators; and (4) to spread concepts learned in school to the home and wider community with the aim of "enlightening society."[21]

What specific policies did students and teachers have in mind to secure these objectives? First, in honor of the earthquake disaster, the school established "moral cultivation classes" on the first day of each month. In this class teachers read the November 10 imperial rescript and stressed that students adopt the values of "simplicity and fortitude, humanity, fairness, philanthropy, coexistence, loyalty, filial piety, and courage" in their daily lives. One novelty highlighted by the editors of Kyōiku was the school's program to "correct moral trends" by "providing students with appropriate criticisms and interpretations of current ideological and social problems" in Japan.[22] In this class held once a month, students were required to each raise a question or submit a comment to their teacher in relation to a social problem in Japan. The teacher would then copy each question and distribute them to every student. After discussing the questions at home with parents and siblings, children—usually with other students in small groups—would be responsible for explaining possible causes behind the social problem in question and be encouraged to work with the teacher and other students in devising possible solutions to rectify it.

A final aspect of the "Kobayashi plan" worth mentioning was its emphasis on physical education. The drafters of this plan argued that physical education "not only helped improve the physical strength of students" but also embedded discipline, illustrated the importance of cooperation, and fostered spirit in its practitioners. Each day after lunch and after school every Thursday, teachers and students performed team exercises or held athletic competitions. The school also established an "early risers" club to encourage morning exercises and athletics. In addition, the school introduced nature walks, mountain-climbing excursions, sporting competitions, and community events throughout the year. The school also organized two ten-day excursions during the "dog days of summer" and the "midwinter break" to foster a sense of community and to bolster the industrious spirit of students and teachers.[23]

Miyake Setsurei (Yūjirō), journalist, former bureaucrat, and social commentator, believed that the state needed to do more than just use schools and focus on children to guarantee the moral and spiritual renewal of society. In a speech entitled "Teito fukkō to seishin sakkō" (Imperial capital reconstruction and spiritual renewal), delivered before an assembly of delegates representing Young Men's Associations of Japan (Nihon seinenkan), Miyake urged the state to expand adult education facilities and increase the number of neighborhood training centers and lecture halls through which the government could reach the masses and instill values of frugality, simplicity, thrift, contribution, and sacrifice on a regularized basis. In Miyake's opinion, Japanese subjects—particularly the working classes who had only a rudimentary education—needed to undergo an almost continual process of directed spiritual maintenance and renewal. Claiming that people had, since the end of the Russo-Japanese War, lost the desire to better themselves and the nation, Miyake argued it was both the responsibility of the state and civic-minded people to "help people make themselves stronger." Similar to many feudal lords and samurai retainers who grew weak during the extended, extravagance-filled peace of the Tokugawa period, Miyake claimed, people of Japan had become soft, material minded, and lazy since 1905. Moreover, people had lost awareness of their potential. Through government intervention with the help of neigborhood volunteer associations, Miyake suggested that "even the good-for-nothings of ordinary times" could be trained "to make great contributions when necessity arose."[24] There was no better time to initiate new policies aimed at spiritual renewal than after the great shock and upheaval associated with the 1923 disaster.

What would happen if people, not just Miyake's "good-for-nothings of ordinary times," did not want to embrace or even follow moralist-inspired prescriptions or overtures for spiritual renewal? What if people shied away from or outright resisted gentle attempts by government officials to persuade society to accept a more moderate path? What could the state do if their best efforts to reconstruct the popular mind and to encourage diligence and thrift fell on deaf ears? A number of bureaucratic elites grappled with this issue, and all came to a similar conclusion: the government should use the heavy hand of state intervention—the iron fist as opposed to the velvet glove—to compel people to adopt new moral, ideological, and economic customs and mores. The postdisaster period in which many people existed in a heightened state of vulnerability and relied on the government for basic sustenance was an ideal moment to implement far-reaching changes to society.

PROPOSING SPIRITUAL RENEWAL THROUGH THE HEAVY HAND OF STATE INTERVENTION

Christian socialist and progressive activist Abe Isoo was one individual who urged the government to intervene in a heavy-handed fashion to facilitate a program of national renovation. Writing in the December issue of *Chihō gyōsei*, Abe rejoiced at the opportunity that he believed finally existed to "initiate a revolution in the materialistic and spiritual lives of all of us."[25] As one of Japan's most outspoken and long-standing advocates for social and moral reform, he argued that the disaster, if used correctly by the government, would allow reformers to achieve their long-sought desire. The disaster not only turned Tokyo upside-down but also severed the "habits of convention" from people's lives and compelled people to reflect on their daily practices. But, he warned, the postdisaster window of opportunity would not last forever, and thus he urged the government to act quickly. As the nation's memory of the disaster slipped away and a state of normalcy returned, so too, he argued, would the bad customs and conventions reassert their insidious hold on people's lives.

What did Abe propose, and how did he suggest the government should intervene? The abolition, not restriction, of alcohol was Abe's first target. "From a national point of view," he claimed, "alcohol is one big evil."[26] The consumption of alcohol had a desultory impact on people's health, wealth, and national productivity. Moreover, the production and consumption of alcohol was, in Abe's mind, a wasteful pursuit. The immediate adoption of prohibition under the guise of national mourning for the earthquake victims would not only improve people's wealth and well-being but also enable the government to pay for a larger reconstruction program. Abe claimed that Japanese spent around ¥1.4 billion per year on alcohol, a figure that was actually an underestimate.[27] "If we stop the consumption of alcohol for one year, we can pay for the reconstruction of the capital," he suggested. But "why stop there?" If alcohol was banned for three years, Japan "could build the most splendid, largest city in the world."[28] After three years, Abe concluded, most people would lose their taste for, and convention of, drinking alcohol, and the nation and its people would be far better off as a result. Abolition of alcohol could then become permanent.

A clear sense of unreality defined this proposition. However impassioned, Abe gave no thought to, or instructions on, how money that was once spent on alcohol would be channeled in the direction of the imperial capital. Would people in Kyūshū or even Tokyo want to give up alcohol to rebuild the capital? Nothing, however, curbed his enthusiasm. One gets the impression upon reading this article that Abe cared first and foremost about abolition of alcohol, a policy he had in fact championed since the turn of the century, and less about actually paying for a grandiose reconstruction plan. Monetary savings secured from abolition and how to collect it for reconstruction were secondary considerations for Abe, but the reconstruction effort served as a convenient justification for adopting prohibition.

Prohibition was only one of many policies that Abe urged the government to adopt in the postdisaster emergency period. He also encouraged the government to pass laws that would restrict people from wearing extravagant clothing. Abe claimed that women in particular had become too fashion conscious and spendthrift on their appearance since the wartime economic boom. Simplicity and functionality should be the form and basis of all apparel, and the government should mandate such a policy. Abe lauded the president of Japan Women's University for proscribing all but simple cotton clothes for its students following the disaster. This was exactly the type of leadership that was needed for the government to truly turn the disaster into a blessing. In this proposal Abe echoed many commentators who likewise urged the government to adopt more restrictions on extravagant clothing and cosmetics. Tawara Magoichi proposed that high-quality silk of *meisen* classification and above be reserved exclusively for the export market. Not only would this help prune spending on everyday luxuries in Japan, but it would also mean higher export earnings for the nation. If policies such as these were not adopted, Japan would likely experience more incidents of people confronting and sometimes accosting overdressed women in public. While Tawara claimed that "such acts were unacceptable," he cautioned that the "imprudent behavior of these gaily dressed ladies was also inexcusable." In this light, he concluded that it was "understandable that some people might feel like punishing these ladies."[29]

After restricting extravagant clothing, the government, Abe argued, should turn its attention toward banning other unnecessary and ultimately wasteful daily habits or practices. First, he argued that the custom of exchanging seasonal gifts for the sake of following convention was absurd and should be abolished. Did people really enjoy giving and receiving these gifts, offerings that were exchanged because convention demanded it? To support his calls for abolition, Abe reprinted a letter he had supposedly received from an unnamed "noble woman" who had long shared his beliefs but resisted adopting his suggestions because society demanded otherwise. The earthquake disaster, which destroyed her family's home, had, she declared, freed her from the prison of "convention." "I did not have the courage to abolish the custom of exchanging seasonal gifts with relatives and acquaintances before," she wrote, "even though I always thought it was ridiculous. Now," she concluded, "since we have lost our property" the custom could be abolished once and for all.[30] The posting of New Year's greetings cards could also be stopped for the sake of saving money. Weddings and funerals, Abe suggested, had become too ornate and extravagant since the turn of the century, and he therefore asked the government to place clear guidelines—and restrictions if necessary—on the conduct and scale of these ceremonies.

Abe's greatest hope was that the government would intervene and outlaw the licensed prostitution system in Japan as part of a larger moral regeneration campaign. "Just as slums in a city were risky from a hygienic point of view," Abe argued, "the existence of licensed prostitution was dangerous from a moral perspective."[31] If the government was committed to a program of spiritual renewal and moral regeneration following the earthquake, Abe argued, it simply could not allow houses of prostitution to be rebuilt in the Yoshiwara and Susaki districts of the capital. Because both districts had been destroyed in the fires, Abe argued that it was now the perfect time to legislate them out of existence permanently. Abe considered the licensed prostitution system as nothing more than slavery for the girls and women who had been sold into service and a moral contagion for men. Rather than accept licensed prostitution because it helped preserve the moral standing and virginity of "good" Japanese girls by giving young men a regulated and state-monitored outlet for their sexual desires—a claim made by supporters of the system—Abe argued that the matter-of-fact transaction of sex tempted young men into immoral premarital sex. Not only was this harmful to young men and women at the time of discretion, but, he argued, the habit of commercial sex that began with a casual visit to a brothel would encourage male promiscuity later in life. Abe concluded that as a modern nation committed to a project of spiritual renewal, Japan could simply not afford to allow licensed prostitution to continue.

Could the changes proposed by Abe be implemented in a nation that was becoming more urban, more pluralistic, and more wealthy? A number of commentators from opposite poles of the political spectrum suggested otherwise. Left-wing writer Yamakawa Hitoshi claimed that neither spiritual renewal nor a physical reconstruction could be achieved in a democratic society in which capitalism was the ruling economic ideology. Private property, he argued, led to selfishness and capitalism led to "excessive profit-making." He claimed that both contributed to extravagance. That people in Japan, even the wealthy capitalists outside Tokyo who had not suffered during the disaster, failed to protest the profit control acts or the emergency requisition orders implemented by the government in September 1923 led Yamakawa to believe that Japanese might finally be willing to jettison capitalism and embrace left-wing statist policies. Observing that people were more vulnerable after a major natural calamity and thus more likely to accept, if not endorse, state intervention, Yamakawa urged the government to seize all private property and to implement greater economic and social controls over society and the economy. A "new Tokyo" and "society" could only be realized, Yamakawa concluded, when "capitalist power" was overthrown and private property abolished.[32]

Army General Ugaki Kazushige advocated many similar policies to Yamakawa's. A general whose career is often defined by his plans for national renovation, Ugaki in 1923 urged government officials to extend the antiprofiteering measures implemented in the disaster region to all areas of the country to curb what he saw as a widening and ultimately destructive division between the classes. Ugaki also suggested that the postdisaster period was an ideal time to implement a series of aggressive new taxes, forced saving, and even excess land and asset confiscation measures targeting upper classes that would allow the state to pay for reconstruction. He also argued that the state should force all people to be more frugal through increased taxation, particularly on luxury items. Ugaki expected fierce resistance to this and other interventionist policies from wealthy landholders and industrialists. Moreover, he suggested that no politician in increasingly democratic, Westernized, and pluralist Japan would have the strength and determination to implement far-reaching reform measures that were necessary to renovate society. Contemporary politicians, particularly those from the political parties, only advanced "popular policies" that would endear them to voters. To Ugaki, these politicians lacked conviction, bravery, and a decisive spirit that was essential "to transcend" the narrow interests they represented for the sake of the nation. Political power, he concluded, needed to be concentrated more effectively.[33]

Progressive-minded bureaucrat Nagai Tōru argued just the opposite and suggested that the heavy hand of government was far too blunt an instrument to move society in the right direction. Nagai claimed that managing society was not the sole responsibility of the state. People in local communities must take a degree of responsibility in maintaining the health, welfare, and ideological well-being of the community. To best facilitate a sense of public mindedness, to counter the growing trend of laxity, and to reinvigorate society, Nagai suggested that Japan needed to undergo a "democratic, evolutionary reform in politics, industry, and society." How could this best be accomplished? Nagai had one answer: "the immediate enforcement of universal suffrage."[34] Universal suffrage, he suggested, would encourage more participation across society and persuade people to work with one another and government officials in fostering "cooperation and harmony between the social classes." Just as some European nations were forced to accept greater democracy after the war to establish a stable and participatory society, Nagai concluded that Japan must follow suit after the trauma of the 1923 disaster.[35]

Economist Fukuda Tokuzō agreed that making Japan more democratic would assist with the overall project of spiritual renewal. "The question of what to do with buildings and city streets was trivial," he argued, in comparison to the real issue of "what the first task of restoration must be." His answer was straightforward: to "realize complete universal suffrage." Fukuda believed that democracy would nurture civic activism and the birth of an autonomous spirit for local government and society. The implementation of universal suffrage would also make government more responsive to the people, their needs, and their desires. Fukuda believed that if people felt part and parcel of a larger project of spiritual renewal from the local level up, greater numbers of citizens would subscribe to it than if it was simply a message delivered from the top down.[36] Given the diverse range of opinions presented in postdisaster Japan on how best to implement a plan of spiritual renewal and economic retrenchment, what concrete policies did government elites adopt to facilitate this end? They focused their priorities on economic retrenchment first and foremost.

A LIMITED WAR ON LUXURIES

Launching a limited war against luxuries was one proposal that found many adherents in postearthquake Japan. As historians have documented, many prewar Japanese political elites saw the purchase of luxuries as not only wasteful, but an economic threat to the nation.[37] Moreover, the desire to own and the actual purchase of luxuries were often seen as morally suspect actions, symptomatic of a larger spiritual deficiency. These notions were amplified in the wake of the 1923 tragedy when many commentators seized on the supposed frivolous, decadent, and luxury-minded habits of Japanese and suggested that such behaviors had contributed to the disaster. Curbing the purchase of luxuries, therefore, appealed to those in and outside of government who supported the interconnected policies of spiritual renewal and economic revitalization. The earthquake and the subsequent calls for reconstruction and rebirth put both policy objectives in clear and immediate focus. As early as April 1924, newspapers across Japan began to publish stories on discussions taking place in relation to the "problem of luxuries."[38] How, newspapers asked, would the government restrict the purchase and consumption of luxuries? The editors of the newspaper *Yamato* questioned whether elites would place a simple tax on luxuries. Or, they asked, would officials take a bolder step and outlaw luxurious items outright? Would taxes and proscriptions be placed on both foreign- and domestic-made items? Newspapers produced by and for foreigners in Japan likewise followed these rumors with intense scrutiny and speculation.

When the forty-ninth session of the Imperial Diet opened in July 1924, the government revealed its multifaceted plan to encourage thrift and diligence to curb what ministers argued was the wasteful and dangerous consumption of luxuries. Finance Minister Hamaguchi Osachi, a career Finance Ministry bureaucrat first elected to parliament in 1915, was the spokesmen for this undertaking. Nicknamed the lion of the Diet, Hamaguchi introduced his government's policy with a stark warning. In a solemn manner and serious tone, the Shikoku native declared that "national finances were confronted by a very grave crisis." It was a well-recognized "fact," Hamaguchi declared, that the "luxurious and frivolous habits formed during the wartime boom" still threatened the nation. Unnecessary spending had led to high inflation, and this trend would only intensify as "large sums for physical reconstruction" were spent in the coming years. Unless people became thrift conscious and adopted a policy of self-denial, the economic crisis would worsen. Hamaguchi professed that his government was united in its belief that one of its chief duties was to "awaken" all Japanese to "the crisis" that was upon the nation.[39]

Hamaguchi thereafter outlined his government's plan to "replenish the country's economic vitality" and to reconstruct the nation. While admitting that a long-term policy of renovation could not yet be implemented, as the current government had just taken power on June 11, the present cabinet would begin the difficult process of moral and economic reinvigoration nevertheless. First, he announced the government's plan to issue savings bonds to "absorb" funds that would flow into the economy through Tokyo's reconstruction. The bonds would be issued, Hamaguchi articulated, in small denominations so that people across Japan could assist with the national tasks of "capital accumulation" and Tokyo's reconstruction. Hamaguchi declared that it was a moral imperative for the government "to establish the foundation for further economic development" and increasing the rate of personal savings, and government capital was a key first step.

Increased rates of savings, however, would only go so far to combat the crisis looming over Japan. The emergency that the nation faced, Hamaguchi argued, was moral as well as economic, and it was now time for decisive steps—overdue in the finance minister's opinion—to curb the insidious grip that luxuries held over people. To restrict the consumption of luxuries, Hamaguchi detailed a 100 percent tariff that his government wished to place on a comprehensive list of luxury items imported from overseas.

The list of items classified as luxuries, as introduced by the director of the Taxation Bureau, Kuroda Hideo, was indeed comprehensive.[40] Some 123 categories and over 400 individual items would, if parliament passed the bill, be subject to the 100 percent tariff. The list of items to fall under the proposed luxury tariff provides crystal clear insight as to exactly what items government leaders in the 1920s saw as luxuries. Items of personal apparel and fashionable accoutrements were one area of focus. Clothing included in the proposed tariff ranged from raincoats, dress shirts, shirtfronts, shirt collars, and shirt cuffs to undershirts and drawers. Gloves, socks and stockings, shawls, comforters, mufflers, neckties, trouser suspenders, trousers, belts, hats, hat bodies, caps, bonnets, hoods, boots, shoes, slippers, sandals, clogs, and even shoelaces likewise made the list. Any article of clothing that had even one ounce of fur, a single feather, or a thread of silk, or was trimmed with precious metals, metal-covered buttons, precious stones, semiprecious stones, pearl, elephant ivory, tortoise shell, or any embroidery would also be hit with the proposed tariff. All jewelry made of precious metal, pearl, ivory, amber, coral, or meerschaum would also be subjected to the tariff. Individual accoutrements such as cuff links, spectacles, monocles, umbrellas, parasols, and walking sticks were also included.

Household items and toiletries deemed luxuries were just as numerous. Anything made of pottery, glass, gypsum, gilt, or silvered metal, including door and window hinges, locks, and keys, fell under the tariff. Mirrors, standing clocks, statuary, curtains, blinds, tablecloths, cutlery, carpets and rugs, bed quilts, mattresses, air cushions, and artificial flowers made the list. Any item, household or otherwise, made of wood—cut, sawn, or split—including brushes and brooms was listed, as were items made of velvet, plush, or precious metal. Perfumes, musk, artificial musk, soaps, tooth powder, any drug or medicine, sandalwood, Borneo camphor, vanillin, joss sticks, and even mosquito netting were likewise included.

Items for sports and leisure were also earmarked for the tariff. These included playing cards, musical instruments, binoculars, picture postcards, Christmas cards, photographs, albums, any item with paper lace, and calendars. Any article associated with billiards, tennis, cricket, hunting (including firearms), fireworks, chess, or other games was included. All film (movie film excepted), photographic, and phonographic instruments and their parts, including anything deemed a "talking machine," was also listed. In case politicians missed anything, customs officials were granted authority to place this 100 percent tariff on any item deemed a "toy" upon closer inspection.

Was anything left? Yes, certain food, beverages, and raw materials were likewise earmarked for inclusion. Anyone who drank cocoa, black tea, fruit juice and syrups of any kind, mineral water, soda water, Chinese liquor, beer, ale, porter, or stout would feel the sting of the 100 percent tariff. If one also consumed honey, jams, fruit jellies, cheese, confectionery items and cakes—including Christmas cakes—and biscuits (not sugared), one must pay for the luxury of consumption. Raw materials that made the tariff list included all furs, all leather products, anything made of feathers or down, animal husks, coral, pearl, or precious or semiprecious stones or metal.

On paper the savings bond and the luxury tariff bills seemed destined to appeal to virtually every group in Japan who had argued against the evils of luxury and the importance of thrift, diligence, and moral regeneration following the 1923 disaster. Those who preached retrenchment from an economic standpoint would support the effort to increase savings and dissuade people from purchasing luxury items. They would also welcome the prospect of less money flowing out of Japan to the pockets of overseas producers. If implemented, the tariff would likely increase Japan's balance of trade, which in 1923 languished in deficit by a figure of ¥646 million; imports totaled ¥2.45 billion while exports amounted to ¥1.81 billion. Those who believed that the purchase of luxuries was both symptomatic as well as a cause of Japan's moral decline would relish the opportunity to restrict their consumption; placing temptations out of reach, many argued, would break the daily habit of luxuries. Parliamentarians who represented rural districts would not criticize the bill but rather embrace it as it excluded any item—luxury or otherwise—produced in Japan. The only quarter from which the Japanese government expected cries of protest was foreign governments and their citizens who lived in Japan. Neither was a deterrent.

Responses to the government luxury tariff proved mixed. Select newspaper editors and commentators such as Tokutomi Sohō applauded the bill. In an editorial published in the *Kokumin no tomo*, Tokutomi claimed that it was the only praiseworthy action undertaken by the new government.[41] Editors of the *Yamato* praised the bill in principle but questioned whether the tariff would actually dissuade people from purchasing luxuries in practice.[42] The "somber mood following the earthquake had lasted

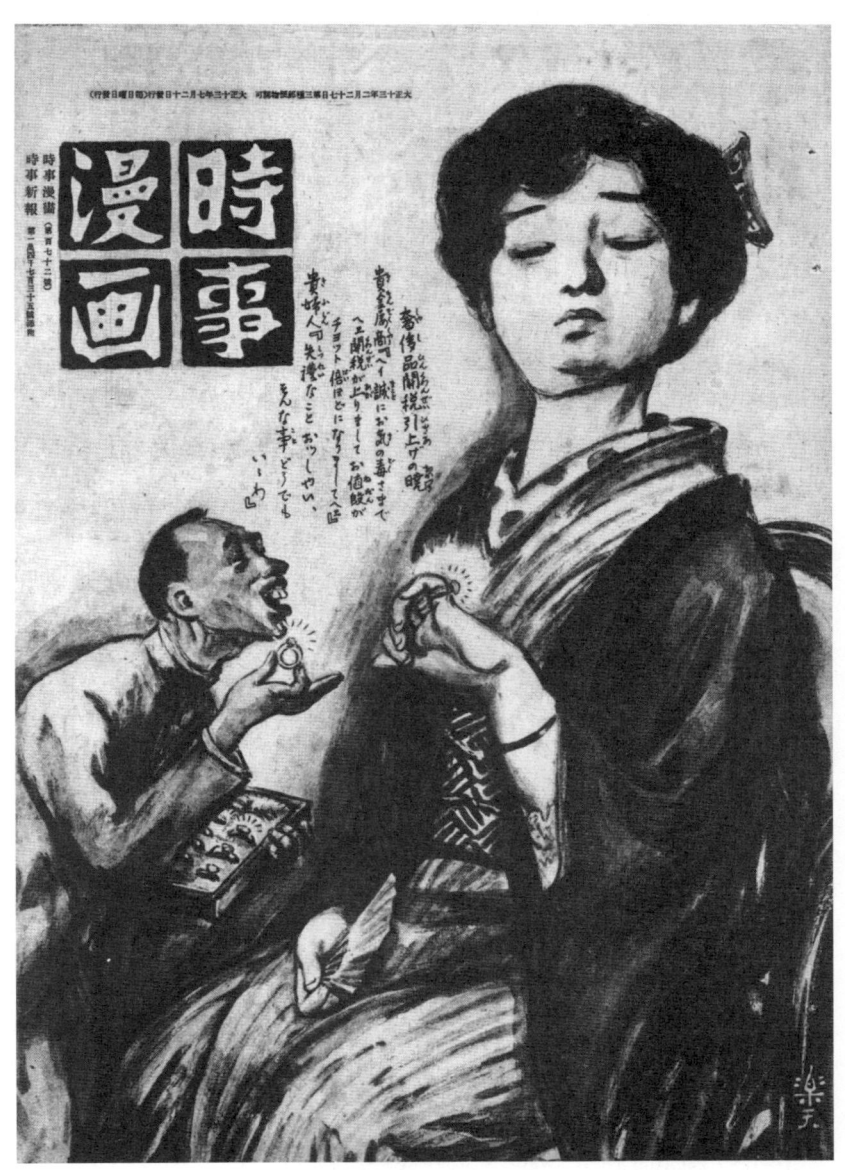

FIGURE 7.1 A discerning customer exclaiming, "The price is inconsequential!" in response to the doubling of prices following the passage of the luxury tariff in 1924

Source: Jiji manga no. 172 (July 20, 1924), cover image. Ohio State University Billy Ireland Cartoon Library & Museum.

only temporarily," they wrote, and the allure of luxuries had returned in postearthquake Japan with a vengeance. In the editors' opinion, people had once more become "wedded to luxuries." Political satirist Kitazawa Rakuten captured this opinion with panache in the July 20, 1924, issue of *Jiji manga*. On the cover of this issue, Kitazawa depicted a well-to-do woman shopping for jewelry. In polite Japanese, the sales assistant apologetically informs her that it is a great pity that due to passage of the luxury tariff, the price of the ring she has placed on her finger will be 100 percent more expensive. The woman replies with insouciance: "Who cares about the fact that the price has doubled; it is inconsequential."

More than a few parliamentarians argued that the tariff did not go far enough. Mutō Kamon (Kenseikai, from Gifu Prefecture) argued that the government had neither cut enough from the budget nor raised the luxury tariff high enough to change behaviors and correct national finances in fundamental ways. To Mutō, the government's actions resembled a "profligate father preaching thrift to his fast son."[43] When questioned on the floor of the Diet, the director of taxation admitted that spending on items that would fall under the proposed luxury tariff was only a very small percentage of the overall amount Japanese people, businesses, and government agencies spent on imports. Foreign-made luxuries, in fact, accounted for ¥34 million in 1921, ¥38.4 million in 1922, and ¥25.2 million in 1923. During these years, the total value of imports ranged from ¥2.1 to ¥2.5 billion. Undeniably, the bill was more symbolic than substantive, but it was just one part of a renewed focus on thrift and diligence in which the earthquake and its memory would play a leading part.

The foreign community in Japan, as the government predicted, proved to be the most vocal critics of the bill. Expatriates in Kobe organized several high-profile protest meetings and sent numerous letters of complaint to the Japanese government. The primary foreign language newspaper in Japan, the *Japan Weekly Chronicle*, contributed to this campaign and published a series of stinging editorials against the bill. Its editors also translated and reprinted numerous Japanese-language newspaper articles that challenged the efficacy of the bill. What was the basis of these critiques?

In the first instance, many newspapers—Japanese as well as English—criticized the bill for *not* including domestic-made luxury items. The *Tokyo asahi* argued that curbing the consumption of luxuries was far more complicated than merely placing a tax on foreign luxury goods entering Japan from overseas.[44] Any measure that stopped short of placing a tariff on domestic-made luxury items, the editors concluded, would neither be consistent nor have the desired effect of decreasing the amount spent on the consumption of luxuries. The *Japan Weekly Chronicle* provided the most spirited response to the tariff and asked if "the finance minister and his senses had parted company" during the drafting of this bill. "Does Mr. Hamaguchi really mean what he says," the editors wrote, "when he declares that articles on his list cannot be manufactured in Japan?" "Either he has not read his own bill," they concluded, "or else he is profoundly ignorant of the products of his home country."[45] Opponents of the luxury tariff bill argued that placing a tariff only on imported luxury items would have two negative consequences. First, it would lead people to buy domestic-made luxury items rather than forsake them altogether, and this in turn would cause a proliferation of domestic producers. Second, sellers of domestic-made luxury items would increase their prices to a level just below the cost of foreign-made luxury goods; goods whose prices would increase by 100 percent if parliamentarians passed the tariff as law. The bill would not fundamentally alter people's spending patterns but rather shift their focus toward domestic goods and drive up prices and ultimately inflation.

Other newspaper editors and Diet members questioned how Hamaguchi and his advisers in the Finance Ministry and taxation office decided on which items were "luxuries." At a July 10 meeting of the special parliamentary committee on the luxury tax, Maeda Yonezō (Seiyūkai, from Tokyo) asked why sporting equipment was included on the tariff list when the government's stated policy was to encourage physical fitness among its people.[46] Sakaguchi Takenosuke, author of a 1924 tome on the merchandising of high-end products in Japan, wrote the most animated and passionate challenge to the luxury tax bill. In an extended essay published in the *Jiji shinpō* and reprinted in the *Japan Weekly Chronicle*, Sakaguchi lamented that in subjecting "shirts, collars, cuffs, neckties, trousers, suspenders, belts, hats, boots, shoes, slippers, other foot ware [sic], soaps and perfumes," the government would tax "whatever one wears" on "whatever part of the body." He then ridiculed the government, stating: "If one were to dispense with the use of these luxuries, one would be forced to return to the primeval mode of living—a little far more uncivilized than that of Formosan aborigines. Can the people of a first-rate power adopt such a life without a loss of dignity?" He then, in an equally tongue and cheek style, condemned the categorization of velvet and plush as luxuries. "As these are mainly used in making seats in train and tram cars," he wrote, "the authorities probably consider it a luxury as a habit for people to sit down" while riding these vehicles. He warned all Japanese that unless one traveled on the "Tōkaidō without sitting down, one cannot claim that one has resisted luxurious habits successfully."[47]

Despite spirited debate and calls that the bill was either too sweeping or too narrow in scope, parliamentarians passed the antiluxury tariff during the July session. A 100 percent ad valorem duty was enacted on July 31, 1924. Even goods shipped from overseas ports prior to passage of the bill were hit with the surcharge upon importation to Japan if they had not cleared customs by the end of July. Arrival of the NYK cargo vessel *Mitomaru* on July 23 led to widespread rejoicing by the owners of its cargo. Buried within its holds were 3,000 tons of items deemed luxuries that escaped the tariff, including precious metals and stones, perfumes, woolen yarn, and fifty grand pianos destined for Yokohama. Had the ship arrived just eight days later, its cargo would have been subjected to the 100 percent duty. The tariff, however, was neither the first nor the last act taken to curb the consumption of luxuries or to advance a course of moral regeneration.

EMPLOYING THE EARTHQUAKE FOR THRIFT AND DILIGENCE

The Great Kantō Earthquake, or more precisely the memory of this tragedy, played an important role in the central government's launch of the 1924 Campaign for the Encouragement of Diligence and Thrift (Kinken shōrei undō). The Katō cabinet implemented the campaign to encourage people across Japan to become moderate, thrift conscious, diligent, and selfless in economic as well as social life.[48] Its genesis was the 1919 Campaign to Foster National Strength (Minryoku kan'yō undō), which was aimed at boosting savings and decreasing the flow of capital overseas as a result of foreign purchases. Together, the 1919 and 1924 initiatives aimed at promoting a sense of frugality, sacrifice, and hard work. Both programs, moreover, used new technologies and mediums to reach people; the two-year 1924 campaign incorporated over two million emotive posters, numerous other public advertising displays, radio broadcasts, and motion pictures.[49]

A major difference between the 1919 and 1924 campaigns, however, was the government's evocative use of the 1923 tragedy to emphasize the urgency and importance of the later program. The Katō cabinet launched the Campaign for the Encouragement of Diligence and Thrift on the one-year anniversary of the Great Kantō Earthquake. This was followed by seven well-choreographed "Diligence and Thrift Weeks" held between 1924 and 1926. Often thrift weeks were held in conjunction with important anniversaries associated with the disaster, as was the case with the one initiated on the three-year anniversary in 1926. Not only were the citizens of Tokyo encouraged to save for the nation's benefit, they were also asked to make donations on twenty street corners throughout the capital of goods as well as money that charitable organizations could use to assist those in need.[50]

On September 1, 1924, Prime Minister Katō issued a formal statement that encompassed his thoughts on the past and his vision for the future. He asked all Japanese to refresh their memories and to use the anniversary of Japan's earthquake calamity as a "starting point" for national reconstruction. The "prosperous national capital had been destroyed in one day" Katō wrote, and only "hard work and endurance" will enable its return to prosperity. Katō compelled the people to "mark this day" as the day that old habits of laxness were broken and spendthrift practices ended and a new moderate course was charted. Katō admitted that this would not be an easy task as the "extravagant and wasteful habits" of the wartime boom era were so entrenched in peoples' minds. It was easy, Katō argued, to forsake a humble and diligent life once tempted by extravagance, decadence, and laxness, but such an existence was hollow. He ended his speech by revealing that his "most fervent desire" was to have people reflect on the earthquake disaster and to embrace concepts of simplicity, moderation, and fortitude in all endeavors. He then declared that people across Japan must "willingly share responsibility for saving the nation from its [current] difficult position." To Katō, sharing responsibility meant saving money and adopting a lifestyle of diligence and thrift.[51]

Home Minister Wakatsuki Reijirō, whose ministry was responsible for the 1924 campaign, was even more direct in evoking the memory of the 1923 catastrophe. In a well-publicized speech on September 1, 1924, at the Tokyo Prefectural Hall of Commercial and Industrial Promotion, Wakatsuki told his listeners that "Japan must never forget" the extraordinary crisis of the unprecedented earthquake disaster and the trials and tragedy it created. But, with passion and conviction, he stated that Japan's "finances and economy" faced an equally "great crisis" in 1924. Unless the "whole nation—the government and the people—summoned the courage to make a united, untiring effort" the home minister predicted that Japan would be unable to surmount its current crisis.[52]

Thereafter the home minister expended considerable effort articulating the specific nature, causes, and potential solutions to Japan's economic crisis. Many of the statistics, themes, and slogans he employed during his speech later appeared on posters, were broadcast via radio programs, and were published in all types of mediums during subsequent Diligence and Thrift Weeks. One of the key factors behind the economic crisis, in Wakatsuki's opinion, was debt. By the home minister's calculations, Japan's national debt in 1913 was ¥2.58 billion; in 1924 it stood at ¥4.68 billion. During this time the national budget had gone from just over ¥500 million to ¥1.6 billion. Moreover, the trade surplus of ¥1.4 billion that Japan had amassed between 1913 and 1919 had evaporated: between 1919 and 1924 Japan suffered a trade deficit of nearly ¥2.7 billion. Since 1913, Wakatsuki claimed, government spending, both local as well as national, had rocketed out of control, and people had become too consumer oriented and enamored of foreign-made luxury items. Such behavior had placed Japan at the edge of a treacherous precipice. Could Japan avert a financial disaster that was larger than the earthquake? Wakatsuki suggested that salvation was possible but would require sacrifice.[53]

Japan, Wakatsuki argued, could look to Britain's wartime and immediate postwar experience to find solutions to its current economic crisis. The British people, he claimed, had "devoted themselves wholeheartedly to the nation" during the war by "restricting consumption." Moreover, they had also purchased government bonds amounting to £240 million between 1916 and 1919 and £205 million between 1919 and 1922. Apart from people spending less, the government also led by example and adopted a retrenchment-minded philosophy of austerity after the war concluded. In each year since 1920, Wakatsuki proclaimed with no little envy, Britain had reduced its national expenditures and was likewise able to reduce its tax intake. Whereas the British government spent £1.65 billion in 1919 and received only £1.39 billion of income, the following year it claimed £1.42 billion of revenue while spending only £1.19 billion. Expenditures totaled a meager £893 million in 1922 while income had receded to £910 million. Both the British people and their leaders, Wakatsuki concluded, had adopted the wonderful mindset and practice of frugality and retrenchment.[54]

Given Britain's successes, the home minister argued that both the people and the Japanese government must unite and work together to reorient Japan away from its dangerous immoderate path. The earthquake had provided an undeniable shock to Japan, Wakatsuki claimed, and it was now up to the government and the people to use this "trial from heaven" for good results. The "government must reorganize and retrench its finance and administration," he argued, while the "people must create a national custom of diligence and thrift." The Katō cabinet would do its part and reduce expenditures, but the people must follow suit and combat luxury, save, and purchase reconstruction bonds. "Diligence and thrift," he argued, were the "mother of austerity, fortitude, and true heartedness." Moreover they were an elixir that if taken by society could serve as a panacea to combat Japan's economic *and* social problems. "Diligence and thrift," he argued, "brings a stable economic life to individuals and leads to increases in productivity and the accumulation of capital to a society. From a moral point of view, they are the individual's driving force for the cultivation of independence and self-respect; and, for society, they are the simplest and most common form of social service. In other words, diligence and thrift are the key to bringing economy and morals together."[55]

In the first instance, the Katō government implemented a policy of fiscal retrenchment in an attempt to set an example for the nation to follow as it charted a moderate path of diligence and thrift. In preparing the national budget for fiscal year 1925, the government trimmed ¥83 million from the budget, lowering it to ¥1,542,989,000 from the ¥1,625,024,000 that was spent in 1924.[56] While this reduction of roughly 6.6 percent reversed the expansionary trend of budgets since 1914, it was not enough of a reduction to avoid criticism. During the preparation of the budget, longtime champion of retrenchment and sworn enemy of luxuries Horie Ki'ichi accused the government of being reckless with finance and predicted that only a small amount would be reduced from the budget. He asked, "Is the government really qualified to conduct propaganda for thrift . . . when it is a great spendthrift in practice?"[57]

When the Katō cabinet's budget was released in November 1924, the editors of the *Japan Weekly Chronicle* took this argument further and claimed that the government had no legitimacy to orchestrate its first Diligence and Thrift Week to coincide with the one-year anniversary of the November 10, 1923, rescript on the invigoration of the national spirit. Moreover, they argued that bombarding people with pronouncements to "abstain from fish on Monday, flesh on Tuesday, and sake on Wednesday, etc.," was too simplistic to succeed.[58] This echoed a piece written earlier by Yoshino Sakuzō and published by the *Japan Weekly Chronicle* under the headline "Thrift: How Can the Poor Save?" In this article, Yoshino lauded the government for producing banners that exclaimed "Let us live frugally" and "Let us work hard and save much." However, he intimated that society was "too complex" for simple mottos of diligence and thrift "to be acted on by an obedient public meekly and unquestioningly." He concluded that "it is now in the army and navy alone that a large number of men can be moved about by order of their superior authorities as one moves one's hands and feet."[59] Given Yoshino's reservations about the potential efficacy of the diligence and thrift campaign, how did people respond to this and other initiatives geared toward reducing consumption for both economic and moral reasons?

THE ALLURE OF THE MODERN CONSUMER CULTURE AND THE NEVER-ENDING SUASION CAMPAIGNS

Did the individuals who championed progressive-minded interventionist approaches after the 1923 earthquake succeed in either encouraging the government to adopt their recommendations or in their overall efforts to facilitate spiritual renewal? The government certainly became more sophisticated from the 1920s onward in employing novel ways of reaching the public. Movies, radio broadcasts, publications, and well-orchestrated suasion campaigns became part of everyday life in interwar and wartime Japan. It would have been next to impossible for an urbanite not to see some type of government campaign for diligence and thrift, moral regeneration, or spiritual renewal.

It is impossible to write with certainty whether or not the government's campaigns for diligence, thrift, and moral reconstruction had any direct impact on Japanese society. Examining data related to savings and spending, however, provides some clues. Between 1924 and 1926, the years of the diligence and thrift campaign, withdrawals from postal savings accounts decreased from ¥936 million in 1924 to ¥844 million in 1926; a decline of 9.8 percent.[60] Deposits fell by only ¥4 million yen to ¥931 million. The fiscal year following the completion of the campaign, 1927, saw an extraordinary increase in deposits by the amount of ¥635 million yen; this equated to a 68 percent increase over the previous year's deposits. The overall balance of postal savings in 1927 rested at ¥1.6 billion. It is impossible to tell whether this was a direct result of the government's diligence and thrift campaign but Japan's postal savings balance increased each year from 1924 to 1931. In doing so, it more than doubled from ¥1.1 billion in 1924 to ¥2.8 billion in 1931. People across Japan also purchased over ¥200 million in earthquake reconstruction bonds between 1924 and 1929, though the purchase of these bonds was slower than government officials hoped and expected.[61] Japan's national budget decreased during this same time from ¥1.6 billion in 1924 to ¥1.4 billion in 1931.[62] In short, people were saving more while the national government spent less.

As people saved more and the government practiced thrift, was there a decline in the level of spending on items that officials deemed luxuries? Yes and no. If one looks only at the years immediately following the in-

TABLE 7.1 Spending on phonographs and records in Japan, 1912–1930

YEAR	AMOUNT SPENT (YEN)	INDEXED TO 1912 (1912 = 100)	CONSUMER PRICE INDEX FOR ALL ITEMS (INDEXED TO 1912)
1912	2,997,000	100	100
1919	14,652,000	489	211
1920	21,561,000	719	221
1921	14,261,000	475	202
1922	13,214,000	441	199
1923	13,134,000	438	197
1924	17,150,000	572	199
1925	11,715,000	391	201
1926	11,052,000	369	192
1927	9,792,000	327	189
1928	9,434,000	315	182
1929	10,134,000	338	178
1930	11,482,000	383	160

Source: Ōkawa, Chōki keizai tōkei, 6:250–51.

TABLE 7.2 Spending on cameras and camera parts in Japan, 1912–1930

YEAR	AMOUNT SPENT (YEN)	INDEXED TO 1912 (1912 = 100)	CONSUMER PRICE INDEX FOR ALL ITEMS (INDEXED TO 1912)
1912	50,000	100	100
1919	482,000	964	211
1920	2,056,000	4,112	221
1921	2,459,000	4,918	202
1922	3,435,000	6,870	199
1923	1,221,000	2,442	197
1924	1,674,000	3,348	199
1925	593,000	1,186	201
1926	347,000	694	192
1927	1,513,000	3,026	189
1928	1,917,000	3,834	182
1929	1,777,000	3,231	178
1930	2,336,000	4672	229

Source: Ōkawa, *Chōki keizai tōkei*, 6:250–51.

troduction of the luxury tariff, a decrease in spending on a range of so-called luxury items is apparent. For instance, spending on phonographs and records dropped precipitously after the imposition of the luxury tariff in July 1924. Spending levels on these items did not recover to the level of 1920 until the mid-1930s. Spending on cameras and camera parts, another item classified as a luxury in the 1924 tariff, likewise fell dramatically after the tariff was introduced. Spending recovered quickly, however, and superseded the 1920 level in 1930. Cosmetics, on the other hand, showed no significant decline. By 1930, people were spending 30 percent more money on cosmetics than they had one decade earlier. In terms of alcoholic beverages, spending on sake fell from ¥1.16 billion in 1920 to ¥913 million in 1930. Sales of shōchū, on the other hand, increased from ¥68 million in 1920 to ¥73 million in 1930.[63]

Too many variables likely influenced why people spent, what they purchased, or how much they saved to make any singular conclusion about

TABLE 7.3 Spending on cosmetics in Japan, 1912–1930

YEAR	AMOUNT SPENT (YEN)	INDEXED TO 1912 (1912 = 100)	CONSUMER PRICE INDEX FOR ALL ITEMS (INDEXED TO 1912)
1912	836,000	100	100
1919	11,898,000	1,423	211
1920	29,230,000	3,496	221
1921	21,661,000	2,591	202
1922	28,249,000	3,379	199
1924	31,429,000	3,759	199
1925	32,936,000	3,940	201
1926	29,289,000	3,504	192
1927	35,961,000	4,302	189
1928	36,340,000	4,346	182
1929	39,162,000	4,684	178
1930	41,517,000	4,959	160

Source: Ōkawa, Chōki keizai tōkei, 6:250–51.

whether these declines were tied to the government's diligence and thrift campaigns. The fact that savings increased and the purchase of certain luxuries decreased temporarily after 1924, reversing trends that had developed over the previous eight years, is instructive but not conclusive. We know beyond any doubt that the economic impact of the earthquake and concerns held by elites about the inflationary influence reconstruction spending would have on the economy encouraged government officials to adopt retrenchment-oriented budgetary policies after 1924. In the 1920s this was the orthodox economic policy to follow.

It is likewise impossible to answer with certainty whether any of the government's suasion policies geared toward reorienting society away from the perceived trajectory of moral decline had any lasting effect on society. Had the government taken heavy-handed authoritarian measures such as abolishing licensed prostitution, proscribing the construction or reconstruction of dance halls, cafes, and department stores, or banning the sale of luxury goods outright, historians today could conclude with relative ease that the earthquake had a fundamental, transformative impact on state-society relations. None of these developments occurred as a result of the 1923 earthquake. The government did not launch such interventionist policies in 1924 because Japan was not an authoritarian state with a unified group of policy makers who believed that this was an appropriate path to follow. While voices from the left and the right urged radical, sometimes revolutionary solutions to Japan's perceived crises, leaders in 1924 chose a far more moderate course to effect change within society. The government pursued moral suasion and economic tariffs rather than coercion and proscription with mixed results. But even the path of suasion was restricted because many Japanese leaders were enchanted by the belief that the crisis of economy necessitated reduced spending. Funds pruned from government expenditure could have been used to build more lecture halls, public cafeterias, schools, community centers, parks, and other facilities where officials could inculcate values and practices in the pursuit of reforming if not perfecting society.

Like those planners who sought a sweeping transformation of Tokyo's built environment, people who desired a radical regeneration or economic reorientation of society found themselves disappointed in the years following the 1923 calamity. Many claimed that the purchase of luxuries and the practice of morally questionable pursuits had increased. These individuals worried that the resilience of luxury and the allure of materialism, consumer conveniences, and modern entertainment associated with the city might render all future attempts at renewal failures. However destructive and dislocating the earthquake disaster was, many feared that it might amount to nothing more than a momentary interruption on Japan's otherwise steady post–World War I decline. According to commentator Nose Yoritoshi, the Japan of 1925 was still facing a national crisis because the hoped-for spiritual reconstruction had simply not materialized. Nose in fact suggested that people were becoming more extravagant and decadent than at any previous time in the history of the nation. This "fact" was all too obvious, Nose argued, if one saw the holiday displays at Mitsukoshi and Matsuzakaya department stores. More to the point, he claimed that "the unhealthy, luxurious, idle lifestyles to which we became accustomed in prosperous times during the war—before the earthquake—were still dominating our popular mind." It was, he wrote, "just like a snake coiling around our necks." He argued that "public morals were going back to the deplorable state of the pre-earthquake times." Added to this, increased calls for expanded democracy, the elevation of individualism, proposals supporting equal rights for men and women, and the increased reports of unconstrained sexual behavior and juvenile delinquency all led Nose to conclude that the Japanese people had learned nothing from the earthquake disaster.[64]

In a dramatic visual fashion, Kitazawa Rakuten expressed the dissolution that many, including Nose, felt at the lack of progress in the overall campaign for spiritual renewal and physical reconstruction on the two-year anniversary of the disaster: transformation he suggested was chimeric. On the cover of the August 31, 1925, edition of *Jiji manga*, Kitazawa placed an angry-looking catfish—the creature in Japanese folklore supposedly responsible for unleashing earthquakes—rising over Tokyo. The catfish declares: "It's as if you've had your eyes closed. Perhaps I should shake things up again."[65] Before the catfish, and no doubt one reason behind its palpable exasperation, are a man and woman dressed in luxurious Western attire. The woman's umbrella reads *keichō fuhaku* (frivolous and thoughtless). Kitazawa critiqued the slow process of reconstruction planning by placing a seemingly bored bureaucrat dressed in yellow sitting at a desk staring uninterestedly at a banner, which reads *kukaku seiri* (land readjustment). A corpulent plutocrat in a green suit stuffs yet more money into his pockets, having gained wealth through reconstruction. A Japanese lumber dealer, sitting on a pile of unused timber, laments the decline in commodity prices that occurred as a result of the abolition of import tariffs on construction goods. At its essence, Kitazawa's cover image highlights the fact that the earthquake changed neither peoples' attitudes and behaviors nor how the city was rebuilt following the 1923 disaster.

Nose and Kitazawa expressed concerns held by many commentators and officials that the nation had missed a unique opportunity following 1923. Both believed that the nation had failed to use the disaster and the subsequent reconstruction project to prune out luxurious and extravagant habits and behaviors and dangerous ideologies that sprang from urbanization, the growth of modern consumer culture, and larger global trends of a social, political, and ideological nature. Other commentators and government officials lamented the fact that by 1926, people were spending more at cafes, restaurants, theaters, and movie houses than ever before. Metropolitan police officials announced publicly that Tokyoites had spent ¥3.2 million more at these places of business during the first ten days of January 1926 (¥8.5 million in total) than they had one year earlier.[66] As many economic and social historians have suggested, Japan,

FIGURE 7.2 Two years after the earthquake, the catfish exclaiming, "Perhaps I should shake things up again!"

Source: Jiji manga no. 228 (31 August 1925), cover image. Ohio State University Billy Ireland Cartoon Library & Museum.

particularly urban Japan, was becoming a wealthier, consumer-oriented society, and people were spending more on all types of items.[67] This longer-term trend was not arrested because of the 1923 earthquake or subsequent calls for moderation. If the "civilization-upturning" disaster could not wake people from their dreamlike existence and compel people to reflect and reform, Nose and others wondered what, if anything, could.

Throughout the 1930s, and particularly after Japan and China went to war in 1937, government officials asked similar questions with increasing frequency. Initially the government emphasized suasion and urged greater numbers of middle-class urbanites to join neighborhood associations, particularly district commissioner (*hōmen iin*) networks.[68] More than simply provide or distribute welfare, these middle-class participants served as local leaders in moral regeneration and suasion campaigns launched to encourage reform. Numbers illustrate their growth: the number of district commissioners rose from 102 in 1928 to 3,320 in 1937.[69] Perhaps one of the most important longer-term legacies of the 1923 disaster was that it convinced many bureaucratic elites that maintaining and strengthening urban Japanese people in mind and body required as much attention and effort—if not more—as those programs launched to invigorate and revitalize people in rural communities. In postearthquake Japan, state officials began to invest more time and effort in organizing, training, and eventually mobilizing ward and neighborhood associations, district commissioners, schools and youth groups, and other quasi-governmental vigilance groups in Tokyo and other major urban centers. None of these changes took place overnight. Moreover, none was undertaken exclusively because of the earthquake disaster.

If state officials needed any convincing as to the difficult but necessary task they faced after 1923, however, they had to look no further than the writings of Tokyo novelist Uno Kōji, whose moving evocation began the first chapter of this study. Uno ended his reflection on the catastrophe with a poignant passage that illustrates both the transitory impact of the earthquake on his thoughts and the resilient allure of the consumerism and self-indulgence often associated—correctly or not—with 1920s Japanese urbanism. Two weeks after *"that day,"* Uno bluntly told a group of friends that he "missed this café and that café in the Ginza" and "the cinema, the theatres, and entertainment of Asakusa." He confessed that he desperately wanted to drink as he had before the disaster, "sitting under the bright lights" of the city. For a man who had, over the first two weeks of September 1923, survived an extraordinary calamity, witnessed unprecedented destruction, and reflected on the tens of thousands of corpses that littered the remains of Japan's capital, Uno concluded his discussion by telling his friends: "Two hundred kilometers, four hundred kilometers, and six hundred kilometers from Tokyo, Shizuoka, Nagoya, and Osaka [still] exist. In these cities, the bands are playing in the cinemas, and people are bustling in the busiest areas of town as usual, just like we did in our town before September 1. This thought makes me want to fly to those places, if only I had wings."[70]

Uno would not have to grow wings and fly to other cities in Japan to reacquaint himself with the pleasures of urban, consumer-oriented life. Despite calls from officials to reconstruct the capital to reflect and reinforce new values, Tokyo reemerged from the reconstruction project with many of the old places of entertainment, vice, consumerism, and economic disadvantage that existed before and so worried countless commentators on the eve of calamity.

Eight
READJUSTMENT: REBUILDING TOKYO FROM THE ASHES

We the people of Tokyo are now standing on the world stage, performing a play called reconstruction. Every single move of ours represents the honor of the Japanese people.... At this critical time I strongly believe that you will carry out this great project with endurance, diligence, and a smile.
—Nagata Hidejirō, 1924

In every neighborhood, ugly arguments and fights are destroying the peace and friendship among the residents. The fine tradition of neighborhood cooperation is being ruined. This is because of land readjustment.
—Imperial Capital Reconstruction Institute, 1925

In May 1924 Nagata Hidejirō began one of his most important public relations campaigns as mayor of Tokyo with a tour. Unlike previous trips taken in late 1923, Nagata did not focus his energies on surveying destruction or consoling victims. Rather he spoke about the future. He stressed the importance of all Tokyoites working together to rebuild the capital. Nagata's objective was to convince his fellow citizens to embrace the idea of land readjustment (*kukaku seiri*) as a way to rationalize plots of land, if not whole neighborhoods in Tokyo. Land readjustment, or so Nagata claimed, could enable the city to widen roads, create sidewalks, establish small parks, build more and larger bridges, and create fire-prevention zones in the capital. The people of Tokyo faced a moment of decision. People could either embrace the "absolutely necessary" and "ultimately beneficial" program of land readjustment or they could "look on with indifference" as the old city of Tokyo reasserted itself and the city's golden opportunity vanished.[1]

Fearing ambivalence, if not outright resistance, from Tokyoites, Nagata used evocative language, emotive themes, and practical reasoning in selling the concept of land readjustment, a technique he argued would reorder the city's devastated wards. The mayor and his entourage eventually brought models, maps, and posters of what "new neighborhoods" might resemble once completed.[2] If "old Tokyo" reappeared with its winding streets and back alleys, narrow, fire-prone bridges, and few open spaces, Nagata suggested, "our descendents might suffer the same terrible pain as we did in September." If this were to occur, all 100,000 who perished "would have lost their lives in vain." Nagata confessed that he could not live with such a scenario and thus declared it was a duty to future generations to forge a new Tokyo. With only a little shared sacrifice, a spirit of hard work, and a willingness to vanquish "conservative thoughts and temporizing actions," a splendid capital could be forged that would not only improve the city for current residents but also guarantee "the happiness of our descendants."[3]

What did Nagata mean by "shared sacrifice"? The mayor informed his listeners and the readers of the pamphlet that accompanied his speech that most people would be asked to sacrifice roughly 10 percent of their land without compensation for the betterment of the city. He took great pains to tell Tokyoites that any loss of land over 10 percent would result in "proportionate compensation" and that no one was going to lose their land in its entirety. Everyone would equally suffer short-term pain through the land readjustment process, but everyone would receive long-term benefits by living and working in a splendid new Tokyo.

Residents responded to these overtures with confusion, ambivalence, and in many instances outright anger. Initially many residents remained confused over the complexities associated with land readjustment. Others worried that their lives would again be upturned as land confiscation and forced relocation took place. Rumors that the compensation money would be miniscule and paid only in bonds rather than in cash also caused disquiet among residents. Other rumors were given a degree of legitimacy by elite-level critics of land readjustment, including Egi Kazuyuki, who encouraged residents to resist its implementation without fair compensation.[4] While debates over the reconstruction budget and plans for spiritual renewal and fiscal retrenchment ignited contentious debates at the elite political level, plot by plot readjustment, removal, and relocation of people's homes and businesses proved every bit as contentious. Virtually every aspect of Tokyo's reconstruction, from the planning of roads and sidewalks and the readjustment of irregular land plots to the ultimate design of the earthquake memorial hall, attracted differing types of protest and disputes. An examination of these local level disputes captures what "reconstruction" meant for countless citizens, and how and why so many individuals voiced criticisms of, if not open resistance to, readjustment and official reconstruction.

LAND READJUSTMENT IN POSTDISASTER TOKYO: REALITIES AND RUMORS OF RUIN

Well before the Yamamoto cabinet had decided on a reconstruction budget in November 1923 or bureaucrats had begun to conceptualize land readjustment, Tokyoites began the taxing and emotional task of rebuilding their homes, small businesses, and lives. The rapidity and scale of individual reconstruction amazed commentators and government bureaucrats.[5] But it also raised concerns. Bureaucrats worried that they would find themselves in the unenviable position of tearing down people's makeshift homes and shops when a thorough reconstruction plan was implemented. The government dealt with this potentially difficult problem through Imperial Ordinance 414, issued on September 15, 1923. It classified all buildings constructed between September 15 and the end of February 1924 as "temporary structures" that would have to be removed by the end of August 1928. Owing to the large numbers of residents who began to reconstruct temporary dwellings, the government extended the deadline for the completion of structures classified as temporary to the last day of August 1924. By the middle of 1924 over 170,000 small-scale building projects had commenced or been completed, causing a severe spike in the price of basic building materials, particularly lumber.[6] Tokyoites still wondered as late as December 1923 when and how formal reconstruction would begin, and who would direct its implementation.

A series of laws passed and further imperial ordinances issued between late December 1923 and mid-April 1924 outlined the techniques that would be employed and who would oversee reconstruction. The first was the Special Urban Planning Law (Tokubetsu toshi keikaku hō) promulgated by the Imperial Diet on December 24, 1923.[7] The key features of this law were as follows. First, it superseded the existing 1919 Urban Planning Law in Tokyo and Yokohama. Second, it allowed the government to use land readjustment as a technique to alter the boundaries of residential property lots within the urban, built environment. Through land readjustment, the government gained the ability to claim private property and turn it into public space in the form of roads, sidewalks, parks, schools, or other infrastructure. Moreover, under the 1923 law up to 10 percent of anyone's land could be confiscated by the state without monetary compensation. If the area of land confiscated exceeded 10 percent, compensation would be paid at a rate determined by a land readjustment board.[8]

While land readjustment had been employed regularly in rural or suburban fringe areas as a way to facilitate road construction, it had not been used in built-up, urban areas within Japan. Moreover, when it was used in rural or suburban areas, agreement by at least two-thirds of the landowners owning at least two-thirds of the land in the designated project area was required to implement land readjustment. The special law of 1923 required no similar majority endorsement. Finally, the government gained the ability to remove an owner, leaseholder, or tenants from one plot of land and relocate them to a new location if the land vacated was deemed necessary for the public good.[9]

From the conceptual map in figure 8.1 designed by urban planner Takeuchi Rokuzō, one can see the aims of land readjustment.[10] On the

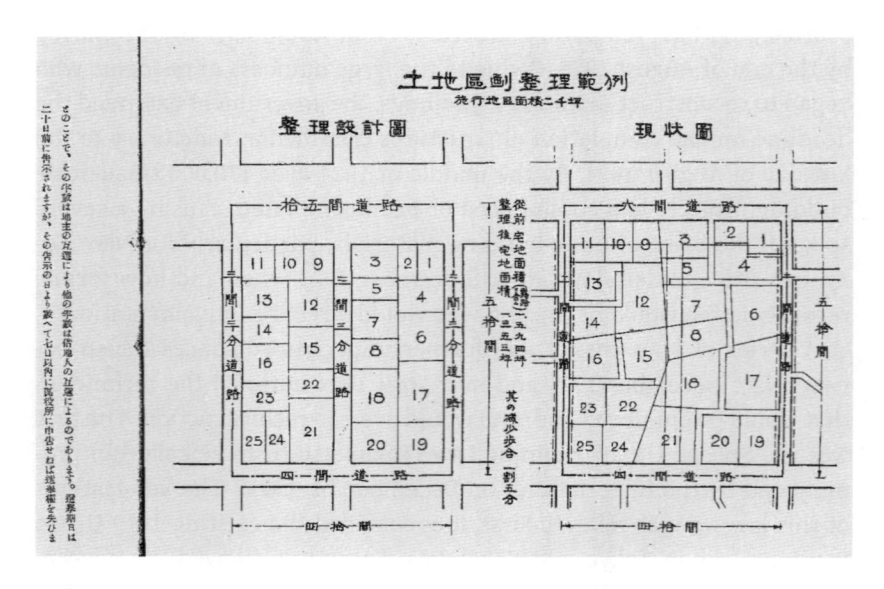

FIGURE 8.1 The aims and objectives of land readjustment: a before-and-after comparison of a hypothetical block in Tokyo

Source: Takeuchi Rokuzō, "Kukaku seiri no sekkei narabi ni shikō," in Gotō Shinpei, ed., *Teito tochi kukaku seiri ni tsuite*, 41. Reproduced with permission of Tokyo Metropolitan Archives.

right side of the diagram is a representative city block of Tokyo with its irregular-shaped plots of land and winding back alleys. On the left sits an idealized block that is smaller overall and comprises rational, more uniform allotments. The reduction in private land has enabled planners to widen the major road this block fronts as well as place an auxiliary road between both halves of the blocks. From a block-level perspective, the neighborhood has become more rational and user friendly.

Clearly, however, there are some winners, some losers, and those who experienced little loss. For example, lots 4, 5, 6, and 17 lose significant street frontage, and their owners could reasonably be expected to complain. Lots 12, 15, and 22 gain more significant street frontage, even if it means a loss of overall land size. These owners would be more likely to accept readjustment. Lots 3, 9, 10, and 11 remain pretty much unchanged except for a loss of about 10 percent of their total land, a loss more than likely made up for by an overall general improvement in the neighborhood through land readjustment.

On March 15 Imperial Ordinance No. 49 detailed how land readjustment would take place under the special urban planning law. Over the following week, the Reconstruction Bureau (Fukkō kyoku), led initially by Naoki Rintarō, announced that roughly 30 million square meters of once devastated land would be divided into sixty-six readjustment districts. The Reconstruction Bureau designated that land readjustment in fifteen districts would fall under the jurisdiction of the Home Ministry, while the City of Tokyo took responsibility for land readjust within the other fifty-one districts. The Reconstruction Bureau then directed landowners and businesses that held leases within the readjustment district—but not tenants—to elect a land readjustment committee within their respective areas. Less than a month later the number of members that were to be elected for each district was announced. Ballots began in early May and finished by August 19 in each of the districts administered by the Home Ministry. All but two of the districts administered by the City of Tokyo had elected their committee representatives by December 31, 1924.[11]

The responsibility these local committees assumed was extensive. Using survey data collected by government officials and army engineers, elected representatives immersed themselves in virtually all decisions related to land readjustment. How individual parcels of land would be readjusted to make plots and entire neighborhoods more rational, how much compensation money was to be paid to landholders who experienced a loss of land greater than 10 percent, and the size and location of land substitution lots all fell under their jurisdiction. Committee members also

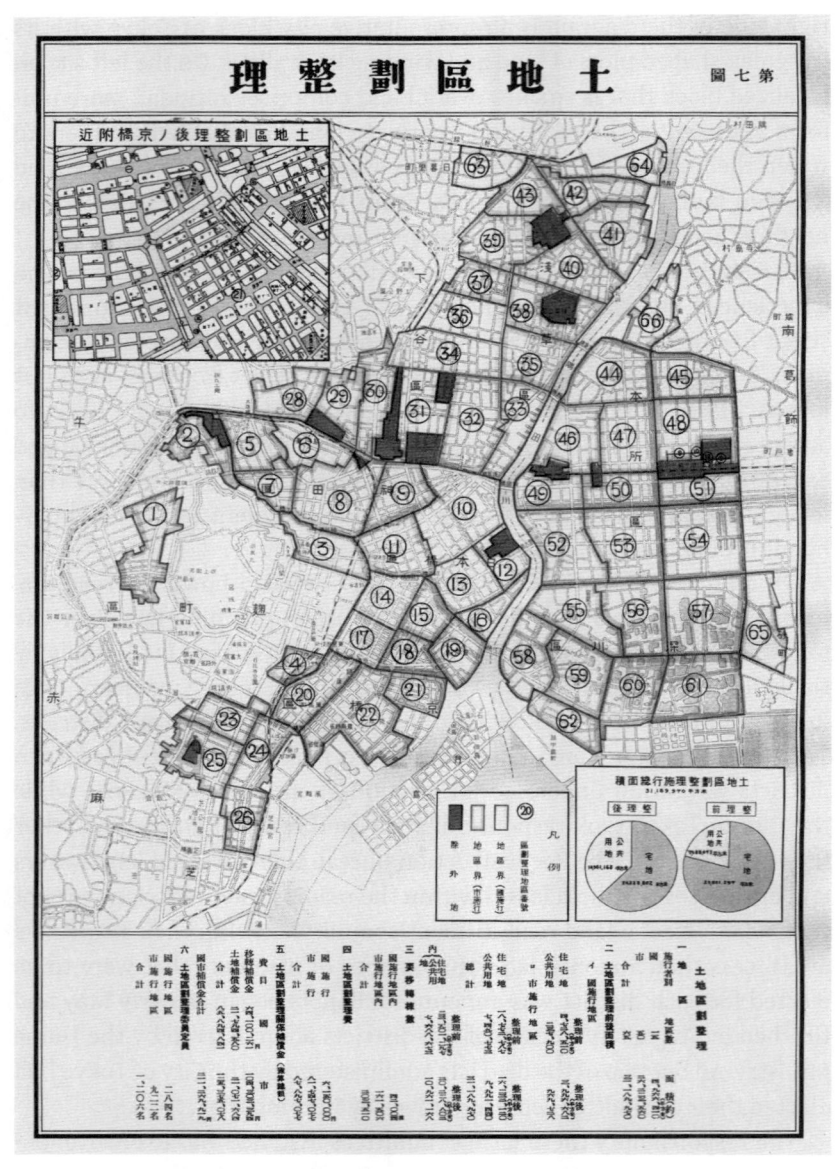

FIGURE 8.2 Land readjustment districts in Tokyo, 1924–1930

Source: Tokyoshi, *Teito fukkō jigyō zuhyō*, 7. Reproduced with permission of Tokyo Metropolitan Archives.

found themselves on the receiving end of hundreds of petitions and complaints and were forced to adjudicate the resultant disputes.

Even before local committee members were chosen, and only two days after the Reconstruction Bureau announced the parameters by which representatives would be elected, residents across Tokyo expressed anger over possible land readjustment. On March 17 residents gathered at the Nishikanda Club in Kanda Ward and established an association to halt land readjustment. The following day the newly formed group sent a petition to the mayor of Tokyo outlining their concerns.[12] Their petition highlighted a number of key points. The first argument made by the Kanda petitioners echoed claims advanced by elites such as Egi Kazuyuki and Itō Miyoji, who argued that uncompensated confiscation of land was unconstitutional. The signatories argued that landowners must receive compensation for all land that was confiscated, not merely confiscated land that exceeded 10 percent of the total area of land owned. Second, they argued that the implementation of land readjustment be postponed until August 1928 to allow landowners to more fully "recover from the economic losses suffered in the great earthquake disaster." Finally, they argued, "tenants must have the right to be involved in the selection of readjustment committee members." Mayor Nagata's response, if any was given, is not known.

The issue of tenants' rights to participate in the selection of land readjustment committee members was taken up by hundreds of groups across Tokyo. Residential protection leagues formed in each of the sixty-six land readjustment districts, and many sent petitions to the Reconstruction Bureau. "For the sake of residential peace and security," read one submission, "we would like tenants to be included in the election of readjustment committees." In a April 9, 1924, petition, residents from Nihonbashi Ward declared that "it was careless beyond words" that "hundreds of thousands of tenants were not given the right to participate in the election of land readjustment committees."[13]

More than two thousand residents from Asakusa Ward met on April 29, 1924, to draft an appeal to the mayor, the Reconstruction Board, and the Home Ministry. While claiming that restricting tenants from the selection process of land readjustment committees was unjust, they went further and claimed that land readjustment must not be allowed to proceed in any form whatsoever. Asakusa had already begun to recover "faster than any other ward," they wrote, and although it still looked "unsightly" in places, residents by themselves had already begun to rebuild as they desired. If the government proceeded with land readjustment and "destroyed the fruits of our hard work," the people will feel "as if we had been struck by yet another great earthquake and fire." Even residents from Honjo Ward echoed many of these concerns and suggested that land readjustment was a "betrayal" of the people who, despite losing everything in the great calamity, had forged ahead and begun to reconstruct Tokyo. "All of our efforts will have been a total waste," they declared, if the government's plan was implemented.[14]

Despite the vocal outcry of local residents, the Reconstruction Bureau, the City of Tokyo, and the Home Ministry refused to alter the overall plan for land readjustment. Moreover, they rejected calls to slow its implementation or to allow tenants the right to participate in the selection of land readjustment boards. Government officials understood that reconstruction would take at least five years to complete, and any postponement would only make resistance grow stronger. In relation to the issue of tenants' participation, only landowning males were allowed to vote in elections. Officials believed it would thus be inappropriate for a large mass of nonlandowning tenants to select people who would have a significant say in how land would be readjusted and lots rationalized throughout the capital.

IMPLEMENTING THE GRAND PROJECT FROM THE BIRD'S-EYE PERSPECTIVE

The readjustment and rationalization of over 30 million square meters of land was an extraordinarily complex undertaking. Even before landowners and businesses had selected local land readjustment committees, government officials had begun the process of surveying every meter of Tokyo set for land readjustment. The aim of land readjustment was to rationalize plots of land and neighborhoods, particularly in relation to the vast new road network that was to emerge within the capital. Whenever possible, Reconstruction Bureau officials hoped to eliminate the irregular land plots and meandering streets and alleys that defined many areas of Tokyo. Though they were not always successful, lots within many of Tokyo's wards became more standard in shape, as can be seen from the before-and-after maps of Land Readjustment Area No. 12 in figures 8.3 and 8.4. This area, which bordered the Sumida River, was trumpeted by the city as one of the most successfully readjusted wards in the capital.

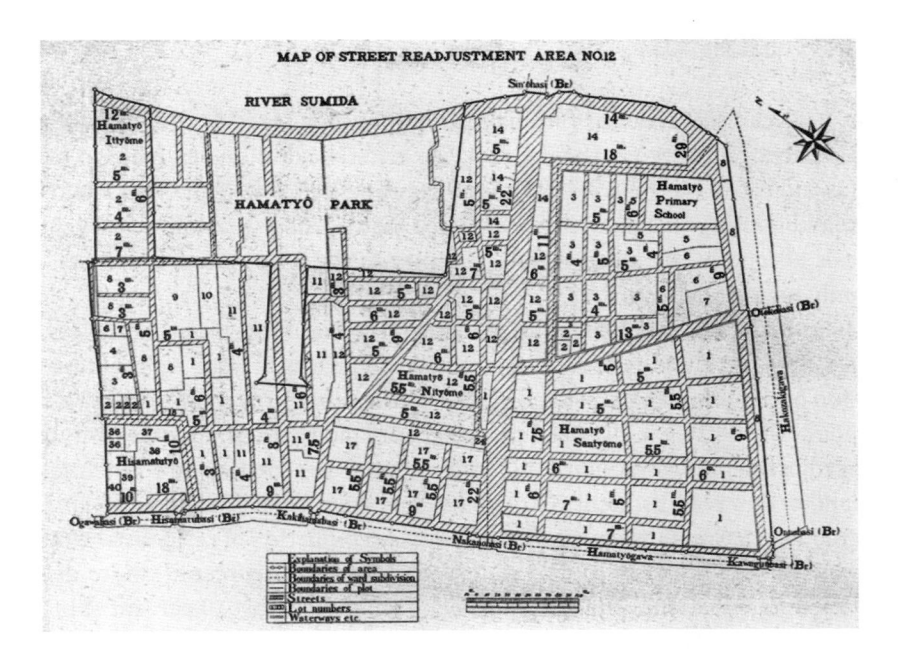

FIGURE 8.3 Readjustment area no. 12 before land readjustment

Source: Tokyo Municipal Office, *Tokyo Capital of Japan*, 70. Reproduced with permission of Tokyo Metropolitan Archives.

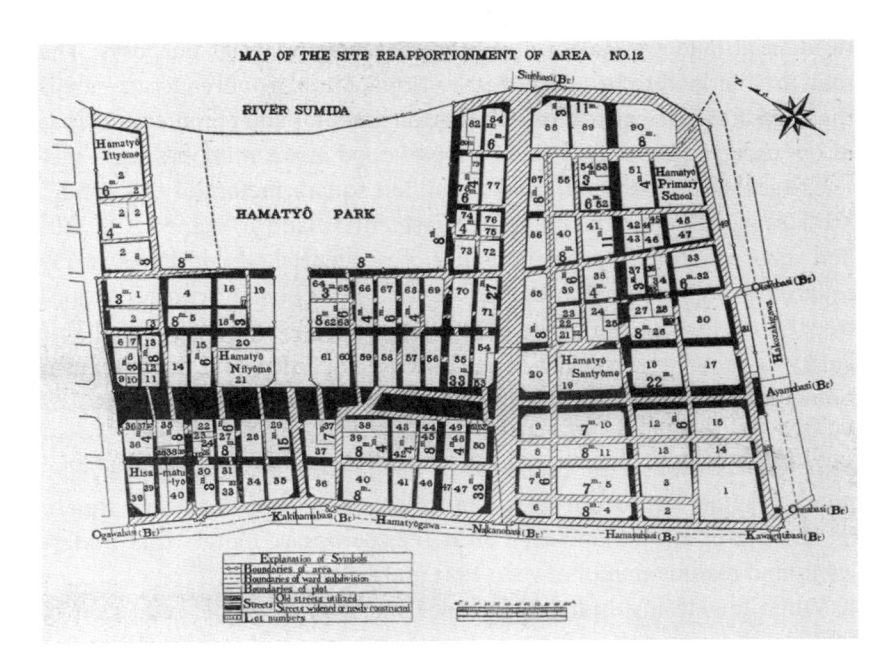

FIGURE 8.4 Readjustment area no. 12 after land readjustment

Source: Tokyo Municipal Office, *Tokyo Capital of Japan*, 71. Reproduced with permission of Tokyo Metropolitan Archives.

One can see the clear rationalization that took place directly beneath Hamachō (Hamatyô in the figures) Park and the addition of a major road (as indicated by the solid dark line) running below it. The other major road bisecting this district led to Honjo via the Shin Ōhashi Bridge. Parcels of land in the lower right quadrant, on the other hand, underwent only minor readjustment.

In pursuit of rationalization and greater uniformity among land plots, residents rarely, if ever, suffered a catastrophic loss of land. An examination of the total land areas readjusted by the City of Tokyo illustrates this point. Of the 24.9 million square meters of land that fell under the city's readjustment jurisdiction, 18.7 million were designated as residential land before readjustment. The postreadjustment total of residential land stood at 15.8 million square meters: the total reduction of residential land was thus 2,861,696.89 square meters, or roughly 15 percent.[15] More detailed statistics show that virtually every landholder lost 10 percent of his or her land.[16] The government was required to pay roughly ¥12.6 million for residential land that was taken beyond the 10 percent compensation-free threshold. The readjustment district that lost the most residential land was Area 43, which lost a total of 23 percent of its residential land.

In land readjustment areas under Home Ministry control, even less residential land was taken and used for nonresidential purposes. The total area of land to be readjusted by Home Ministry bureaucrats—areas that had a higher concentration of governmental and commercial areas as opposed to homes and small shops—stood at 6.3 million square meters. Residential land comprised 4 million square meters of this total.[17] After readjustment, the amount of residential land remained 3.99 million square meters. The total amount of residential land lost was 76,359 square meters. Land taken that exceeded the 10 percent compensation-free threshold was a miniscule 263,572 square meters. However small, this land was valuable and resulted in ¥21.8 million being paid in compensation. The Home Ministry–controlled readjustment district that lost the highest percentage of residential land was Area 31; it suffered a loss of 21 percent. In short, the total area of public land increased in Tokyo by roughly 12 percent, from 7,656,884 square meters to 11,282,560 square meters out of a total land area of 31,166,265 square meters that underwent land readjustment between 1924 and 1930.

While few, if any, individuals experienced catastrophic reduction of their land, many suffered major inconveniences and upheavals during the process of readjustment. Most disruptions stemmed from the removal and relocation, or the disassembly and reassembly, of buildings and structures either to new lots or to new placements on their existing lots that were forced to undergo readjustment. Even when land was not exchanged for a different plot, virtually every building on every plot of readjusted land required some degree of relocation. Takebe Rokuzō, a high-ranking official involved in land readjustment within the Relief Bureau, confirmed that after the initial survey of land in 1923 and early 1924, he believed that roughly 70 percent of all structures within the sixty-six readjustment districts would have to be moved.[18] In reality reconstruction officials moved 99 percent of all structures, a figure that represented 203,510 buildings.[19] In many cases buildings were moved only meters; other times they were moved to nearby lots but never outside of a readjustment district.

The process of relocation was every bit as complex and taxing as the readjustment of individual property lots. Once land had been surveyed and readjustment planning was completed, groups of thirty to fifty buildings were earmarked for relocation at one time. The first relocation order issued by the Home Ministry occurred on October 23, 1924, in Readjustment Area 12. The City of Tokyo gave its first relocation orders to residents in Readjustment Area 42 in April 1925.[20] Relocation ended on November 27, 1929, when the last earmarked structures were relocated within Land Readjustment Area 20.

Initially citizens throughout Tokyo protested forced removal and relocation as much as they did the overall policy of land readjustment. Takebe recalled that as he toured neighborhoods on survey duty, he saw that people had placed numerous signs and posters expressing their outrage at possible removal. People complained not about the removal process itself, but primarily because they assumed that compensation money for removal would be too low, averaging roughly ¥27.50 per tsubo of land that a building occupied.[21] In reality, nearly 50 percent more was spent on compensation per tsubo. "Every day and night," Takebe reminisced, "Reconstruction Bureau members stood at street corners" trying to convince people to embrace removal and relocation. By 1928, he concluded, most did, in large part because compensation proved adequate to meet all expenses. Even businesses and small shop owners that were forced to close temporarily to accommodate moving were given monetary compensation based on an intricate formula. Some buildings were dragged on purpose-built giant rollers, others were disassembled and reassembled, while still others were completely demolished and constructed anew. All told, relocation expenditures constituted the lion's share of land readjustment costs. A total of ¥155,318,997 was spent on the relocation of structures.[22] Takebe estimated that over two thousand employees were engaged in relocation activities and that the total man-hours worked on this project surpassed 2 million.[23]

IMPLEMENTING THE GRAND PROJECT FROM THE WARD LEVEL UP IN KANDA, HONJO, AND NIHONBASHI

Seen from the bird's-eye perspective, land readjustment and relocation might seem straightforward and logical—almost a sterile process that residents eventually endorsed for the betterment of each neighborhood. While it is true that residents accepted the alterations to their neighborhoods, this does not mean that such changes were implemented without considerable local-level contestation. Landowners and shopkeepers from every readjustment district submitted petitions to their local readjustment committees and the Reconstruction Bureau in the hope of securing greater monetary compensation, less drastic alterations to their property, or more time to implement relocation. An examination of citizen submissions from three distinct wards paints a human picture of land readjustment that elite-level urban history does not provide.

Land Readjustment District No. 8 in Kanda Ward is a good place to begin an examination of readjustment from the ward level perspective up. Virtually the entire district had been destroyed by fire in the immediate aftermath of the Great Kantō Earthquake, but residents had returned "promptly, and renewed its appearance" with a high degree of reconstruction spirit. With 164,679 tsubo of residential land, District 8 was the sixth largest district out of the total sixty-six to be readjusted. Some 17.3 percent of private residential land was transformed into public land during the readjustment process. When the project was completed on January 20, 1929, a total of 6,062 buildings had been relocated within this district.[24] Twenty men and representatives of two businesses made up the local Land Readjustment Board: its chairman was Akikusa Ai'ichi; its vice chairman was Toriyama Shūsuke; and the two businesses with representation on the committee were Hakushin-sha Inc. (a paper merchant) and Mineshima Co. Ltd.

Readjustment District 8 in Kanda comprised many small shops and businesses, giving it a mixed residential and commercial feel. As in present-day Kanda, many new and secondhand booksellers and stationery shops dotted the area. Its reputation in 1922 as a "town of students" meant that numerous small restaurants and secondhand clothing stores often bustled with activity. When packaging the merits of land readjustment to its residents, Reconstruction Bureau officials pointed out that District 8 was blessed with many transportation networks that could, if expanded on, guarantee its commercial success for generations. Along with numerous intercity train lines, the Yamanote and Chūō lines also snaked through the district. Moreover, two major roads bisecting this district were set to be upgraded to major highways through the readjustment process—a process that would see both roads widened and connected to an extensive local road network. "The ward's traffic network," Reconstruction Bureau officials suggested, "is the center of people's hopes for the future development of the ward as a students' quarter as well as commercial center." As such, they expected people to embrace land readjustment with considerable enthusiasm. They were mistaken. Officials admitted in 1930 that residents and the members of local readjustment boards "clashed emotionally and cast a dark shadow over the entire project."[25]

Business owner Igarashi Makita was one individual who Reconstruction Board members believed would support the aims and objectives of land readjustment. Igarashi's business, a billiard hall that also sold drinks and snacks at 38-banchi, Ogawachō, Kanda, was popular with residents. On paper, Igarashi's business was set to gain under the draft land readjustment proposal. He was to be rewarded with a longer street frontage from which to advertise his business and hopefully attract more customers. Reconstruction Board members believed that this would more than make up for the 1.8 meter reduction in overall width of his property lot and thus his place of business. Igarashi thought otherwise and made his opinions known in two terse, straightforward petitions. "My hall," Igarashi wrote, contained "rows of billiard tables between which there was sufficient room for players to move about freely."[26] He argued that reducing the size of his business by 1.8 meters would cause "great damage" because two billiard tables would have to be removed so that people could move freely enough to play. Had authorities thought about such a scenario, Igarashi asked, when devising his new allotment?

Igarashi followed this question with a straightforward request: that his allotment be expanded by 1.8 meters so that he could operate as many tables as before. Alternatively, he asked the local Readjustment Board to compensate him more generously. Igarashi wanted the board to not only purchase the two billiard tables that would have to be removed but also provide compensation for future losses based on the earning potential of those two lost tables. Both requests were too much for the Local Readjustment Board or the Reconstruction Bureau to stomach. "We can accede to neither request" was the short and resolute reply. A unique corollary to this story is that the owner of the building in question was Dr. Itō Chūta, a Tokyo University professor of architecture who was a vocal proponent of land readjustment and a radical reconstruction of Tokyo.[27]

Takahashi Hisako, a female landowner who had run a very successful business at 1-banchi, Yanagichō, Kanda Ward, for more than twenty years, likewise protested when the details of her new allotment were announced in September 1925. Writing what the Local Readjustment Board described as one of the "most detailed and convincing petitions," Takahashi pointed out that the shape of her "old shop," which she had rebuilt with her own funds in 1924, was unique precisely because it was irregular and did not conform to the rectangular-shaped property that she was set to inhabit after land readjustment was completed. Her prosperity over the years, Takahashi argued, was because of a 10.8-meter protrusion of her shop frontage that incorporated what would have been the sidewalk, had one existed. Since sidewalks existed on both lots adjacent to hers, people walking down the sidewalk on which her business was located would have to walk out of their way to avoid walking through her shop. She had even built large, open entrances in line with where a sidewalk should have existed, to encourage people to enter her shop. In essence, the front of her shop was a pedestrian thoroughfare that Takahashi described as an "easy to enter, easy to buy" feature. It was invaluable to her business.

From the rational urban planner's perspective, this was exactly the type of lot that land readjustment was to rid Tokyo of forever. From Takahashi's perspective, this irregularity was her business's lifeblood. Adding insult to injury, the lot to which she was to relocate had a smaller shop frontage than the one she currently occupied. Not surprisingly, Takahashi did not give up without a long, protracted fight. Over the course of two years, she sent a number of detailed, well-written, and rational petitions to the local Land Readjustment Board. Board members would not budge on the shape of her new property: no similar protrusion would be allowed. But, they did recognize that "the substituted land was disadvantageous to the party that appealed" and therefore gave Takahashi a much higher compensation payment that took into consideration the special features of her previous property.²⁸ Through the sheer force of money, an equitable agreement was reached in 1928.

It was not just individual shopkeepers or landowners who complained about land readjustment and what it would mean for residents in Land Readjustment District 8. In 1925 the chairman of the local assembly came to the Reconstruction Bureau office on behalf of his residents with a simple request. He asked officials to allow "our neighborhood" to retain a sacred tree that resided at 1-banchi, Awajichō, Kanda Ward. According to local lore, the large ginkgo tree that measured at least two arm spans in diameter at its base was magical. As the chairman relayed to a slightly amused Reconstruction Bureau official, "any insulting or disrespecting act [against this tree trunk] would receive divine punishment."²⁹ When the bureaucrat replied that the tree must come down, the chairman warned the official that no individual in the ward would dare cut down this tree, no matter how much money was offered. The chairman knew the people he represented well. Only after repeated attempts to secure a contractor within the district failed did the Reconstruction Bureau find an outside party to remove the tree.

Even then, removal of the tree proved difficult and ultimately costly to the contractor. Upon seeing the tree in person and the faces of the frightened residents who gathered round, the contractor decided to be safe rather than sorry. He therefore enlisted the help of a Shinto priest, who informed the contractor that an elaborate and costly ceremony and subsequent festival would have to be held in order not to anger the spirit that inhabited the tree. Indeed, after the tree was successfully felled, a large-scale festival was held at the contractor's expense. It comprised copious amounts of sake, sweets, and other sacred offerings. Straw festoons were then hung around the trunk and an elaborate prayer was given so that no harm would come to anyone present or the neighborhood. A purification ritual was held and, after more sake was consumed, the trunk was taken by ox cart to Kanda Myōjin Shrine. Residents followed behind, singing a Tokugawa-era woodcutter's song. The contractor later admitted to Reconstruction Bureau officials that it was a difficult task to remove the tree from such a built-up neighborhood. Moreover, he complained that he gained no money through this job after paying more than ¥50 for the ceremony and subsequent festivities. When asked by Reconstruction Bureau officials if he wanted to lodge a formal complaint and to request more funds for his services, the contractor replied, "I'd better not complain: it is after all a sacred tree."³⁰

Land Readjustment District 52, which comprised parts of Honjo and Fukagawa Wards, is another good example to explore. Similar to District 8 in Kanda, the land of Honjo and Fukagawa had been annihilated by postearthquake fires. Unlike Kanda, however, this district's pre-earthquake character was defined by industry, both heavy and medium. This trend continued with rapidity after reconstruction work began. In fact, the city zoned this district and much of the surrounding land as industrial. This did not mean, however, that the area was devoid of residents. On the contrary, Reconstruction Bureau officials described it as an "industrial area where many residents lived in poverty." This was an understatement, betrayed in part by the large number of social welfare institutions that dotted this district. They included the Kōtōbashi milk distribution center, the Kōtōbashi childcare center, the Oshiagechō childcare center, the Hōonji playground, the Narihirabashi employment agency, the Kōtōbashi employment agency, the Honjo public pawnshop, the Oshiage public pawnshop, and many other social welfare institutions. Reconstruction Bureau officials tasked with mapping and surveying this district stopped listing all the social welfare institutions present in the district because they suggested that "there were too many to enumerate."[31]

Similar to District 8 in Kanda, District 52 in Honjo and Fukagawa was large. With 157,703 tsubo of residential land, it was the tenth largest district out of sixty-six to be readjusted. Sixteen percent of residential land was lost to "public land" during the readjustment process. A total of 5,680 buildings were relocated, and the last one settled on its new lot on November 22, 1929. Some 377 land owners and 2,039 business leaseholders elected twenty representatives to sit on the local Land Readjustment Board. Shimada Tōkichi served as its chairman, Miyamura Kameichi served as its vice chairman, and one individual represented the Nippon Ice Manufacturing Company.[32]

Even though Reconstruction Bureau officials proclaimed proudly that "there were few instances of ugly disputes over readjustment," the process still took considerable time. Readjustment began on April 17, 1924, but was not completed until November 1929. Why did so few disputes erupt? Reconstruction Bureau officials praised the work of the local committee, claiming that they had formed many small groups that worked directly with the people almost on a resident-by-resident basis.[33] A more logical and perhaps truthful explanation is that the large number of poor tenants who had no property rights to defend meant that few residential-related disputes emerged.

One dispute, however, warrants discussion in part because the Readjustment Board deemed it selfish and frivolous. From the petitioner's standpoint, the decrease in property size would cause an inconvenience, and he wanted this known to authorities. In describing his potential loss, however, he proved to be flippant and ineffective. The petitioner, Oguro Ishimatsu, lived at 41-banchi, Nishirokkenboricho, Fukagawa Ward, where he owned a fairly large building that he divided into nine parts. He lived in one section and rented the other eight subunits to tenants. This was his sole source of income. This well-to-do landlord spent his ample spare time collecting and breeding birds. The petition read:

> I am known in the neighborhood as a bird lover. Before the earthquake disaster, I bred seventeen or eighteen species of birds with much love. Since the earthquake, I developed an interest in chickens and am currently keeping seven or eight of them. But my substitution land is too small to build a sufficient pen for my chickens for which I care so much. There is no roof for them to avoid rain and there is no room for them to bask in the sun. I feel very sorry for the chickens.
>
> Before readjustment, my land was large enough not to have to make the chickens suffer such misery. But the size of land was reduced and I now can no longer enjoy raising my beloved chickens to satisfaction. I feel truly sorry for them and would like the authorities to take measures with particular sympathy so that I can take care of my chickens without putting them in misery.[34]

One wonders if Oguro cared as much for his tenants and if he had, whether a petition lodged on their behalf would have received any different response. Ultimately the Readjustment Board simply refused to respond.

A dispute more typical of this heavy industrial ward was the one that arose from a petition submitted by Kawasaki Sadakichi. Kawasaki owned a lead pipe factory at 1-banchi, Nishimotocho, Fukagawa Ward. The dispute did not involve substitution land, removal, or relocation, but rather compensation for damage his factory suffered due to reconstruction work adjacent to his factory. Kawasaki had been told that a contractor employed by the Reconstruction Bureau would carry out extensive river embankment work near the Onaki River and that this work would require 6-meter trenches to be dug near his factory. When Kawasaki questioned what impact this excavation would have on his property, the Reconstruction Bureau told him not to worry because "the work would be carried out by specialist engineers and a skilled contractor." Furthermore, they gave a "guarantee" that "his factory's operation would not be affected."[35] These were famous last words.

After digging large trenches, the contractor pounded numerous pine logs into the ground to strengthen the embankment work in preparation for laying a permanent foundation. The vibrations that resulted from this pounding caused large fissures to form throughout Kawasaki's neighborhood. To make matters worse, a freak storm deposited a sizable amount of rain, which filled many of the fissures with water. Both led to considerable ground movement and, as a result, the foundation of Kawasaki's factory subsided. With an uneven floor, Kawasaki found that the pipes his factory produced were no longer uniform or satisfactory. They were, he claimed, "worthless." He therefore requested that the Reconstruction Bureau pay ¥1,741 in compensation for the temporary closure of his factory. He also asked for ¥5,215 as reimbursement to cover the repair costs of his pipe-making machine, a lead-melting furnace, and his factory's foundation.

The Local Land Readjustment Board refused to adjudicate, as the issue in question did not relate directly to land readjustment. It did, however offer to mediate the dispute. The Reconstruction Bureau acknowledged that serious damage had been inflicted on Kawasaki's factory but initially refused to accept any responsibility. It argued that further investigation was needed. If rainwater seeping into the fissures caused the foundation to move, the Reconstruction Bureau argued, this would constitute an act of nature for which no one would be held responsible. If the pounding of pine logs into the ground ruined Kawasaki's foundation, then it was the contractor's fault and settlement money would be arranged. But, the bureau argued, the fault may well have been the poor construction of Kawasaki's foundation, for which only he would be responsible. After an extended investigation, the Reconstruction Bureau ruled that rainwater had led to the movement of Kawasaki's foundation. Therefore no money would be paid. Kawasaki challenged this verdict, and after a heated exchange between all three parties, the contractor and Kawasaki reached an undisclosed yet amicable solution.

A final district that warrants attention was Land Readjustment District 10 in Nihonbashi. Unlike in Districts 8 and 52, the Home Ministry took responsibility for land readjustment in this area. It did so in part because this was a high-profile area of commerce that also included numerous national government buildings and facilities. Though the district was heavily damaged throughout and annihilated in a number of areas by the postearthquake fires, Reconstruction Bureau officials proclaimed that "the ward's position in the commerce and industry world was too firm to rock." Home to iconic department stores Mitsukoshi and Shirokiya as well as the Tokyo Rice Exchange, the Tokyo Market for Sugar Wholesalers, the Tokyo Stock Exchange, and the Bank of Japan, to name just a few, this district "reached the height of prosperity in the Taishō era," as another official described it, and emerged as "the center of business and trade" in Tokyo. While Reconstruction Bureau officials claimed that there were too many social welfare facilities to enumerate in Honjo District 52, they declared that there were "too many scenic spots and places of historic interest to mention" in District 10. It also had some of the highest real estate valuations in Japan. Reconstruction Bureau officials used a local idiom to describe the value of land in the district: "Here, a jug-full of dirt is worth a jug-full of gold."[36]

Along with commercial and industrial clout, District 10 was also big in a geographic sense. Its 185,601 tsubo of residential land made it the third largest readjustment district in Tokyo. Some 16.6 percent of its private land became public land during the readjustment process, and 6,730 buildings were relocated within this district by November 13, 1929. This district was also home to many well-established community associations that reflected and enhanced civic pride: the Army Reservists Association had 5,000 branch members, the Youth Association had 6,600 members, and more than 146 neighborhood associations from this area had registered with the Home Ministry. A total of 624 landowners and 4,994 business leaseholders selected 24 members to serve on the Local Land Readjustment Committee: Kikuchi Chōshirō served as chairman; Okamoto Hiroshi served as vice chairman; and one individual represented the Star Metal Company.

Given the high valuation of land, the strong sense of community identity that existed within this district, and its overall size, Reconstruction Bureau officials anticipated many significant disputes and in fact warned the Local Land Readjustment Board of just such a scenario. This prediction proved prescient. The Tōyō Car Company, located at 4-banchi, 2-chōme, Yonezawachō, Nihonbashi Ward, found itself in a remarkably ironic position. This passenger transport (taxi), car rental, and service station business was not asked to move to substitute land. Moreover, it was not set to lose any significant amount of land, as had occurred with virtually every other landowner in Tokyo. Most individuals and businesses that lost roughly 16 percent of their land for the public good would have envied a similar fate. Not so for the owners of Tōyō. The road that their business fronted was set to narrow in size by 3 meters: the width would decrease from 11 meters to 8 meters. Such a development, Okada Ryū and Sekimoto Usaburō, Tōyō's managing directors proclaimed, would devastate their business.

In a well-written petition submitted to the Local Land Readjustment Board, both company leaders articulated their concerns effectively. One point both men made was that far fewer vehicles would travel down a road that was going to narrow, particularly since a parallel road would be 11 meters in width. "For a company that [in part] sells petrol," their petition read, "reduced traffic" could gravely weaken business. They also expressed concern about the effects of their business on their neighbors. The Reconstruction Bureau had earmarked the park directly opposite Tōyō's business to become a sizable residential area after readjustment was completed. "If our company's large vehicles come in and go out, both day and night," the petition read, "it will cause residents a serious inconvenience."[37] The Local Land Readjustment Board agreed and decided to relocate Tōyō to a site that fronted a major thoroughfare. Everyone was happy with the outcome.

A woman who was ordered to move her teashop to 38-banchi, Hisamatsuchō, Nihonbashi, was not nearly as successful. The unnamed proprietress simply did not wish to move to this location, writing, "No matter how hard I try, I don't feel like building a house there. If you can, please give me any different plot of land." When asked why she had such reservations by a curious Reconstruction Bureau official, she declared that she felt the site, formerly the location of the Meijiza Theater, was haunted. Her foreboding focused on the theater's large trap door that existed under where the stage once stood. Even though officials planned to fill this cellar with earth and remove the trap door, she simply didn't want to build a house over this site "no matter what." The woman even "burst into tears" in front of Reconstruction Bureau officials and pleaded not to be moved. Neither her tears nor her petition had any impact. The local Land Readjustment Board recommended that she call a Shinto priest to "purify the location." The woman agreed, and a site visit at the end of the readjustment process reported that the woman was happy with the new location and that no strange occurrences had hampered her business. As one Reconstruction Bureau official noted, "after the disaster . . . it was a wonderful time for Shinto priests to gain quick riches."[38]

These microlevel studies of reconstruction tell us much about postdisaster Tokyo. First, they highlight something that the grand planners who debated reconstruction plans in autumn of 1923 rarely acknowledged or perhaps even considered. Reconstruction was ultimately a very local phenomenon that influenced Tokyo's inhabitants on a plot-by-plot basis. It was also a process that residents took with extraordinary seriousness, sometimes as if their future prosperity rested on how the city was reconstructed. In many cases it did. Moreover, it shows how people who owned land or a business possessed the willingness to voice their concerns, and in many cases the desire to shape the built environment of their neighborhoods through an intricate, bureaucratic process. It is wrong to suggest that activism evaporated over the course of reconstruction; people submitted petitions and raised concerns up to the end of the readjustment process. While many landowners may have harbored misgivings or outright resentment against losing—on average—15 percent of their private land for public good, it did not cause catastrophic damage to anyone, and residents accepted the finality of land readjustment. But did this sacrifice make Tokyo more livable, more modern, or more disaster resistant? Put another way, what precisely did the government do with the new public land that it secured from landowners during the process of readjustment?

SUCCESSES IN CONSTRUCTING NEW TOKYO: TRANSPORTATION INFRASTRUCTURE

One of the most notable features of "new Tokyo" to emerge as a result of the reconstruction program was an extensive, modern transportation network that focused on roads. City officials and residents had complained about Tokyo's roads for years prior to the Great Kantō Earthquake with little avail. The criticisms were straightforward: Tokyo's roads were too few, and those that existed were too narrow. Sometimes unpaved, particularly in the poorest areas of the city, many roads lacked sidewalks and thus created headaches for pedestrians as well as motorists or those who used horse carts, pushcarts, rickshaws, or bicycles. Finally, the overall network was often described as serpentine in layout and difficult to navigate. These were valid observations. After an in-depth examination in 1920, municipal officials concluded that nearly 15 percent of Tokyo's roads required paving. The city created a seven-year plan set to begin in 1921 to pave Tokyo's unpaved streets at a cost of ¥40 million.[39] The problem of Tokyo's unpaved streets was so visible in 1920 that the imperial family contributed ¥3 million to initiate this project and to guarantee that politicians would follow suit and support it through to completion. While the project began in 1922, it was interrupted by the 1923 calamity, leaving over 80 percent of the road surfaces earmarked for paving unfinished.

Constructing roads and an integrated transportation infrastructure network, however, became a central focus of the reconstruction project. The Home Ministry took responsibility for constructing and widening all roads that, when finished, would measure 22 meters or more in width. These were designated as major trunk roads and were designed, wherever possible, to follow the existing road network.[40] To facilitate their expansion, government officials earmarked these roads to be widened through the process of land readjustment. The total length of these roads measured 115,816 meters and comprised 3,261,927 square meters. The City of Tokyo, on the other hand, took responsibility for constructing what were classified as "auxiliary" roads, namely, all roads that when completed would be less than 22 meters in width. The length of these roads came to 124,159 meters, comprising a total of 1,747,147 square meters.

Many roads also became noticeably wider. The total area of roads in Tokyo increased from 9,283,867 square meters on the eve of the Great Kantō Earthquake to 13,551,923 square meters in 1930. That equated to a 45 percent increase over the 1923 figure. Roads, however, did not get much longer. The length of roads increased by only 136,207 meters, from 1,011,350 in 1923 to 1,147,557 in 1930, a 13 percent increase. Increases were not uniform throughout Tokyo. Rather, almost all expansion occurred within the devastated parts of eastern Tokyo, where the land readjustment process had taken place. In areas where land readjustment occurred, the length of roads increased from 606,100 meters in 1923 to 729,029 meters in 1930, a 20 percent increase. In areas where it did not take place, the length of road increased by only 3.3 percent.

A more striking figure surrounds the expansion of the total area of roads in the districts that underwent land readjustment. The total road area in such districts increased from 5,873,259 square meters in 1923 to 9,992,058 square meters in 1930. This represented an increase of just over 70 percent. In nonreadjusted land areas of Tokyo, the increase was only 149,266 square meters, from 3,410,608 square meters in 1923 to 3,559,874 square meters in 1930 (4.4 percent). One final statistic worth quoting surrounds the percentage of road space in relation to the overall built environment of Tokyo. Roads accounted for 11.6 percent of the capital's built environment in 1923 and just over 18 percent in 1930.

FIGURE 8.5 The wide and impressive Shōwa Dōri after reconstruction

Source: Postcard, author's private collection.

What did this increase in the road network mean for the citizens of Tokyo? Beyond doubt, it led to greater, more efficient mobility. Though the population of Tokyo had not yet returned to its pre-earthquake level by 1930, there was a virtual explosion in the number of passenger trips on all forms of public transportation within the capital. The total number of trips on public transport taken in greater Tokyo rose from roughly 700 million in 1922 to 1.06 billion in 1928.[41] The new road network also led to changes in the way people moved within Tokyo. In 1922 there were roughly 480 million trips taken on public trams, while buses accounted for approximately 20 million trips. By 1928 the number of tram trips had decreased to approximately 450 million, while bus trips had risen to 60 million.

The bustling nature of Tokyo's transportation infrastructure also meant that there were more traffic related accidents, injuries, and deaths in the capital. In 1922, 10,246 accidents involving vehicles of some type were reported to the police.[42] A total of 6,405 injuries were also reported, and 122 fatalities occurred. In 1928, 25,946 accidents were reported to the police. Injuries totaled 14,466, and fatalities jumped to 229. In short, more people were on the move in new, faster, modern ways than before, and this likewise translated into an increased number of vehicle-related accidents.

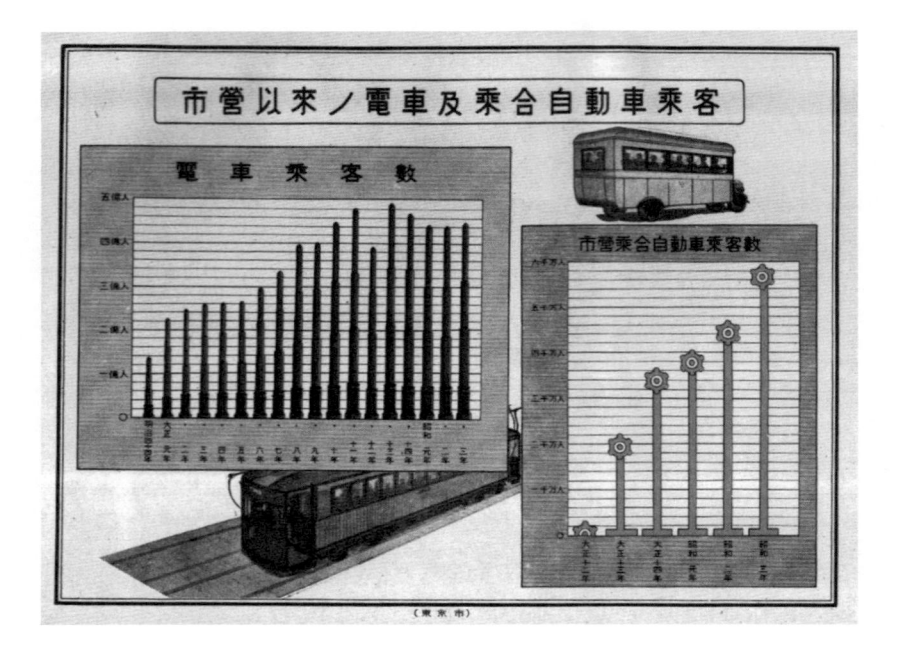

FIGURE 8.6 Chart indicating the number of bus and tram trips in Tokyo

Source: Nihon Tōkei Fukyūkai,*Teito fukkō jigyō taikan*, 1:8. Reproduced with permission of Tokyo Metropolitan Archives.

Bridges likewise assisted the movement of ever-increasing numbers of Tokyoites following reconstruction. The Home Ministry oversaw the construction of 96 bridges connected to large trunk roads at a cost of ¥32,272,000.[43] Fifteen further bridges were readjusted as part of Tokyo's canal improvement plan at a cost of just over ¥2 million. The City of Tokyo constructed 129 bridges connected to auxiliary roads and upgraded 57 existing bridges as part of the road-paving scheme associated with land readjustment. These projects cost the city ¥14,522,746. While adding to the functionality of Tokyo's transport system, 6 of the key bridges constructed by the Home Ministry that spanned the Sumida River served as visual icons of modernity. Five remain in place today with the same structural shape with which they were constructed between 1926 and 1931: Eitai-bashi (1926); Kiyosu-bashi (1928); Kuramae-bashi (1927); Komagata-bashi (1927); and Kototoi-bashi (1931); while the Aioi-bashi had its original design changed. The Kiyosu Bridge was designated as one of the sights of new Tokyo featured in an official 1930 postcard set.

FIGURE 8.7 Kiyosu Bridge spanning the Sumida River

Source: Postcard, author's private collection.

A final piece of the transportation infrastructure that was improved markedly during Tokyo's reconstruction was its river and canal system. On the eve of calamity, sixty-seven rivers and canals were used daily by traders to transport goods throughout the city, but particularly in the southern and eastern wards that experienced considerable damage in 1923. Tokyo's water-based infrastructure network measured 86.4 kilometers in length and comprised a total area of 4,297,520 square meters. As with the capital's road network, traders had complained for years about the sorry state of Tokyo's water transport system. At low tide boatmen found it difficult to use the canals because many were often too shallow to navigate. At high tide transportation proved just as difficult because many bridges had been constructed too low to permit underneath passage. Both afflictions were cured through the postearthquake reconstruction project.

As with virtually every other transportation-related reconstruction project, the national government and the City of Tokyo divided the canal and river improvement undertaking. The Home Ministry assumed responsibility for the improvement of eleven canals that measured 14,590 meters in length and had an average width of 40 meters and depth of 1.8 meters. Along with widening and deepening canals, the national government also strengthened embankments throughout Tokyo. In total, the Home Ministry spent just over ¥28 million on canal and river transportation improvement. The municipal government of Tokyo, on the other hand, spent just over ¥3 million on smaller canal projects and work on the Arakawa drainage canal.

Roads, bridges, and canals, all expanded through the process of land readjustment and were undeniably the most successful projects completed under the reconstruction program.[44] Not surprisingly, these projects received the largest amount of reconstruction appropriations from both the national government and the city. In 1930 the City of Tokyo and the national government published their final budgets associated with the reconstruction project to coincide with the celebration of new Tokyo in March of that year. Including ¥194,124,374 in loans and subsidies to the municipal government to cover city reconstruction projects, the national government appropriated ¥545,362,233.[45] The subsidies to Tokyo amounted to 35.6 percent of the overall national budget. Spending on land readjustment, streets, bridges, canals, and river embankments totaled ¥295,087,465. This figure amounted to 54.1 percent of the overall budget.

Tokyo government's spending mirrored these percentages. The municipal government spent ¥350,287,214 on reconstruction. Appropriations for land readjustment, streets, bridges, and canals equaled ¥188,669,259, or 53.8 percent of overall spending. If one includes the ¥40,939,096 spent on reconstructed primary schools and the ¥40,211,321 spent on sewage treatment facilities—often considered the two other successes of reconstruction—the total spent on all of these projects equated to 77 percent of the city's total reconstruction budget. This emphasis on transportation infrastructure, sanitation, and early childhood education left only a small amount of money available for two items that many residents and social bureaucrats believed were needed dearly in new Tokyo: parks and social welfare facilities.

IMPROVEMENTS, BUT STILL LEFT WANTING: PARKS AND SOCIAL WELFARE FACILITIES IN NEW TOKYO

Tokyo was anything but a "green city" in the years before the Great Kantō Earthquake. In 1922 just 1.7 per cent of Tokyo's urban space was classified

FIGURE 8.8 Sumida Park in 1930

Source: Postcard, author's private collection.

as parkland. Not only was this miniscule in real terms, but also in a comparative context. Parks comprised 20 percent of the urban space of Paris, 14 percent of the urban space of Washington, D.C., and 9 percent of the total environment of London.[46] The Home Ministry thus designed and created three large parks in the capital: Sumida Park in Asakusa and Honjo Wards; Hamachō Park in Nihonbashi Ward; and Kinshi Park in Honjo Ward. When completed, these parks added nearly 290,000 square meters of park space to Tokyo. More than just open spaces, however, these parks included a variety of sports and exercise facilities, such as tennis courts, playgrounds, sand pits, baseball diamonds, and even a swimming pool at Hamachō Park. But three large parks, however well positioned and equipped, could not by themselves transform Tokyo into a green city or even one that had as much park space as London.

Municipal authorities therefore undertook a more extensive park-building program after the 1923 disaster. City planners created ninety-nine small parks in Tokyo with more than half attached to primary schools. Even with the addition of these new parks, however, Tokyo remained bereft of green space. At the completion of the reconstruction in 1930, parks constituted just 3.7 percent of Tokyo's overall urban space. Within a scaled-back reconstruction project, parks had fewer bureaucratic champions than schools, bridges, or new and widened roads, which many elites viewed as economic and social necessities. Moreover, an expansion of parkland would have required further land confiscation or purchase, thus making parks expensive and contentious.

Similar factors hindered a grandiose expansion in the city's social welfare infrastructure. While a sizable cadre of bureaucratic elites advocated a complete transformation of the social welfare system, budgetary factors tempered these ambitions. Out of the total ¥350,287,214 spent by the city on reconstruction, social welfare works received only ¥4,525,000: this equated to 1.3 percent of the city's overall reconstruction budget and only 0.5 percent of the total spent on reconstruction if all national, prefectural, and municipal expenditures are added together.[47]

Though the amount spent was small in comparison to other reconstruction expenditures, city officials appropriated the money wisely. The municipal government constructed most of the new structures in Tokyo's poorest and most disadvantaged areas where they would have the greatest impact. Officials added a total of sixty-six social welfare facilities to the city's built environment with the vast majority built in the wards of Fukagawa, Honjo, Asakusa, and Kyōbashi, and on the northwestern boundary of Koishikawa Ward.

Pawnshops became an important component of the urban safety net that city officials constructed in the aftermath of the Great Kantō Earthquake. The 1923 disaster caused considerable disruption to finances across Tokyo, and officials desired to create a formal and well-regulated system of financial organizations for the capital's poorest residents. Pawnshops were viewed as more than just banks for Tokyo's poor. Rather, they were meant to keep many of Tokyo's poorest workers from borrowing money from private lenders at extremely inflated interest rates. The city established pawnshops to provide financial services and educational classes on such topics as finance, interest, savings, debt, and insurance. By 1930 the city had constructed seven municipal pawnshops at a cost of ¥152,250. Supplementing the city's two pawnshops that existed in 1923, the financial institutions were located in Asakusa (three), Fukagawa (two), Honjo (two), and one each in Kyōbashi and Koishikawa. Upon completion, these institutions replaced the eleven temporary pawnshops created in the immediate aftermath of the earthquake through a grant from the Home Ministry.

Rather than just assisting the capital's poorest residents to better manage their money, the City of Tokyo also built facilities that would help

FIGURE 8.9 Interior of a public dining hall in reconstructed Tokyo

Source: Tokyo Municipal Office, *The Reconstruction of Tokyo*, 319. Reproduced with permission of Tokyo Metropolitan Archives.

people find jobs. Over the course of the reconstruction project the municipal government created fourteen permanent employment matching agencies and located them in Koishikawa (four), Honjo (three), Asakusa (two), Shiba (two), and one each in Yotsuya, Kōjimachi, and Shitaya. These agencies supplemented the three that survived the earthquake. When operational, these fourteen new institutions became more than just centers where workers looked for jobs and employers shopped for cheap laborers. City officials provided job seekers and their families with an array of services including wage advances and traveling expenses for individuals willing to move to secure employment. Employment-matching agencies also provided training in basic vocational skills for select professions and rudimentary financial services for workers and their families that supplemented those offered by municipal pawnshops.

Public dining halls, cheap lodging houses, and public baths were three further types of social welfare facilities that the municipal government created to cater to the needs of the city's poorest inhabitants.[48] The city constructed eleven new dining halls that supplemented both public cafeterias that existed before the disaster. Municipal authorities spent

¥463,510 on these facilities that provided free or subsidized meals seven days a week. City officials placed eight of the thirteen public dining halls in the wards of Fukagawa, Honjo, and Asakusa. The city also constructed ten cheap lodging houses for male workers of scant means and placed them near public cafeterias and employment-matching agencies. Providing beds in shifts that corresponded to factory working hours, these facilities afforded lodgers austere, dormitory-like accommodation. Though spartan, they were undeniably more comfortable and conducive to rest than sleeping in public places. In addition to meals and shelter provided by these two types of facilities, the city also constructed and managed three public baths, two of which were located in Fukagawa and one in Kyōbashi.

The City of Tokyo also created social welfare initiatives geared toward women and children. Municipal authorities constructed ten childcare facilities during the reconstruction project. When opened, these crèches enabled mothers to place children in state-sponsored care facilities during the hours of their employment. Officials at municipal crèches did more than just supervise playtime activities. Rather, they used these institutions to monitor the physical and mental health and well-being of children from the poorest parts of the city. The city operated four day-care facilities in Fukagawa, and one each in Honjo, Asakusa, Shitaya, Koishikawa, Kyōbashi, and Yotsuya. For unemployed women seeking to learn vocational skills, the city opened five women's workhouses. The facilities were created so that women from poorer families could learn to sew, knit, and produce other handicrafts that could be sold to supplement their family's income. On average, two hundred women per day attended these institutions in 1930.[49]

What can one make of the city's expansion of social welfare facilities and programs following the 1923 calamity? From Nagai Tōru or Abe Isoo's perspective, the city's undertakings were not extensive enough to ameliorate the new and accumulated social ills that emerged with Tokyo's modern development. When compared to what existed prior to 1923, however, it is clear that the city made important progress. The structures themselves—all assembled with reinforced concrete and built to reflect architectural modernity—were constructed with permanence in mind so that they could serve as a foundation on which to expand in the future. Could more have been done? Certainly. Could it have been accomplished at a cost the city could have realistically afforded? Probably, if social management had been the city's chief priority. It was not. A greater expansion of social welfare facilities, as with an expansion of all other types of municipal infrastructure, moreover, would have required more money with

FIGURE 8.10 Children's day-care facility in Ryūsenji, Shitaya Ward

Source: Tokyo Municipal Office, *The Reconstruction of Tokyo*, 319. Reproduced with permission of Tokyo Metropolitan Archives.

which to purchase land. Securing appropriations as well as land would have resulted in greater contestation from elites with competing visions and resistance from residents, many of whom desired a quick return to a pre-1923 state of normalcy. Municipal officials and national political elites who desired more facilities comforted themselves with the belief that the conclusion to Tokyo's reconstruction in 1930 was the beginning of a new era of social management, not its culmination. Before that future arrived, however, city officials wanted to make certain that no one would forget the tragedy of 1923. To facilitate this end elites from both the city and the national government endeavored to create a memorial unlike any built before in Japan. As with many aspects of Tokyo's reconstruction, the design and construction of this hall was soon enmeshed in controversy.

THE EARTHQUAKE MEMORIAL HALL CONTROVERSY

Throughout the autumn of 1923, the site of the former Honjo Clothing Depot became a place of reflection, mourning, and solemn remembrance.

Virtually every politician, surviving relative, or curious onlooker who ventured to this location realized that it was too sacred for anything but a monument. Though the City of Tokyo devised initial plans to transform this location into a park, Mayor Nagata rejected the idea less than one month after the earthquake. He decided, with the support of national elites, that a monument must be constructed at the hallowed site. By mid-October a collection of political and religious elites discussed precisely what the memorial should entail. While they reached no definitive conclusions during 1923 or the first half of 1924—as other reconstruction issues took precedence—a general consensus emerged that the site must contain an impressive memorial hall, a park that could accommodate large gatherings of people, and a charnel house for the remains of the cremated who once filled the locale.

When the pressing and immediate issues of relief and recovery, the reconstruction budget, and the plan for urban land readjustment had been decided upon, officials took concrete steps to create a fitting earthquake memorial. In June 1924 the municipal government created a public-private partnership institution known as the Taishō Earthquake Disaster Memorial Project Association (Taishō shinsai kinen jigyō kyōkai) (hereafter referred to as the Memorial Association) to oversee the project. The association included national political elites such as Gotō Shinpei, local leaders such as Nagata Hidejirō, and private, albeit well-connected, citizens such as Shibusawa Eichi. Claiming that it was the "duty of those who survived the disaster to console the spirits of the dead," the society's prospectus advocated developing a solemn memorial hall and sacred repository for the remains of the dead.[50] In June 1924 the leaders of this organization stated that the memorial must not just encourage people to reflect on the past calamity but must also educate future generations so that no similar catastrophe would ever again befall the residents of Tokyo.

As with virtually every aspect of reconstruction, the issue of money soon dominated construction discussions. Memorial Association members debated among themselves how much money would be needed to build a fitting memorial. Moreover, they questioned from what sources money would be made available. As to the former question, the members estimated that roughly ¥1 million would be needed to clear the site and to construct the memorial. In relation to the latter question, Tokyo municipal officials agreed to provide funds to cover half of the construction costs. Memorial Association members decided that the other half of construction funds would have to be raised privately through donations and subscriptions.⁵¹ In this endeavor, they hoped to enlist the support of Buddhist associations throughout Japan to raise roughly ¥500,000.

Memorial Association members concluded that raising this sort of money—on top of the money raised previously for relief—would require a concerted, well-choreographed publicity offensive. To foster excitement and to launch the national campaign drive, Memorial Association leaders took an imaginative course of action. In December 1924 they opened the design of the earthquake memorial complex to a national competition. Advertised in newspapers, academic journals, and flyers sent to architectural firms across Japan, the association asked interested architects, engineers, and designers to create a unique memorial that in the first instance paid tribute to the victims. The overall design, moreover, needed to take into consideration a number of other points. First, numerous religious ceremonies would be held on the site each year, and thus the judges declared that the building should present a reflective, dignified presence. Second, they articulated that the building must be built to the highest earthquake and fireproof standards available at the time of construction. Finally, they decreed that all building materials must, unless absolute necessity dictated otherwise, originate from Japan.

The contest fostered extensive interest within Japan. When the deadline of February 28, 1925, was reached, 220 applicants had submitted entries. While the ¥3,000 yen first prize was significant, the status and acclaim secured by being selected as the designer of the earthquake memorial was immeasurable. For two weeks, seven judges—politician Okada Tadahiko; engineers Itō Chūta, Tsukamoto Yasushi, Sano Toshikata, and Satō Kōichi; Tokyo Art School principal Masaki Naohiko and Tokyo Municipal Park Section director Inoshita Kiyoshi—deliberated over the entries.⁵² How the judges voted individually or what influenced their decision is not known, but on March 14 they selected engineer-cum-architect Maeda Kenjirō's entry and awarded it first prize. Maeda was an engineer for Daiichi Bank who had an illustrious history of winning competitions. One newspaper reported that he had won similar competitions to design select railway stations for the South Manchurian Railway Corporation and the Kobe Public Hall.⁵³ As part of the publicity-oriented undertaking, the Memorial Association displayed Maeda's winning entry at the Tokyo Municipal Hall in Ueno Park from March 21 to 23, 1925.

As can be seen from the picture of Maeda's winning design (fig. 8.11), the central feature of his entry was a 53.33 meter tower with a diameter of 8.79 meters. Made of reinforced concrete, the tower was an impressive

FIGURE 8.11 Maeda's winning entry for the earthquake memorial hall

Source: Tokyo shinsai kinen jigyō kyōkai hen, *Taishō daishinsai kinen kenzōbutsu kyōgi sekkei zushū*, 2. Reproduced with permission of Tokyo Metropolitan Archives.

design. It included ten stories and a basement. The total building area comprised 1,093 square meters. The basement housed a large, sacred chamber where the remains of victims would be entombed. Directly above the charnel house and in the middle of the large tower, Maeda placed a sacred white marble pillar that would represent the spirits of the dead. People could, the architect suggested, pray to the central pillar from around the radius of the memorial monument. Above the pillar at the top of the 53 meter tower, a flame would burn for the souls of the dead. The tower would be made up of circular staircases so that nothing would pass over the central pillar.

Maeda suggested that numerous features of his design would draw people to the central sacred pillar and create a solemn atmosphere of reflection. First, eleven black marble pillars would surround the central white marble pillar. The round roof above the main part of the tower would be composed of opulent stained glass. Colored rays of "electric light," Maeda wrote, would illuminate the main hall and create a mysterious, contemplative mood. Two semicircular rooms positioned at ten o'clock and two o'clock from the main hall would be designated as a worship room and an exhibition room, respectively. The worship room was designed with two floors; one for general visitors and the other for priests. The two-story exhibition room would be used to display artifacts taken from the disaster area as well as art displays, survivor accounts, and photographs taken of devastated and reconstructed Tokyo. Apart from these two rooms, Maeda designed other, separate buildings that would house an office and lounge for priests, a public park, a park caretaker's office, a bell tower built of reinforced concrete to house a bell donated by Buddhist associations from China, and a public toilet.

When the design was displayed in public, it took only days for cries of protest to filter back to the Memorial Association judges. The *Japan Weekly Chronicle* reported that the judges had been confronted with claims that Maeda's design was a "reproduction" of a Prussian triumphal tower. When presented with images purported to be of the German original, the judges began a preliminary investigation into the matter. At the end of April 1925, they declared that "similarity in design" was "insignificant and unworthy of further attention."[54] Neither this decision nor the passage of time, however, made Maeda's design any more palatable to those opposed to its appearance.

Disquiet over the memorial design surfaced regularly over the following year and intensified in conjunction with religious ceremonies held at the site. Criticism became strongest following a vernal equinox ceremony held in March 1926.[55] All thirteen Japanese Buddhist sects united in opposition to Maeda's design.[56] The Japan Buddhist Federation (Nihon bukkyō rengōkai), located at 36 Shinbori-chō, Shiba Ward, Tokyo, took the lead role in fomenting dissent against the planned memorial. It likewise organized a sophisticated public relations campaign against the memorial's design.[57]

Writing on behalf of "all Buddhists," the federation outlined its reasons for opposing Maeda's design in numerous petitions submitted to the Memorial Association selection committee and to the government of Tokyo. Suggesting that the memorial hall would become one of the most sacred sites in Japan, the federation stated that "the building must conform to people's religious values." The selected design, it argued, did not. Moreover, the federation described the structure as offensive in that it was "a mere imitation of Western architecture." It failed to take into account people's beliefs or those of the majority of victims." The design, the Buddhist Federation concluded, must be changed and a new design that "adopted [Japan's] unique national spiritual culture" be built in its place. They declared that there was no room for compromise.[58]

Buddhist monks from all of Japan's thirteen recognized sects did more than endorse petitions written by the Japan Buddhist Federation. They also took to the streets to rally people and local politicians to their cause. Buddhist priest Shudō Kambayashi encouraged monks to gather signatures and compile petitions against the design of the memorial hall. Buddhists also used the service held for the three-year anniversary of the disaster to secure even greater support for an alteration of the proposed memorial design. Buddhists had created so much disquiet in Honjo Ward that Sugino Zensaku, chairman of the Honjo Ward Council, sent a petition to the memorial hall judges and pleaded with them to revisit their decision and to change the hall's design. He echoed his residents, who argued that the design was nothing more that a "Western-inspired tall tower" that held little outward significance to the people of his ward. "The design of the memorial hall," he concluded, must "reflect what the majority of people want," and few apart from the stubborn judges supported Maeda's design.[59]

Both Sugino and the leaders of various Buddhist organizations made one final point that resonated with the Memorial Association judges: fund raising. Takizawa Sondo, secretary of the Japan Buddhist Federation, told the editors of the foreign newspaper *Trans-Pacific* that his organization received more than 8,000 postcards or letters a day in the spring of 1926 from all parts of the country that critiqued Maeda's design.[60] While this was likely an exaggerated figure, he used it to make a clear point: it would be virtually impossible for Buddhists to raise the necessary ¥500,000 to complete the project when so many people opposed the design.[61] In December 1926 the judges agreed that the design would have to be changed so that the final memorial hall would respect the wishes of the majority of the population. It is not known whether the judges required Maeda to return the prize money, but the notoriety seems to have done little to damage his career. He continued to design buildings and gained fame for such icons as the Shiseido Cosmetics Building in 1928, the Kyoto Municipal Museum of Art in 1933, and Shiseido's famous Ice Cream Parlor, which operated from 1935 to 1940.

Rather than introduce another competition or select the runner-up at such a late date, the judges decided to appoint an architect to "redesign" the memorial hall. Engineer Itō Chūta assumed responsibility for creating the new design and submitted his draft proposal on November 7, 1927. Itō's new hall was entirely "Japanese" in outward appearance. Moreover, its design was influenced heavily by Buddhist architecture. Itō boasted that when completed, the memorial complex would house the country's fourth tallest pagoda.[62] Moreover, he stressed that the exterior was simple yet evocative, traditional yet unique, and a building that would compel visitors to reflect on the great tragedy that had struck the nation.

The building housed a worship hall, a ceremony hall, and anterooms that comprised roughly 984 square meters in total floor space. The pagoda held a charnel house and various other small antechambers for religious

FIGURE 8.12 The completed earthquake memorial hall in 1930

Source: Postcard, author's private collection.

services and affairs. As with Maeda's plan, the final design included a bell tower, an exhibition hall, an attendant's residence and a public toilet. The cornerstone laying ceremony was held on November 27, 1927, and over the course of three years the Toda Construction Company completed Japan's earthquake memorial. At the foundation-laying ceremony, Home Minister Suzuki Kisaburō urged a speedy completion of the project. He also suggested that when completed, the site would guarantee that no individual would ever forget the tragedy that people suffered in 1923. His words were prophetic. The earthquake memorial hall remains today, one of the few buildings in eastern Tokyo to survive the man-made firestorms of March 9–10, 1945, intact. Though this was one of eastern Tokyo's largest open spaces, residents refused to seek shelter there in 1945. The people of eastern Tokyo were all too aware of the catastrophe that had visited this site just twenty-two years earlier and believed that it would again be incinerated on that fateful night in the closing months of the Second World War. It survived intact, though its appearance was not as grand then as it stood on the day of its completion on September 1, 1930. Many of the statues, memorial plaques, and metal fittings had been seized by government officials and melted into scrap iron to support Japan's futile war effort.

The story of the earthquake memorial hall is a fitting microcosm of the ambitious hopes and contested realities associated with Tokyo's reconstruction. Though many elites and municipal officials believed that the design and construction of a memorial hall would unite the nation and be devoid of the local and elite-level contestation that afflicted so many other aspects of reconstruction, it was not. The novel idea of holding a national competition for the memorial hall provoked the ire of local residents, who wanted a greater say in how the memorial would be designed and constructed. They were joined in opposition by Buddhist associations, who never felt entirely appreciated for the work they had conducted after the calamity. However national the memorial and reconstruction process had been packaged by elites, it was a very real and local phenomenon for countless residents of eastern Tokyo, every bit as real as the calamity itself had been years earlier.

Nine

CONCLUSION

It is like a dream to look back on what happened at the time [of the earthquake calamity] and afterwards, seeing you at such an occasion today. In those days, it was too embarrassing to stand in public wearing silken attire. To quench our thirst, we had to eat pears. If we remember the situation of those days and if we are prepared to endure like we did, we can accomplish anything.
—Nagata Hidejirō, 1930

Compared to urban reconstructions that have taken place anywhere in the world, our project is unparalleled. . . . Comparing the spirit of Tokyoites today with that at the time of the earthquake disaster, however, how much has it lifted? The enhancement of the national spirit is something we must promote continuously.
—Horikiri Zenjirō, 1930

On March 24, 1930, more than a million Tokyoites participated in the opening act of what became a weeklong series of events held to celebrate Tokyo's rebirth. Beginning just after sunrise, throngs of people gathered near the imperial palace hoping to position themselves for a glimpse of the Shōwa Emperor as he embarked on a well-choreographed inspection tour of new Tokyo. Still more well-wishers packed the 35-kilometer tour route to wave flags and shout "banzai" as the emperor's maroon motorcade drove past. Wherever they could, people congregated at more than half a dozen locations where the imperial entourage stopped to inspect a site classified as a significant reconstruction accomplishment. Tokyo and its people were proud to celebrate and share what they had achieved since the bleak days of September 1923.

The route that the emperor took and the sites selected for inspection provide many clues as to what municipal authorities believed were the meritorious reconstruction successes.[1] The emperor's motorcade traversed the Sumida River over four of the six new bridges—Kototoi-bashi,

FIGURE 9.1 The emperor's motorcade as it left the earthquake memorial hall
Source: Postcard, University of Melbourne Japanese collection.

Kuramae-bashi, Kiyosu-bashi, and Eitai-bashi—heralded as great engineering achievements and exemplars of architectural modernity. The avenues selected for the motorcade included the widest, most magisterial new streets that Tokyo could offer any resident or visitor, including Shōwa dōri, Yasukuni dōri, Asakusa dōri, Kiyosumi dōri, Eitai dōri, and Hibiya dōri. The places where the emperor stopped were no less significant symbols of new Tokyo. He inspected Sumida Park, the largest new park constructed in the capital, and drove by three others including the Hamachō Park in Nihonbashi Ward and two parks in Fukagawa Ward. He also visited a modern, reinforced concrete hospital constructed in Tsukiji.

Three locations visited by the emperor, however, stood out in significance. To coincide with the moment the earthquake struck Tokyo, he made a personal inspection of the unfinished Taishō Earthquake Memorial Hall in Honjo Ward. After examining the building, charnel house, and grounds of the project site, he offered condolences to the spirits of the dead in a solemn, private ceremony. From death and remembrance he next turned to life and the future of Tokyo by visiting the newly constructed Chiyoda Primary School in Nihonbashi Ward. During his daylong tour, the emperor spent more time at the state-of-the-art school than at any other location. He met with teachers and students and spent nearly an hour inspecting the three-story, reinforced concrete symbol of modernity. He took great interest in the exterior of the building, its classrooms, its public facilities, and a number of visual and material displays placed in its spacious auditorium.

Perhaps the most memorable sight of the day, however, was the view afforded the emperor from the bluff at Ueno Park directly beneath the statue of Saigō Takamori. From this vantage point, the emperor was given an unobstructed view of the reconstructed city. The scene before him in 1930 stood in stark contrast to the ruined city he saw from this very same location on September 15, 1923, when, as crown prince, he toured the city on horseback guarded by armed military personnel. Looking at the devastated city from Ueno Park just days following the calamity, he might have shared journalist Karl Padek's opinion that the disaster might "remove Japan from the list of great powers for a decade."[2] Tokyo of 1923 looked as if it had been struck off the face of the planet by forces almost beyond comprehension. Given the contrast presented in 1930, it was clear that Tokyo had not just recovered but in many places looked modern, once again bustling, and, in a few places, even awe-inspiring.

The roughly ¥744 million that the City of Tokyo and the national government spent on reconstruction by 1930 had done much to improve the city's infrastructure and its outward appearance. Roads in most places of the previously destroyed sections of the city were wider, straighter, and paved. Virtually all had been reconstructed with modern sidewalks. Land plots in much of eastern and central Tokyo had been rationalized, though by no means standardized. Many new government buildings, from schools to ministry offices, emerged from reconstruction as impressive, and in many cases imposing, symbols of authority, modernity, and seeming permanence. The city also gained modern water and sewer systems that, while virtually invisible and thus unheralded, were nevertheless significant accomplishments that benefited large numbers of residents.

If one examines how the reconstruction budget was spent, it is not surprising that rebuilt Tokyo looked more modern and rational and emerged as a more functional city. The lion's share of spending had gone to pay for roads, canals, bridges, and the process of land readjustment. Specifically, ¥487,906,815 million was spent on these endeavors, equating to 66 percent of the total reconstruction budget (see tables 9.1–9.3).

While the final budget was larger than the ¥468 million agreed upon in December 1923, it was still much less than many hoped. Numerous

FIGURE 9.2 Two exemplars of New Tokyo: the wide avenue of Nihonbashi dōri in Ginza and the Kiyosu Bridge

Source: Postcard, author's private collection.

TABLE 9.1 National government reconstruction expenditures

EXPENDITURE ITEMS	COST (YEN)	PERCENTAGE OF OVERALL EXPENDITURES
Repair and improvement of trunk roads, including bridges	257,458,400	34.6
Repair and improvement of canals and creeks	28,879,065	3.9
Subsidy for construction of buildings in fire prevention zones	18,000,000	2.4
Construction of new parks	11,900,000	1.6
Land readjustment	8,750,000	1.2
Total	324,987,465	43.7

Source: Tokyo Municipal Office, *The Reconstruction of Tokyo*, 214.

TABLE 9.2 Tokyo Prefecture reconstruction expenditures

EXPENDITURE ITEMS	COST (YEN)	PERCENTAGE OF OVERALL EXPENDITURES
Repair and improvement of prefectural roads	18,754,036	2.5
Educational facilities	3,250,000	0.44
Total	22,004,036	2.94

Source: Tokyo Municipal Office, *The Reconstruction of Tokyo*, 214.

individuals believed both in 1923 and 1930 that the government should have budgeted far more on what was described as a once-in-a-lifetime opportunity to transform Tokyo into a modern imperial metropolis that embodied and projected new values and promoted Japan's position as a leader in East Asia. Horikiri Zenjirō, Tokyo's mayor in 1930, was one such individual. While acknowledging many accomplishments associated with reconstruction at the city's reconstruction festivities, he also stated publicly what many people felt privately but were too diplomatic to articulate: that much more could have been achieved. He evoked the memory of Gotō Shinpei, who died the previous year a broken man, compelling a large gathering of dignitaries to remember his spirit of optimism in the face of calamity. Horikiri then lamented that "the reconstruction plan that is to be completed today is much smaller than Mr. Gotō intended."[3]

Horikiri understood that most parts of western Tokyo that had escaped the postearthquake conflagrations experienced little change in relation to land plots and did not have their road networks altered, straightened, or widened in revolutionary ways during the seven-year reconstruction project. Moreover, Gotō's dream of turning Tokyo into a greener city through the creation of extensive new parks and a well-integrated system of open space fire breaks withered. Tokyo of 1930 was not a brand new city that emerged from the ruins of its former self. Rather, it was a reconstructed city that possessed many of the same urban vulnerabilities that existed on the eve of calamity in August 1923. American military planners hoping to coerce submission through urban area bombing

TABLE 9.3 City of Tokyo reconstruction expenditures

EXPENDITURE ITEMS	COST (YEN)	PERCENTAGE OF OVERALL EXPENDITURES
Land readjustment	93,951,000	12.6
Repair and improvement of auxiliary roads	80,114,314	10.8
1st- and 2nd-term repair, restoration, and expansion of electrical enterprises	56,920,000	7.6
Sewage system	41,811,321	5.6
Primary schools	41,056,583	5.5
Restoration of water system and expansion of pipes and mains for water supply	25,000,000	3.4
Construction of central wholesale market and branch establishments	21,400,000	2.9
Small parks	13,752,175	1.8
Local bridges	13,684,221	1.8
Social welfare	4,525,000	0.6
Sanitation system	3,100,000	0.4
Rubbish disposal	1,850,000	0.2
Total	397,164,614	53.4
Reconstruction grand total	744,156,115	100

Source: Tokyo Municipal Office, *The Reconstruction of Tokyo*, 214–15.

Note: Figures include ¥213,374,768 in loans and subsidies from the central government, which were approved as part of overall spending during the forty-seventh, forty-ninth, fiftieth, and fifty-sixth sessions of the Imperial Diet.

during the Second World War were aware of Tokyo's vulnerabilities and specifically targeted areas that they knew had burned during the 1923 catastrophe: the results were equally devastating in March 1945.[4] A wartime propaganda brochure (fig. 9.3) illustrates American awareness of Tokyo's vulnerabilities.

FIGURE 9.3 World War II–era American propaganda image referring back to the Great Kantō Earthquake

Source: Photo, author's private collection.

Years later, on the sixtieth anniversary of the Great Kantō Earthquake, when the war was a somewhat distant, yet still painful, memory for many Tokyoites, the Shōwa Emperor discussed the firebombing of Tokyo. He stated: "Gotō Shinpei designed a massive reconstruction plan for Tokyo, a plan that if implemented, might have reduced considerably the wartime fires of 1945 in Tokyo. I now think that it was very unfortunate that Gotō's plan was not put into action."[5] Many Tokyoites eventually shared the emperor's sense of regret at what had failed to materialize between 1923 and 1930.

Though the municipal and national governments spent only a meager amount constructing parks and other open green spaces, they budgeted even less money on social welfare facilities. Dreams that a reconstructed, well-designed, and well-managed city could ameliorate the social ills and side effects of urban, industrial modernity proved wildly ambitious. Many of the social problems and disadvantaged areas that had existed before the earthquake returned almost as quickly as reconstruction commenced. None of these areas was highlighted during the 1930 celebrations. It is just as instructive for historians today to examine which districts the emperor's motorcade avoided in March 1930 because city officials did not want to tarnish the celebration. The emperor did not travel through what many considered to be the new or reemergent slum areas of eastern Fukagawa or Honjo. In fact, city officials were so worried that impoverished and unemployed people from the poorest areas of Tokyo might mar the 1930 celebrations that they decided to employ, albeit temporarily, virtually everyone who sought employment to assist with numerous aspects of the weeklong festivities, from street cleaning and garbage collection to decorating the city's streets and sidewalks.[6] Likewise, the emperor did not inspect a dance hall or café, though he had ample from which to select. Tokyo housed more dance halls, cafés, department stores, restaurants, cabarets, and movie theaters in 1930 than existed in 1922, much to the chagrin of bureaucratic elites and concerned social commentators who believed the popularity of such places were symptomatic of a spiritual degeneration of society. Not unsurprisingly, the emperor also did not visit the entertainment quarters of Asakusa or the licensed prostitution quarters of Yoshiwara or Susaki that reemerged and flourished in the seven years after 1923.

People who experience a megadisaster such as the 1923 Great Kantō Earthquake and those who interpret and construct such occurrences often believe that a catastrophic event of this magnitude will change everything. The devastation caused, the intense reflection that follows, and eventual calls for unity, resolve, and sacrifice in the aftermath inveigle us to think that lasting change will result. We comfort ourselves with the notion that cities and societies will be rebuilt better than before and that people will change their thoughts and behaviors in the wake of a personal, local, or national calamity. Perhaps this is a human survival mechanism triggered when faced with catastrophic destruction and seemingly immeasurable loss. Perhaps it also betrays our faith in modernity and progress. Ian Buruma, professor of human rights at Bard College, endorsed wholeheartedly the disaster-as-agent-of-change interpretation in the wake of Japan's most recent earthquake and tsunami calamity of 2011. He wrote, "Natural disasters have a way of changing everything, and not just in material terms."[7]

Rarely, however, if history serves as a judge, have disasters imparted change in societies, behaviors, or systems in such a sweeping manner. Many parts of the marginalized Lower Ninth Ward of New Orleans are still a wreck after Hurricane Katrina battered the American Gulf Coast in 2005. The collapse of numerous schools in Chengdu as a result of the 2008 Sichuan earthquake has not led to greater local-level political transparency or successful parent-led involvement in educational affairs but rather resulted in greater attempts to silence and suppress dissent. In tsunami-ravaged communities along Japan's northeastern coast, such as Babanakayama and Ōfunato, attempts to resettle homes, communities, and villages to higher tracts of land have been plagued with difficulties and riddled with contestation since the March 11, 2011, disaster. "We should all be working together," declared Yoshihiro Miura, a forty-six-year-old fisherman from Babanakayama. "But even in this little village, there's this kind of wrangling. It's just human nature."[8]

Anthropologists, social scientists, and historians who have examined disasters and the reconstruction processes that follow have found numerous instances in which the contestation, resistance, and resilience—not to mention competing visions of what should be reconstructed and how—replace the optimism and opportunism that often spring forth from devastated, postdisaster landscapes. Sometimes people defy authority and attempt to rebuild where and how their homes and lives existed prior to calamity. Whether in Babanakayama in 2011, Port-au-Prince in 2010, Banda-Aceh in 2004, or Fukagawa Ward in 1923, resilience and a desire to return to normalcy often trump or at least hinder grand plans of a revolutionary new future. Virtually always, competing visions over reconstruction plans provoke fierce resistance and political struggles. "Disasters," as Susanna Hoffman reminds us, "set a critical stage bringing out and igniting arenas of contestation within a society."[9]

Apart from inspiring opportunism and unleashing contestation, disasters and the reconstruction processes that follow also reveal much about a society. As Hoffman provocatively asked, "When catastrophe upends a culture's cards, are matters altered or merely disclosed?"[10] Often they reveal far more than they ever alter. Perhaps this is their most important, utilitarian quality to historians and people who study disasters. Interconnections among politics, society, nature, ideology, economics, religion, and the built environment can be observed in the wake of catastrophe. Moreover, when one explores a disaster as more than just a cataclysmic event, examining it within the social, political, cultural, economic, and historical contexts in which it took place, one finds that it often amplifies preexisting concerns, trends, and patterns of behavior rather than transforming existing orders in radical ways. This was certainly true of Japan following the Great Kantō Earthquake.

What did the earthquake and the subsequent reconstruction projects reveal about interwar Japan? At the most basic human level, the land readjustment and reconstruction process illuminated the remarkable resilience of Tokyoites. Many residents returned to the locations of their former lives and began the challenging tasks of rebuilding their homes, often without government support or direction. When given a chance to voice their concerns and raise objections to plans that had been crafted by elites with little input from local residents, Tokyoites lobbied and petitioned local authorities with resolve. While residents were not able to fundamentally shape the overall plans for reconstruction, they found creative ways to voice opinions and utilized with aplomb all political and legal avenues open to them.

On a broader governmental level, the fractious debates over the scope and scale of reconstruction illustrated the unwillingness and inability of many elites and institutions to put differences aside at a moment of national emergency and to work together effectively. Individuals from different government agencies fought bitterly over what to reconstruct, how to rebuild, and, most important, how much to budget for reconstruction. Attempts to sooth disputes, tame jealousies, temper bureaucratic rivalries, and reach consensus under the aegis of numerous committees accomplished little: often they calcified disagreements and institutionalized antagonisms. Parliamentarians angered that they had not been consulted in reconstruction planning voiced outrage over the bureaucratic structures put in place to plan and implement reconstruction. They were joined by other elected elites representing constituents outside the disaster zone who believed that large monetary outlays directed toward the capital would endanger the economic vitality of the nation. Japan's quasi-democratic polity comprised many diffuse groups who held only limited powers yet guarded them with ferocity. It thus proved slow, fractious, and weak in responding to the complex task of rebuilding Tokyo and reconstructing the nation. That Japan's political system lacked an active and engaged arbiter of power more dominant than the other actors further complicated attempts to implement transformative policies after 1923.

The earthquake disaster also revealed—and amplified in dramatic ways—elite-level concerns about the state of Japanese society at this agonizing moment of modern revelation. The capital-destroying calamity helped galvanize elite opinion behind the notion that Japan was mired in an era of acute moral decay and social degeneration. In the opinion of many commentators, Japanese people—particularly the emergent middle classes of Japan's expanding cities—had become too consumer oriented, wedded to luxuries, and wasteful. In the political sphere, they had become enamored of dangerous, if not ultimately subversive, ideologies. In the social sphere, many elites complained that people displayed flippancy and laxness, whereas before they had exhibited respect for authority and resolve not only in the face of adversity but also in everyday situations. Commentators suggested that urbanites had become bewitched by the temptations of a frivolous lifestyle and beguiled by the hollow allure of hedonistic pursuits. Numerous individuals who interpreted and constructed the earthquake calamity as an act of heavenly punishment to admonish Japanese society hoped that this disaster would serve as *the* event that would shake people loose from the clutches of regress. In a new Japan, society would again embody frugality, austerity, diligence, decorum, resolve, sacrifice, and restraint.

Many Japanese people, however, held different opinions on 1920s urban modernity. Rather than see it as a threat or as an insidious disease that weakened the nation, they often embraced it willingly and enjoyed much of what modernity had to offer. Tariffs on luxury items made only a small and short-lived dent in the overall consumption of consumer goods. Dance halls, coffee shops, and department stores proliferated in reconstructed Tokyo, much to the chagrin of disaster opportunists who wished to use the disaster to assist with moral regeneration. Social commentator, journalist, and philosopher Miyake Setsurei lamented that people had quickly lost sight of the earthquake's meaning. In autumn 1923 he hoped that the Great Kantō Earthquake would be not only the event that led to a complete refashioning of Tokyo but also the crisis that triggered a social, moral, and ideological transformation of Japanese society. Later historians, he hoped, would divide Japan's modern history with 1923 as a clear demarcation line akin to what the Meiji Restoration of 1868 had been for earlier generations. Neither happened. By 1924 Miyake worried that few meaningful changes, apart from a more rationally planned city, had emerged from the ruins of Tokyo. "This earthquake came too suddenly," he declared, "and went very quickly, leaving us with only a dreamlike impression."[11] Nose Yoritoshi was even more damning in his criticism of postdisaster society, writing that "danger past, gods forgotten" had become the touchstone phrase of reconstruction.[12]

The political tensions that the reconstruction project exacerbated as well as the elite-level anxieties that the disaster amplified and people's seeming unwillingness to embrace a moral reorientation elicited two distinct responses. On the one hand, individuals from across the political spectrum, including socialist intellectual Yamakawa Hitoshi, General Ugaki Kazushige, and progressive Christian social reformer Abe Isoo, advocated the adoption of more heavy-handed, intrusive, top-down approaches to address the perceived regression of society and decline of the popular mind. They likewise implied that Japan needed a more unified polity in which fewer actors wielded more political power. No unanimity existed, however, within Japan's burgeoning pluralistic polity to accept such authoritarian, high modernist prescriptions. Moreover, no one in Japan's highly bureaucratic and politically diffuse governmental structure had the power to unite, let alone browbeat, elites with competing visions to accept radical reconstruction and political regeneration initiatives as well as political reorganization.

Far more elites embraced a moderate path. These individuals saw the potential for manifold rewards if government could harness technocratic modernity and tap into the ample reservoir of personal resilience exhibited by countless Tokyoites at the neighborhood level during reconstruction.[13] Future conflict, Japanese elites understood, would require the mobilized efforts of civilians and the economy as much as the military. Home Ministry bureaucrats and City of Tokyo officials concluded that the government needed to find new and more effective ways to persuade, manage, organize, and mobilize urbanites. Imaginative suasion campaigns continued throughout the interwar period, but officials put increased emphasis on expanding the number of neighborhood associations (*tonarigumi*) and increasing their status and role in urban society.[14] Between 1923 and 1927 more neighborhood associations were formed across Tokyo than at any other point in the capital's history.[15] Government officials hoped that neighborhood associations would serve as more than just conduits of authority downward, however.[16] They envisaged that these organizations would play leading roles in assisting with the delivery of services, from firefighting to civil defense, and from street cleaning to organizing local festivals: in essence, neighborhood associations would forge and strengthen community bonds among individuals families as well as state-community ties.[17] Moreover, Home Ministry bureaucrats hoped that neighborhood leaders could exert subtle social pressures to encourage the adoption of numerous ideological as well as economic regeneration campaigns ranging from renewed antiluxury drives to war bond purchase initiatives.

Commentators and political elites who reflected on the loss of governmental authority and the murderous violence committed in the aftermath of the 1923 earthquake argued that greater training and physical and moral discipline were required to keep order from evaporating in a future emergency situation. Many saw neighbor associations as key components in this undertaking as well. Educator Matsushita Senkichi wrote in 1924 that proper military drills and integrated physical and mental training in schools and society could help foster a "sense of loyalty, courage, service, devotion, and sacrifice" among practitioners.[18] Once learned in school and reinforced at the neighborhood level, Matsushita claimed, military drills could lead to a "more orderly society" in which people embraced a "moderate and disciplined life." Moreover, he concluded that future training would better prepare people to face and "persevere in any kind of hardship." Only "daily training" and a focus on mental and physical "discipline" embedded at an early age and reinforced throughout life, Ugaki Kazushige concurred, would keep panic from sweeping the nation and order from evaporating in a future emergency contingency.[19]

However successful these nascent urban-based neighborhood associations were, governing elites seemingly always wanted more from the people of Japan. This was particularly true after urbanities exhibited general ambivalence to the introduction of air raid and fire prevention drills across urban areas in the late 1920s. Beginning in 1928 and continuing throughout the interwar period, military officials working in conjunction with retired servicemen's associations, youth groups and young men's associations, neighborhood associations, and metropolitan governments sought to make people aware of the potential dangers of air attack and to prepare them for just such a contingency. These efforts took many forms, from the mundane publication of a journal entitled *Bōkū* (Air defense) and numerous pamphlets and posters to more extravagant air defense exercises and blackout drills. The first large-scale air raid drill was carried out in Osaka between July 5 and 7, 1928; more drills would later be held in Tokyo and other major cities across Japan. The events were choreographed to mimic future air raids—both night and daytime attacks. The army and navy each flew aircraft over sections of the city, dropped fake bombs to correspond with fixed explosions on the ground, and released smoke from canisters that represented poison gas. The Osaka government mobilized 139,764 officials to direct this three-day extravaganza, with the lion's share of supporting officials (121,000) coming from youth group members.[20]

It is nearly impossible to determine how urbanites really felt about these exercises, but what is possible to understand is how government officials perceived of their response. In a word, officials claimed that people were generally unenthusiastic. As one municipal official wrote, "In this international climate of disarmament, the prerequisite idea that the city will be subjected to air attack in the near future is somehow missing."[21] They lamented that the residents of Osaka were far more excited about and engaged with the seasonal fireworks display that took place at the end of July. Tokyo city officials realized this point as well. To counter what they predicted would be a similar lack of enthusiasm for air defense drills, they directed officials and volunteers who organized these yearly events to emphasize that fire prevention training had practical, nonmilitary efficacy: it could help, they argued, in case Japan was struck by another large earthquake. Drills, exercises, and mock air raids continued throughout the 1930s.

As perceived threats and concerns expanded from natural disasters and moral degeneration to encompass foreign enemies and what was packaged as a holy war against China after 1937, anxieties heightened about whether Japanese people could cope with and surmount a period of national emergency. Terada Torahiko, physicist at Tokyo University, reflected on the perceived state of emergency that existed in the late 1930s and conflated Japan's postdisaster experience of 1923 with what the nation and the people faced in 1938: an extraordinary challenge that would require people to work together with resolve and discipline. Writing in his evocatively titled 1938 book, *Tensai to kokubō* (Natural disasters and national defense), he expressed concern that people had "forgotten what it was like when the last disaster occurred." While disasters of the past caused local or regional devastation, the threat Japan now faced was on a national scale. He suggested, "at a deep layer of our national consciousness there is some vague, nightmare-like shadow of uneasiness lingering these days." "Japan," he continued, was situated at the "center of a whirlpool of uneasiness" and a country "surrounded by enemies."[22] It was therefore incumbent upon leaders to take more active steps to prepare society for the exigencies of a national emergency.

From 1937 onward, Japanese leaders implemented a number of laws aimed at preparing the economy and society for national economic and spiritual mobilization.[23] Differing agencies within Japan's polity gained new powers to better control manpower, resources, wages, prices, production, and distribution. Businesses, people, and elites with competing views again met such elite-level overtures with responses that were far less enthusiastic than government officials hoped. People continued to purchase luxuries in quantities that were larger than governing elites believed healthy or necessary.[24] Campaigns that encouraged thrift, diligence, frugality, sacrifice, temperance, and moderation became laws that restricted, then outlawed, luxuries and the sale of items deemed nonessential. Dance halls and other entertainment venues considered frivolous or harmful to national morale were closed in 1940, and increasing levels of conformity and sacrifice were demanded of Japan's population.[25] Attempts to unify Japan's polity—launched concurrently with the adoption of ever increasing economic regulations—proved every bit as difficult to achieve. The Imperial Rule Assistance Association, inaugurated in October 1940, did little to tame bureaucratic rivalries and political jealousies that had accumulated over the previous fifty years of Japan's political and economic development. Japan's economy, which was the smallest and most resource poor of all major combatants in World War II, remained, arguably, the least well organized and efficient.

Japan's leaders understood that more sacrifices were needed from the populace and that institutional rivalries had to be put aside for their country to survive the greatest challenge of the twentieth century: total war. Regardless of this awareness, Japanese elites did a poor job securing both objectives. Catastrophes, whether natural, as was the case in 1923, or manmade, as Japan experienced between 1937 and 1945, were not as easy to manipulate and use to forge lasting, efficient, and popular new orders. While Japan's prewar experience with modernity heightened anxieties and suggested to many the disappearance of old attitudes and behaviors that had enabled Japan to emerge as an independent nation state and imperial power, it also offered new opportunities. Many believed that well-managed technocratic modernity held out the tantalizing prospect that national reconstruction was attainable and that a new order could be fashioned from the ruins of catastrophe or from the clutches of a national emergency. Both beliefs were overly optimistic. Calls for sacrifice, renewal, and regeneration therefore continued unabated throughout the interwar and wartime periods in Japan. They were as prevalent, and became as quotidian, as the tremors that shook Japan's seismically vulnerable archipelago. The attainment of national reconstruction, however, remained as illusive and illusory as a mythical chimera.

NOTES

PREFACE

1. "Admiral Anderson, U.S.N. Says No Thread [sic] of War on Pacific," November 5, 1923, C405, Document 036/2798, RG 38 Records of the Chief of Naval Operations, Office of Naval Intelligence, Confidential "Suspect" and General Correspondence Files, 1913–1924, Entry 78, Box 11, File 91036–2793.
2. *Japan Weekly Chronicle*, September 27, 1923, 436.

INTRODUCTION

1. Anonymous, "Shikai no toshi o meguru: saigai yokuyokujitsu no dai Tokyo" (Great Tokyo two days after calamity: Walking around a dead city), in *Taishō daishinsai daikasai* (The Taishō great earthquake and conflagration), ed. Dainihon yūbenkai kōdansha, 37 (Tokyo: Dainihon yūbenkai kōdansha, 1923).
2. Takashima Beihō, "Shin Tokyo no kensetsu to Tokyokko no iki" (Construction of new Tokyo and the spirit of Tokyoites), *Chūō kōron* 38:11 (October 1923): 52–53.
3. Ibid., 53.
4. Shimamoto Ainosuke, "Seishinteki fukkō saku" (Measures for spiritual restoration), *Teiyū rinrikai rinri kōenshū* 36:254 (1923): 102.

5. Soeda Azenbō, "Taishō daishinsai no uta" (A song of the Taishō Great Earthquake), in *Kantō daishinsai: Dokyumento* (The Great Kantō Earthquake: Documents), ed. Gendaishi no kai, 210–12 (Tokyo: Sōfūkan, 1996).
6. Fukuda Tokuzō, "Fukkō Nihon tōmen no mondai" (The urgent questions confronting Japanese reconstruction), *Kaizō* 5:10 (October 1923): 8–9.
7. The April 18, 1906, San Francisco earthquake and fires destroyed just over 12 million square meters of land, with an estimated loss of life at roughly 3,000. The September 1, 1923, Great Kantō Earthquke and fires destroyed just over 33 million square meters of land in Tokyo, with an estimated loss of life at just over 100,000.
8. Fukuda, "Fukkō Nihon tōmen no mondai," 8–9.
9. Ibid., 13–14, 8–10.
10. Ibid., 13–14, 9, 4.
11. Anan Jō'ichi, "Toshi seikon to daieidan" (The reconstruction of the city and a resolute decision), *Toshi kōron* 6:11 (November 1923): 82.
12. Ichiki Kitokurō, "Fukkō shigi" (Personal opinion on reconstruction), *Chihō gyōsei* 31:11 (November 1923): 7–8.
13. Abe, "Teito fukkō ni yōsuru ōnaru gisei," 21.
14. Shimamoto, "Seishinteki fukkō saku," 105.
15. Rebecca Solnit, *A Paradise Build in Hell: The Extraordinary Communities That Arise in Disaster* (New York: Viking, 2009).
16. Anthony Oliver-Smith and Susanna Hoffman, "Introduction: Why Anthropologists Should Study Disasters," in *Catastrophe and Culture: The Anthropology of Disaster*, ed. Anthony Oliver-Smith and Susanna Hoffman, 3–22 (Santa Fe: School of American Research Press, 2002); Anthony Oliver-Smith, "Theorizing Disasters: Nature, Power, and Culture," in Oliver-Smith and Hoffman, *Catastrophe and Culture*, 23–47; Susanna M. Hoffman, "The Monster and the Mother: The Symbolism of Disaster," in Oliver-Smith and Hoffman, *Catastrophe and Culture*, 113–41; Susanna Hoffman and Anthony Oliver-Smith, "Anthropology and the Angry Earth: An Overview," in *The Angry Earth: Disaster in Anthropological Perspectives*, ed. Anthony Oliver-Smith and Susanna Hoffman, 1–16 (London: Routledge, 1999); Anthony Oliver-Smith, "What Is a Disaster? Anthropological Perspectives on a Persistent Question," in Oliver-Smith and Hoffman, *The Angry Earth*, 18–34; Hoffman, "After Atlas Shrugs: Cultural Change or Persistence after a Disaster," in Oliver-Smith and Hoffman, *The Angry Earth*, 302–25; Anthony Oliver-Smith, "Disaster Context and Causation: An Overview of Changing Perspectives in Disaster Research," in *Natural Disasters and Cultural Responses*, ed. Anthony Oliver-Smith, 1–38 (Williamsburg, Va.: College of William and Mary, 1986).
17. Oliver-Smith and Hoffman, "Introduction," 6.
18. Oliver-Smith, "Theorizing Disasters," 24.
19. Miura Tōsaku, "Hijōji ni arawareru kyōiku no kōka ni tsuite" (Concerning the effectiveness of education in a time of crisis), *Kyōiku jiron* 1384 (1923): 9.
20. Oliver-Smith, "Theorizing Disasters," 26.
21. Marc Bloch, *Feudal Society*, 2 vols. (Chicago: University of Chicago Press, 1961), 1:152.

22. Hoffman, "After Atlas Shrugs," 310; Hoffman, "The Monster and the Mother," 139.
23. Hoffman, "After Atlas Shrugs," 310–11.
24. Virginia Garcia-Acosta, "Historical Disaster Research," in Oliver-Smith and Hoffman, *Catastrophe and Culture*, 49–52.
25. Gregory Bankoff, *Cultures of Disaster: Society and Natural Hazard in the Philippines* (London: Routledge/Curzon, 2003); John Barry, *Rising Tide: The Great Mississippi Flood of 1927 and How It Changed America* (New York: Simon and Schuster, 1997); Douglas Brinkley, *The Great Deluge: Hurricane Katrina, New Orleans, and the Mississippi Gulf Coast* (New York: Harper, 2006); Martin Doering, "The Politics of Nature: Constructing German Reunification During the Great Oder Flood of 1997," *Environment and History* 9:2 (May 2003): 195–214; Shelia Hones, "Distant Disasters, Local Fears: Volcanoes, Earthquakes, Revolution, and Passion in The Atlantic Monthly, 1880–1884," in *American Disasters*, ed. Steven Biel, 170–96 (New York: New York University Press, 2001); Jeffrey Jackson, *Paris Under Water: How the City of Light Survived the Great Flood of 1910* (New York: Palgrave Macmillan, 2010); Alessa Johns, "Introduction," in *Dreadful Visitations: Confronting Natural Catastrophe in the Age of Enlightenment*, ed. Alessa Johns, xi–xxv (London: Routledge, 1999); Michael Kempe, "Noah's Flood: The Genesis Story and Natural Disasters in Early Modern Times," *Environment and History* 9:2 (May 2003): 151–71; Matthew Mulcahy, *Hurricanes and Society in the British Greater Caribbean, 1624-1783* (Baltimore: Johns Hopkins University Press, 2008); Matthew Mulcahy, "A Tempestuous Spirit Called Hurri Cano: Hurricanes and Colonial Society in the British Greater Caribbean," in Biel, *American Disasters*, 11–38; John Mullin, "The Reconstruction of Lisbon Following the Earthquake of 1755: A Study in Despotic Planning," *Planning Perspectives* 7:2 (April 1992): 157–79; Christian Rohr, "Man and Nature in the Late Middle Ages: The Earthquake of Carinthia and Northern Italy on 26 January 1348 and Its Perception," *Environment and History* 9:2 (May 2003): 127–49; Stuart Schwartz, "The Hurricane of San Ciriaco: Disaster, Politics, and Society in Puerto Rico, 1899–1901," *Hispanic American Historical Review* 72:3 (August 1992): 303–34; Theodore Steinberg, *Acts of God: The Unnatural History of Natural Disaster in America* (Oxford: Oxford University Press, 2000); Martin Struber, "Divine Punishment or Object of Research? The Resonance of Earthquakes, Floods, Epidemics and Famine in the Correspondence Network of Albrecht von Haller," *Environment and History* 9:2 (May 2003): 171–93; Donald Worster, *Dust Bowl: The Southern Plains in the 1930s* (New York: Oxford University Press, 1979).
26. Alex Bates, "Catfish, Super Frog, and the End of the World: Earthquakes (and Natural Disasters) in the Japanese Cultural Imagination," *Education About Asia* 12:2 (Fall 2007): 13–19; Janet Borland, "Makeshift Schools and Education in the Ruins of Tokyo, 1923," *Japanese Studies* 29:1 (May 2009): 131–43; Janet Borland, "Stories of Ideal Subjects from the Great Kantō Earthquake," *Japanese Studies* 25:1 (May 2005): 21–34; Janet Borland, "Capitalising on Catastrophe: Reinvigorating the Japanese State with Moral Values Through Education Following the 1923 Great Kantō Earthquake," *Modern Asian Studies* 40:4 (October 2006): 875–908; Greg Clancey, *Earthquake Nation: The Cultural Politics of Japanese Seismicity, 1868-1930* (Berkeley: University of

California Press, 2006); Greg Clancey, "The Meiji Earthquake: Nature, Nation, and the Ambiguities of Catastrophe," *Modern Asian Studies* 40:4 (October 2006): 909–51; J. Charles Schencking, "1923 Tokyo as a Devastated War and Occupation Zone: The Catastrophe One Confronted in Post Earthquake Japan," *Japanese Studies* 29:1 (May 2009): 111–29; J. Charles Schencking, "The Great Kantō Earthquake and the Culture of Catastrophe and Reconstruction in 1920s Japan," *Journal of Japanese Studies* 34:2 (Summer 2008): 294–334; J. Charles Schencking, "The Great Kantō Earthquake of 1923 and the Japanese Nation: Responding to an Urban Calamity of an Unprecedented Nature," *Education About Asia* 12:3 (Fall 2007): 20–25; J. Charles Schencking, "Catastrophe, Opportunism, Contestation: The Fractured Politics of Reconstructing Tokyo Following the Great Kantō Earthquake of 1923," *Modern Asian Studies* 40:4 (October 2006): 833–874; Gregory Smits, "Shaking Up Japan: Edo Society and the 1855 Catfish Prints," *Journal of Social History* 39:4 (Summer 2006): 1045–77; Timothy Tsu, "Making Virtues of Disaster: 'Beautiful Tales' from the Kobe Flood," *Asian Studies Review* 32 (June 2008): 197–214; Gennifer Weisenfeld, "Designing After Disaster: Barrack Decoration and the Great Kantō Earthquake," *Japanese Studies* 18:3 (December 1998): 229–46; Gennifer Weisenfeld, "Imaging Calamity: Artists in the Capital After the Great Kantō Earthquake," in *Modern Boy, Modern Girl: Modernity in Japanese Art, 1910–1935*, ed. Jackie Menzies, 25–29 (Sydney: Art Gallery of New South Wales, 1998). Three works that have focused specifically on the massacre of Koreans are Michael Allen, "The Price of Identity: The 1923 Earthquake and Its Aftermath," *Korean Studies* 20 (1996): 64–93; Michael Weiner, *The Origins of the Korean Community in Japan, 1910–1923* (Manchester: Manchester University Press, 1989); and Michael Weiner, "Koreans in the Aftermath of the Kantō Earthquake of 1923," *Immigrants and Minorities* 2:1 (April 1983): 5–32.

27. Works that come closest but are lacking in detailed analysis or, in the case of one, utilization of Japanese-language materials are Noel Busch, *Two Minutes to Noon: The Story of the Great Tokyo Earthquake and Fire* (New York: Arthur Barker, 1962); Edward Seidensticker, *Tokyo Rising: The City Since the Great Earthquake* (New York: Knopf, 1990); Edward Seidensticker, *Low City, High City: Tokyo from Edo to the Earthquake* (New York: Knopf, 1983).
28. Sheldon Garon, *Molding Japanese Minds: The State in Everyday Life* (Princeton: Princeton University Press, 1997).

1. CATACLYSM

1. Uno Kōji, "San byaku nen no yume" (Dream of three hundred years)," in *Taishō daishinkasai shi* (Documents of the Taishō great earthquake and conflagration), ed. Kaizōsha (Tokyo: Kaizōsha, 1924), 38.
2. Kawatake Shigetoshi, "Sōnan ki" (Meeting with disaster), in Kaizōsha, *Taishō daishinkasai shi*, 27; Miura, "Hijōji arawareru kyōiku no kōka ni tsuite," 9; Yamaka Hitoshi, "Teito fukkō to shakai shugiteki seisaku" (Imperial capital reconstruction and socialist policy), in *Yamakawa Hitoshi zenshū*, ed. Yamakawa Shinsaku (Tokyo: Keisō shobō, 1968), 275.

3. Uno, "San byaku nen no yume," 39.
4. Takashima Heizaburō, "Daishinkasai to kodomo" (The great earthquake conflagration and children), *Jidō kenkyū* 27:2 (November 1923): 49–50.
5. Funaki Yoshie, "Hi de shinauka, mizu o erabauka" (Death by burning or death by drowning), in *Kantō daishinsai: Dokyumento* (The Great Kantō Earthquake: Documents), ed. Gendaishi no kai (Tokyo: Sōfūkan, 1996), 53–57.
6. Tanaka Kōtarō, "Shitai no nioi" (The smell of dead bodies), in ibid., 80–81.
7. *Osaka mainichi shinbun* (English edition), September 6, 1923, 1.
8. Ross Stein, Shinji Toda, Tom Parsons, and Elliot Grunewald, "A New Probabilistic Seismic Hazard Assessment for Greater Tokyo," *Philosophical Transactions of the Royal Society A* 364 (June 2006): 1965–88. In a 2005 publication, a research team headed by Satō Hiroshi claimed to have identified a "new fault line" beneath the Tokyo metropolitan area. See Satō Hiroshi et al., "Earthquake Source Fault Beneath Tokyo," *Science* 309:5733 (July 15, 2005): 462–65.
9. *Osaka mainichi shinbun* (English edition), September 5, 1923, 2.
10. Mononobe Nagao, "Shinsai chihō fukkō ni taisuru kibō" (Hopes for the reconstruction of the earthquake disaster area), *Chihō gyōsei* 32:1 (January 1924): 23–24.
11. Tanaka, "Shitai no nioi," 71.
12. Ibid., 73.
13. Kawatake, "Sōnan ki," 24, 26–27.
14. Funaki, "Hi de shinauka, mizu o erabauka," 54, 56.
15. Kawatake, "Sōnan ki," 29–30.
16. Koizumi Tomi, "Hifukushō seki sōnan no ki" (Meeting with a disaster at the site of the clothing depot), in Kaizōsha, *Taishō daishinkasai shi*, 13.
17. Only a small street separated the 67,000 square meter site of the Honjo Clothing Depot from a 40,000 square meter private garden owned by the Yasuda family. Those seeking refuge thus found 107,000 square meters of "empty land" on the eastern bank of the Sumida River.
18. Koizumi, "Hifukushō seki sōnan no ki," 14.
19. Funaki, "Hi de shinauka, mizu o erabauka," 53–54.
20. For more in-depth analysis of the Korean massacres that followed the 1923 earthquake, see Allen, "The Price of Identity," 64–93; Weiner, *The Origins of the Korean Community in Japan*, 164–201; Weiner, "Koreans in the Aftermath of the Kantō Earthquake of 1923," 5–32; Matsuo Takayoshi, "Kantō daishinsai ka no Chōsenjin gyakusatsu jiken: 1" (Massacre of Koreans at the time of the Great Kantō Earthquake: 1), *Shisō* 471 (September 1963): 44–61; Matsuo Takayoshi, "Kantō daishinsai ka no Chōsenjin gyakusatsu jiken: 2" (Massacre of Koreans at the time of the Great Kantō Earthquake: 2), *Shisō* 476 (February 1964): 110–20; Matsuo Shōichi, "Kantō daishinsai shi kenkyū no seika to kadai" (Results and research tasks of historical study on the Great Kantō Earthquake), *Hosei daigaku tama ronshū* 9 (March 1993): 79–148; and Matsuo Shōichi, "Chōsenjin gyakusatsu to guntai" (The Korean massacre and the army), *Rekishi hyōron* 9 (September 1993): 29–41.
21. Miyao Shunji, govenor of Hokkaidō, reflected on how dismayed he was with earthquake refugees who settled in Hokkaidō and immediately began to spread rumors

of a Korean uprising. Tokyo Shisei Chōsakai, ed., *Teito fukkō hiroku* (Secret notes on the imperial capital reconstruction) (Tokyo: Hōbunkan, 1930), 76.
22. Weiner, *The Origins of the Korean Community in Japan*, 172–74.
23. Suematsu Kai'ichirō, "Daishin ni kansuru shokan" (Impressions related to the great earthquake disaster), *Chihō gyōsei* 31:10 (October 1923): 22–23.
24. Kang Tok-sang and Kum Pyong-dong, eds., *Kantō daishinsai to Chōsenjin* (Koreans and the Great Kantō Earthquake) (Tokyo: Misuzu shobō, 1963), 73. Specific numbers are as follows: Tokyo Prefecture, 1593; Kanagawa Prefecture, 603; Gunma Prefecture, 469; Saitama Prefecture, 366; Chiba Prefecture, 336; and Tochigi Prefecture, 19.
25. References to Koreans were printed as XXX in Tanaka's essay as many, though not all, were censored. Tanaka, "Shitai no nioi," 74–76.
26. Ibid., 76.
27. Matsuo Takayoshi, "Kantō daishinsai ka no Chōsenjin gyakusatsu jiken: 1," 59–61; Weiner, *The Origins of the Korean Community in Japan*, 183–84.
28. Hoashi Ri'ichirō, "Tenrai no kyōkun" (Heaven-sent lessons), *Chihō gyōsei* 31:10 (October 1923): 26.
29. Suematsu, "Daishin ni kansuru shokan," 23.
30. Tawara Magoichi, "Teito fukkō to jinshin" (The imperial capital reconstruction and the popular mind), *Chihō gyōsei* 31:10 (October 1923): 17.
31. Oku Hidesaburō, "Shinsai to dōtokushin no kaizō" (The earthquake and the renewal of a moral spirit), *Kyōiku* 490 (1924): 11–12.
32. Tabuchi Toyokichi delivered this speech in the December sitting of the Imperial Diet. The transcript can be found in Nihon teikoku gikai, ed., *Dai Nihon teikoku gikaishi* (A parliamentary record of imperial Japan), 18 vols. (Tokyo: Dai Nihon teikoku gikai kankōkai, 1930), 14:1468–72.
33. Koizumi, "Hifukushō seki sōnan no ki," 17.
34. Ibid., 15, 18.
35. Ibid., 17–22.
36. Kawatake, "Sōnan ki," 28–31.
37. Tanaka, "Shitai no nioi," 78–80.
38. Nakamura Shōichi, "Hifukushō ato" (The site of the clothing depot), in Gendaishi no kai, *Kantō daishinsai: Dokyumento*, 97.
39. Tanaka, "Shitai no nioi," 80.
40. Takenobu, Yoshitarō, ed., *The Japan Year Book, 1924–25* (Tokyo: Foreign Affairs Association of Japan, 1925), 275.
41. Ibid., 249; *Japan Weekly Chronicle*, April 3, 1924, 473.
42. Shimaji Daitō, "Daijihen no dōtoku oyobi shinkō ni oyobosu eikyō" (The effect of the great disaster on morals and faith), *Nihon kyōiku* 2:11 (November 1923): 114.
43. *Hōchi shinbun*, September 5, 1923, 1; September 6, 1923, 2; September 7, 1923, 1.
44. As of July 31, 1923, the Kōike ginkō had paid-up capital of ¥3 million, deposits of ¥3.8 million, and loans valued at ¥8.9 million. Bank details found in Takenobu, ed., *Japan Year Book, 1924–25*, 461.

45. *Hōchi shinbun*, September 13, 1923, 2.
46. Ibid., September 7 and 15, 1923, 2.
47. *Otaru shinbun*, September 2, 3, and 5, 1923, 1.
48. *Kyūshū nippō*, September 2, 1923, 1.
49. This column began in the September 8 edition of the *Fukuoka nichi nichi shinbun* in which it declared that as of 6 p.m. on September 6, 47,293 bodies had been collected.
50. *Kyūshū nippō*, September 9, 1923, 1.
51. *Fukuoka nichi nichi shinbun*, September 9, 1923, 1.
52. For accounting purposes, families were defined as comprising at least a married couple but often included extended family members. See Tokyo Municipal Office, *The Reconstruction of Tokyo* (Tokyo: Tokyo Municipal Office, 1933), 9.
53. Nihon tōkei fukyūkai, ed., *Teito fukkō jigyō taikan*, vol. 1, part 5, 5.
54. Out of Tokyo Prefecture's overall population of 4 million, 1.55 million were classified as homeless in September 1923. Ibid.
55. Bureau of Social Affairs, Japan Home Ministry, *The Great Earthquake of 1923 in Japan*, 2 vols. (Tokyo: Bureau of Social Affairs, 1926), 1:82–85, 91–92.
56. Takenobu ed., *The Japan Year Book, 1924–25*, 245.
57. Aoki Tokuzō, ed., *Inoue Junnosuke den* (Official biography of Inoue Junnosuke) (Tokyo: Inoue Junnosuke ronsō hensankai, 1935), 195–96.
58. Bureau of Social Affairs, *The Great Earthquake of 1923 in Japan*, 473.
59. Prior to the government's loan intervention, five Japanese insurance companies covered policyholders to a total of 10 percent of any policy's insured value. This figure equaled ¥3.4 million. Foreign insurance companies also paid policyholders a flat figure of 10 percent of total insured policy value, or roughly ¥2 million. See ibid., 485–86, 495–96.
60. Tokyo Municipal Office, *The Reconstruction of Tokyo*, 11–12.
61. The Tokyo Metropolitan Office lists the number of government and public offices that were destroyed by the disaster at 3,563. These include offices such as the Police Training Institute, Forestry Office, Central Meteorological Observatory, Tokyo Rice Exchange Office, Patent Office, Kyōbashi Revenue Office, Tobacco Monopoly Office, various telephone exchange offices, railway offices and stations, and numerous police and fire stations. See ibid., 18.
62. The impact of the 1923 earthquake on education, school buildings, and the urban landscape around schools is discussed in Janet Borland, "Rebuilding Schools and Society after the Great Kantō Earthquake, 1923–1930," Ph.D. dissertation, University of Melbourne, 2008.
63. Borland, "Stories of Ideal Subjects," 23.
64. Tokyo Municipal Office, *The Reconstruction of Tokyo*, 316–18.
65. Bureau of Social Affairs, *The Great Earthquake of 1923 in Japan*, 82.
66. Tokyo Municipal Office, *The Reconstruction of Tokyo*, 20.
67. Ibid., 29.
68. Tsukamoto Seiji, "Teito fukkō no shinseishin" (The new spirit of imperial capital reconstruction), *Toshi kōron* 6:12 (December 1923): 15–16.

2. AFTERMATH

1. As army minister, Yamanashi reduced the size of the army by roughly 2,000 officers and 56,000 enlisted personnel. He believed this reduction could be offset by modernizing the military and emphasizing physical and spiritual training. See Alvin Coox, *Nomonhan: Japan Against Russia, 1939*, 2 vols. (Stanford: Stanford University Press, 1985), 1:18; Leonard Humphreys, *The Way of the Heavenly Sword: The Japanese Army in the 1920s* (Stanford: Stanford University Press, 1995), 91. For a short biography of Yamanashi's career, see Hata Ikuhiko, *Nihon riku-kaigun sōgō jiten* (A comprehensive encyclopedia of Japanese army and navy personnel) (Tokyo: Tokyo daigaku shuppankai, 1991), 152.
2. Yamanashi replaced General Fukuda Masatarō as commander of the martial law headquarters on September 20, 1923. General Fukuda was forced to resign over Captain Amakasu Masahiko's murder of the well-known anarchist Ōsugi Sakae, the radical feminist Itō Noe, and Ōsugi's nephew while in military custody following the earthquake. Humphreys, *The Way of the Heavenly Sword*, 56–57.
3. Hirakata Chieko, Ōtake Yoneko, and Matsuo Shōichi, eds., *Kantō daishinsai seifu, rikukaigun kankei shiryō*, vol. 1: *Seifu kaigenrei kankei shiryō* (Materials related to the administration of martial law) (Tokyo: Nihon keizai hyōronsha, 1997), 460–61.
4. Ibid., 473.
5. Kang and Kum, eds., *Kantō daishinsai to Chōsenjin*, 93.
6. Ibid., 91–93.
7. Hirakata, Ōtake, and Matsuo, eds., *Seifu kaigenrei kankei shiryō*, 460.
8. Susan Weiner, "Bureaucracy and Politics in the 1930s: The Career of Gotō Fumio," Ph.D dissertation, Harvard University, 1984, 36–38.
9. Bureau of Social Affairs, *The Great Earthquake of 1923 in Japan*, 300–301.
10. Takakura Tetsuichi, ed., *Tanaka Giichi denki* (Biography of Tanaka Gi'ichi), 2 vols., (Tokyo: Hara shobō, 1981), 2:344–45.
11. This was the second time in Japan's history that martial law had been declared in the capital. The first time occurred in September 1905 in response to the anti–Portsmouth Treaty riots that broke out in Hibiya following the conclusion of the Russo-Japanese War.
12. Kang and Kum, eds., *Kantō daishinsai to Chōsenjin*, 100–102; Weiner, *The Origins of the Korean Community in Japan, 1910–1923*, 175–76.
13. Yamamoto's recollections are found in Tokyo Shisei Chōsakai, ed., *Teito fukkō hiroku*, 8–9.
14. Ibid.
15. Kang and Kum, eds., *Kantō daishinsai to Chōsenjin*, 101.
16. Hirakata, Ōtake, and Matsuo, eds., *Seifu kaigenrei kankei shiryō*, 461–62.
17. Tasaki Kimitsukasa and Sakamoto Noboru, eds., *Kantō daishinsai seifu, rikukaigun kankei shiryō*, vol. 2: *Rikugun kankei shiryō* (Materials related to the army) (Tokyo:

Nihon keizai hyōronsha, 1997), 2:166–72. From September 2 until October 4, aircraft made 499 flights over Tokyo and Yokohama for reasons of reconnaissance, communication, leaflet dropping, and the transportation of journalists. Total flying time was 2,537 hours and 19 minutes. See Hirakata, Ōtake, and Matsuo, eds., *Seifu kaigenrei kankei shiryō*, 477–78.
18. Hirakata, Ōtake, and Matsuo, eds., *Seifu kaigenrei kankei shiryō*, 1:478–80. Pigeons worked well enough for other military divisions around Japan to request pigeon corps for future contingencies. See *Japan Weekly Mail*, December 6, 1923, 801.
19. Kang and Kum, eds., *Kantō daishinsai to Chōsenjin*, 92.
20. Hirakata, Ōtake, and Matsuo, eds., *Seifu kaigenrei kankei shiryō*, 460–62.
21. Kang and Kum, eds., *Kantō daishinsai to Chōsenjin*, 101–2, 74–75.
22. A summary of these reports can be found in ibid., 104–5.
23. Naimushō keihokyoku, ed., *Shinsaigo ni okeru keikai keibi ippan* (Vigilance units after the earthquake) (Tokyo: Naimushō, n.d). Written in September 1923. Held at Tokyo Metropolitan Archives 093.22 na, 41–43; Kang and Kum, eds., *Kantō daishinsai to Chōsenjin*, 121; Weiner, *The Origins of the Korean Community in Japan, 1910-1923*, 180–81.
24. Kang and Kum, eds., *Kantō daishinsai to Chōsenjin*, 75.
25. Ko Hakushaku Yamamoto kaigun taishō denki hensankai, ed., *Yamamoto Gonnohyōe den* (Official biography of Yamamoto Gonnohyōe), 2 vols. (Tokyo: Ko Hakushaku Yamamoto kaigun taishō denki hensankai, 1938), 2:1054–57; Bureau of Social Affairs, *The Great Earthquake of 1923 in Japan*, 558–59.
26. This figure includes forces drawn from fifty-nine infantry battalions, six cavalry regiments, six artillery regiments, seventeen engineer battalions, two railway regiments, two telecommunication regiments, one aviation battalion, one balloon unit, and one vehicle unit. Hirakata, Ōtake, and Matsuo, eds., *Seifu kaigenrei kankei shiryō*, 1:463–65.
27. Imamura Akitsune, "Shigaichi ni okeru jishin no seimei oyobi zaisan ni taisuru songai o keigen suru kanpō" (Simple methods for reducing the damage to life and property caused by an earthquake in an urban area), *Taiyō* 11:12 (December 1905): 162–71; Borland, "Rebuilding Schools and Society after the Great Kantō Eathquake, 1923–1930," 15–16; Clancey, *Earthquake Nation*, 217–18.
28. *Japan Weekly Chronicle*, September 20, 1923, 412.
29. Details related to the creation of the Emergency Earthquake Relief Bureau and its powers can be found in Kang and Kum, eds., *Kantō daishinsai to Chōsenjin*, 71–73.
30. *Osaka mainichi shinbun*, September 6, 1923.
31. Kagawa Toyohiko, "Shinsai kyūgo undō o kaerimite" (Looking back on the earthquake disaster relief movement), in Gendaishi no kai, *Kantō daishinsai: Dokyumento*, 278–81.
32. Anonymous, "Shikai no toshi o meguru," 40–41.
33. Takenobu, ed., *The Japan Year Book, 1924–25*, 214.
34. Tokyoshi, ed., *Teito fukkōsaishi* (Celebrating the completion of the imperial capital reconstruction) (Tokyo: Tokyoshi, 1932), 571.
35. The following story is recounted in Nagata's memoirs. Ibid., 575–76.
36. Tasaki and Sakamoto, eds., *Rikugun kaigenrei kankei shiryō*, 214–15.

37. Bureau of Social Affairs, *The Great Earthquake of 1923 in Japan*, 129; Weiner, *The Origins of the Korean Community in Japan, 1910-1923*, 164–65.
38. Bureau of Social Affairs, *The Great Earthquake of 1923 in Japan*, 322–23.
39. Den Kenjirō denki hensankai, ed., *Den Kenjirō den* (The official biography of Den Kenjirō) (Tokyo: Den Kenjirō denki hensankai, 1932), 536–46.
40. Bureau of Social Affairs, *The Great Earthquake of 1923 in Japan*, 323.
41. The articles specifically spelled out for requisition included foodstuffs; water; fuel, including oil, gasoline, charcoal, and firewood; houses; building materials; medicine; instruments of conveyance, including automobiles, ships, and railroad freight cars; electrical wiring; and labor. See ibid., 563–65.
42. Ibid., 563–64. Given that the average day laborer in Tokyo made just over ¥2 per day, a ¥3,000 fine was a sizable figure.
43. Takenobu, ed., *The Japan Year Book, 1924–25*, 60.
44. Miyao Shunji's reflections are found in Tokyo Shisei Chōsakai, ed., *Teito fukkō hiroku*, 73–74.
45. Tokyoshi, ed., *Teito fukkōsaishi*, 575–76.
46. Tanaka Masataka and Ōsaka Hideaki, eds., *Kantō daishinsai seifu rikukaigun kankei shiryō*, vol. 3: *Kaigun kankei shiryō* (Materials related to the navy) (Tokyo: Nihon keizai hyōronsha, 1997), 3: 3–11; Takenobu, ed., *The Japan Year Book, 1924–25*, 75–76.
47. Hirakata, Ōtake, and Matsuo, eds., *Seifu kaigenrei kankei shiryō*, 474, 481–82.
48. Ibid., 480–83; Kagawa, "Shinsai kyūgo undō o kaerimite," 280–82.
49. Masuko Kikuyoshi and Sakai Genzō, eds., *Taru o tsukue toshite* (A barrel as a desk) (Tokyo: Seibundō shoten, 1923), 62.
50. Kagawa, "Shinsai kyūgo undō o kaerimite," 281.
51. Ibid., 281–82.
52. Bureau of Social Affairs, *The Great Earthquake of 1923 in Japan*, 325–26.
53. Ibid., 322–23; Kagawa, "Shinsai kyūgo undō o kaerimite," 280–81.
54. Hirakata, Ōtake, and Matsuo, eds., *Seifu kaigenrei kankei shiryō*, 643–44.
55. Ibid., 644–46.
56. Nihon tōkei fukyūkai, ed., *Teito fukkō jigyō taikan*, vol. 1, part 5, pp. 10–11.
57. Ibid.
58. Anonymous, "Shikai no toshi o meguru," 34–35.
59. Namae Takayuki, "Barakku mondai," in Kaizōsha, ed., *Taishō daishinkasai shi* (Documents of the Taishō great earthquake and conflagration) (Tokyo: Kaizōsha, 1924), 31–32.
60. *Japan Weekly Chronicle*, September 6, 1923, 343.
61. *Tokyo nichi nichi*, September 10, 1923, 4.
62. Namae, "Barakku mondai," 30–31.
63. Ibid., 32.
64. *Osaka mainichi shinbun* (English edition), October 5, 1923, 4.
65. *Japan Weekly Chronicle*, September 20, 1923, 396.
66. Namae, "Barakku mondai," 32–35. Namae provides a table with data on the smaller barracks, which eventually housed 39,977 refugees.
67. Takenobu, ed., *The Japan Year Book, 1924–25*, 225.

68. Namae, "Barakku mondai," 32.
69. *Osaka mainichi shinbun* (English edition), September 26, 1923, 1.
70. Yasunari Jirō, "Barakku kara barakku e" (From one barrack to another), in Gendaishi no kai, *Kantō daishinsai: Dokyumento*, 265, 270.
71. *Osaka mainichi shinbun* (English edition), September 27, 1923, 3.
72. Bureau of Social Affairs, *The Great Earthquake of 1923 in Japan*, 341–42.
73. Nagai, "Daishinsai to shakai seisaku," 8–9.
74. The text of Imperial Ordinance 414, which related to laws governing temporary housing, can be found in Bureau of Social Affairs, *The Great Earthquake of 1923 in Japan*, 609–10.
75. *Japan Weekly Chronicle*, September 13, 1923, 371.
76. Ibid., September 20, 1923, 349.
77. Hirakata, Ōtake, and Matsuo, eds., *Seifu kaigenrei kankei shiryō*, 472–73.
78. Tawara, "Teito fukkō to jinshin," 15–17.
79. Shimamoto Ainosuke, "Seishinteki fukkō saku" (Measures for spiritual restoration), *Teiyū rinrikai rinri kōenshū* 36:254 (1923): 102–11.
80. Ichiki, "Fukkō shigi," 5.
81. Matsui Shigeru, "Jikeidan no kaizō ni tsuite" (Concerning the reconstruction of self-defense groups), *Chihō gyōsei* 31:10 (October 1923): 12–13.
82. Suematsu, "Daishin ni kansuru shokan," 20–23.
83. Ugaki Kazushige, *Ugaki Kazushige nikki*, 3 vols. (Tokyo: Asahi shinbunsha, 1954), 3:445–46.
84. Tsuchida Hiroshige, "Kantō daishinsai go no shimin sōdōin mondai ni tsuite: Osaka no jirei o chūshin ni" (Concerning the problems related to the mobilization of citizens after the Great Kantō Earthquake: The case of Osaka), *Shigaku zasshi* 106:12 (December 1997): 63–64.
85. Tokyoshi, *Tokyo shinsairoku*, 3:1633–34.

3. COMMUNICATION

1. Tokyoshi, ed., *Tokyo shinsairoku* (Record of the Tokyo earthquake), 5 vols. (Tokyo: Tokyo shiyakusho, 1926), 2: 430–31.
2. *Tokyo nichi nichi shinbun*, September 9, 1923, 2.
3. Ibid., September 8, 1923, 1.
4. *Osaka asahi shinbun*, September 4, 1923, 1.
5. Ibid., September 5, 1923, 2.
6. Ibid., September 2, 1923, 1; *Osaka mainichi shinbun*, September 2, 1923, 1.
7. *Osaka mainichi shinbun*, September–10. These films were also advertised in this paper's English editions.
8. Ibid., September 13, 1923, 1.
9. Ibid. (English edition), 2.
10. Ibid. (English edition), October 6, 1923, 1.
11. *Tokyo nichi nichi shinbun*, September 6, 1923, 1.

12. *Kyoto hinode shinbun*, September 3, 1923, 1.
13. *Otaru shinbun*, September 6, 1923, 1.
14. *Fukuoka nichi nichi shinbun*, September 9, 1923, 1.
15. Takenobu, ed., *The Japan Year Book, 1924-25*, 249, 275.
16. Tokyoshi, ed., *Teito fukkō jigyō zuhyō* (Imperial capital reconstruction project diagrams) (Tokyo: Tokyoshi, 1930), 3.
17. Tokyoshi, ed., *Teito fukkōsaishi*, 585-86.
18. Tokyofu, ed., *Taishō shinsai biseki* (Beautiful outcomes from the Taishō earthquake) (Tokyo: Tokyofu, 1924). This volume is discussed in Borland, "Rebuilding Schools and Society After the Great Kantō Earthquake, 1923-1930," 101-2.
19. *Yomiuri shinbun*, October 8, 1923, 2. Given that many Koreans were held in so-called protective custody following the disaster, this story seems fanciful.
20. Ibid., October 19, 1923, 2.
21. Ibid., November 17, 1923, 4.
22. Borland, "Stories of Ideal Japanese Subjects from the Great Kantō Earthquake of 1923," 21-34.
23. The stories were published in Monbushō futsū gakumukyoku, ed., *Shinsai ni kansuru kyōiku shiryō* (Education materials related to the earthquake), 3 vols. (Tokyo: Monbushō, 1923).
24. Nihon teikoku gikai, ed., *Dai Nihon teikoku gikaishi*, 14:1455-56.
25. *Kyoto hinode*, October 11 and 12, 1923. Kurushima's tour was detailed in newspapers across Japan. See *Osaka mainichi shinbun*, October 6, 1923.
26. This mirrored the behavior of the Imperial Japanese Navy following the Russo-Japanese War. Throughout October and November 1905, the navy implemented a number of traveling naval pageants in which they displayed picture postcards of victorious battles, captured Russian military hardware, and other memorabilia that highlighted the navy's role in securing victory over Russia. See J. Charles Schencking, *Making Waves: Politics, Propaganda and the Emergence of the Imperial Japanese Navy, 1868-1922* (Stanford: Stanford University Press, 2005), 110-22.
27. Tokyo shinsai kinen jigyō kyōkai, ed., *Hifukushōato: Tokyo shinsai kinen jigyō kyōkai hōkoku* (Remains of the Honjo Clothing Depot: Report by the Tokyo earthquake commemorative project association) (Tokyo: Tokyo shinsai kinen jigyō kyōkai, 1932).
28. Gennifer Weisenfeld discusses one lithograph series produced by Ishikawa Shoten entitled *Tokyo daishinsai gahō* (Great Tokyo earthquake). See Weisenfeld, "Imaging Calamity," 25-29.
29. Andrew Gordon, *Labor and Imperial Democracy in Prewar Japan* (Berkeley: University of California Press, 1991), 44-50.
30. *Japan Weekly Chronicle*, November 8, 1923, 638.
31. Ibid., November 15, 1923, 659.
32. *Hōchi shinbun*, September 5, 1923, 2; *Osaka mainichi shinbun*, September 5, 1923, 1.
33. Tanaka, "Shitai no nioi," 78-80.
34. Anonymous, "Shikai no toshi o meguru," 35-36.
35. Kawatake, "Sōnan ki," 29-30.

36. Soeda, "Taishō daishinsai no uta," 210–12.
37. Takenobu, ed., *The Japan Year Book, 1924–25*, 224.
38. *Osaka mainichi shinbun* (English edition), September 11, 1923, 1.
39. Takenobu, ed., *The Japan Year Book, 1924–25*, 224–25.
40. Nihon tōkei fukyūkai, ed., *Teito fukkō jigyō taikan*, vol. 1, part 13, 26–27.
41. *Tokyo nichi nichi shinbun* (English edition), October 1 and 2, 1923, 1; *Osaka mainichi shinbun* (English edition), October 2, 1923, 1.
42. *Osaka mainichi shinbun* (English edition), October 2, 1923, 1.
43. Nakamura, "Hifukushō ato," 95–97.
44. Anonymous, "Shikai no toshi o meguru," 38.
45. Uno, "San byaku nen no yume," 39.
46. *Hōchi shinbun*, September 10, 1923, 2; September 20 and 28, 1923, 1; *Tokyo nichi nichi shinbun*, September 20, 1923, 1.
47. Soeda Asenbō, "Taishō daishinsai no uta," 212.
48. Kitō Shirō, "Shinda hito" (Dead people), in *Tokyo shiritsu shōgakkō jidō: Shinsai kinen bunshū* (A collection of essays commemorating the earthquake by Tokyo's primary school children), ed. Tokyo shiyakusho, 7 vols., (Tokyo: Baifūkan, 1924), 3:233.
49. Tawara, "Teito fukkō to jinshin," 14.
50. Anonymous, "Shikai no toshi o meguru," 39.
51. *Osaka mainichi shinbun* (English edition), September 25, 1923, 3.
52. Tokyo shinsai kinen jigyō kyōkai, ed., *Hifukushōato: Tokyo shinsai kinen jigyō kyōkai hōkoku*, 313–17; *Tokyo nichi nichi shinbun*, October 20, 1923, 1; *Osaka mainichi shinbun* (English edition), October 21, 1923, 3.
53. Tokyoshi ed., *Tokyo shinsairoku*, 2: 430–33.
54. Nagata Hidejirō, "Teito fukkō to shimin no danryokusei" (The reconstruction of the imperial capital and resilience of the people), *Toshi kōron* 6:11 (November 1923): 25.
55. Sandra Wilson, "The Past in the Present: War in Narratives of Modernity in the 1920s and 1930s," in *Being Modern in Japan: Culture and Society from the 1910s to the 1930s*, ed. Elise Tipton and John Clark, 170–73 (Honolulu: University of Hawaii Press, 2000).
56. Takashima, "Shin Tokyo no kensetsu to Tokyokko no iki," 55, 53.
57. Mizusawa shiritsu Gotō Shinpei kinenkan, ed., *Gotō Shinpei monjo*, reel 56, 21–11 (Tokyo: Yūshōdō fuirumu shuppan, 1979).
58. Hoashi, "Tenrai no kyōkun," 27–28.
59. Tokyoshi, ed., *Tokyo shinsairoku*, 2:430–31.
60. Abe, "Teito fukkō ni yōsuru ōnaru gisei," 17.
61. Okutani Fumitomo, "Kantō no daisaigai wa ikanaru shin'i ka" (In what ways was the Great Kantō disaster divine will?), *Michi no tomo* 401 (November 5, 1923): 10–11.
62. Ōki Tōkichi, "Kokumin shisō zendō saku" (Policies for guiding the nation's ideology), *Chihō gyōsei* 32:2 (February 1924): 20; Tomoeda Takahiko, "Shinsai to dōtokuteki fukkō (The earthquake disaster and moral reconstruction), *Shakai seisaku jihō* 10:38 (November 1923): 66.
63. Shimamoto, "Seishinteki fukkō saku," 102.

4. ADMONISHMENT

1. For descriptions of 1920s Asakusa, see Ishizumi Harunosuke, *Asakusa ritan* (Little-known stories of Asakusa) (Tokyo: Bungei shijō, 1927); Yamada Taichi, *Tochi no kioku Asakusa* (Asakusa local memories) (Tokyo: Iwanami shoten, 2000); Yoshimi Shun'ya, *Toshi no doramaturugi: Tokyo sakariba no shakaishi* (The dramaturgy of the city: The social history of Tokyo entertainment districts) (Tokyo: Kōbundō, 1987); Soeda Azenbō, *Asakusa teiryūki* (Record of the Asakusa underworld) (Tokyo: Kindai seikatsusha, 1930); Miriam Silverberg, *Erotic Grotesque Nonsense: The Mass Culture of Japanese Modern Times* (Berkeley: University of California Press, 2006), 177–202.
2. Seidensticker, *Low City, High City*, 119; Donald Richie, "Forward," in Kawabata Yasunari, *The Scarlet Gang of Asakusa* (Berkeley: University of California Press, 2005), x.
3. Journalist Matsubara Iwagorō described Asakusa as "a vast garbage dump" whose unhealthy elements contaminated the entire city. See David Ambaras, *Bad Youth: Juvenile Delinquency and the Politics of Everyday Life in Modern Japan* (Berkeley: University of California Press, 2006), 36.
4. Okutani, "Kantō no daisaigai wa ikanaru shin'i ka," 15–16.
5. Abe, "Teito fukkō ni yōsuru ōnaru gisei," 19.
6. Ugaki, *Ugaki Kazushige nikki*, 1:452.
7. Ichiki, "Fukkō shigi," 6–7.
8. Hiroi Osamu, *Saigai to Nihonjin: Kyodai jishin no shakai shinri* (Disasters and Japanese people: Social psychology of large-scale earthquakes) (Tokyo: Jiji Tsūshinsha, 1995); Bates, "Catfish, Super Frog, and the End of the World," 13–19; Smits, "Shaking Up Japan," 1050–51; John Milne, *Seismology* (London: Kegan Paul, 1898), 25–26.
9. Agustin Udias and Alfonso Lopez Arroyo, "The Lisbon Earthquake of 1755 in Spanish Contemporary Authors," in *The 1755 Lisbon Earthquake: Revisited*, ed. L.A. Mendes-Victor et al. (New York: Springer-Verlag, 2008), 14; Liba Taub, *Ancient Meteorology* (London: Routledge, 2003), 184–86; Milne, *Seismology*, 25–26; John Gates Taylor, Jr., "Eighteenth-Century Earthquake Theories: A Case-History Investigation into the Character of the Study of the Earth in the Enlightenment," Ph.D. dissertation, University of Oklahoma, 1975, 2–11.
10. Smits, "Shaking Up Japan," 1050–51; Clancey, *Earthquake Nation*, 144–45; Hashimoto Manpei, *Jishingaku koto hajime* (Beginnings of earthquake studies) (Tokyo: Asahi Shinbunsa, 1983).
11. David Milne, "Notices of Earthquake Shocks in Great Britain, and Especially in Scotland, with Inferences Suggested by These Notices as to the Nature and Causes of Such Shocks," *Edinburgh New Philosophical Journal* 31 (1841): 92–93, as quoted in Taylor, "Eighteenth-Century Earthquake Theories," 224.
12. Clancey, *Earthquake Nation*, 6–90.
13. Imamura would be shunned from mainstream Japanese seismology, in part due to publication of a 1905 article in the popular journal *Taiyō* in which he suggested that Tokyo was due for a major earthquake, which would likely be followed by a series of catastrophic fires. For a discussion on the Ōmori-Imamura rift that developed in the early part of the twentieth century, see ibid., 217–20.

14. Charles Davison called Ōmori "our leader in seismology" in Ōmori's obituary published in *Nature*. See Charles Davison, "Fusakichi Omori," *Nature* 113:2830 (1924): 133.
15. Doering, "The Politics of Nature," 195–214; Schencking, "The Great Kantō Earthquake and the Culture of Catastrophe and Reconstruction in 1920s Japan," 294–334; Steinberg, *Acts of God*; Schwartz, "The Hurricane of San Ciriaco," 303–34.
16. For a brief description of this organization, see William P. Woodard, "Interfaith Communication and the Confrontation of Religions in Japan," *Journal of the American Academy of Religion* 32:3 (September 1964): 233–34. The Three Religions Conference is discussed in Anesaki Masaharu, *History of Japanese Religion* (Tokyo: Charles Tuttle, 1963), 388–89.
17. Kiitsu kyōkai, ed., *Shinsai ni kansuru shūkyō dōtokuteki kansatsu* (Moral and religious observations related to the earthquake disaster) (Tokyo: Kiitsu kyōkai, 1925).
18. Shitennō Nobutaka, "Kōkū hōmen yori kantaru teito no seishin teki fukkō ou ronzu" (Discussing the spiritual reconstruction of the imperial capital from the viewpoint of aviation), *Chihō gyōsei* 32:1 (January 1924): 34.
19. Kiitsu kyōkai, ed., *Shinsai ni kansuru shūkyō dōtokuteki kansatsu*, 1–2, 7.
20. Okutani, "Kantō no daisaigai wa ikanaru shin'i ka," 15–16, 13.
21. Excerpts from Okutani's lecture tour are contained in the *Michi no tomo* publication. Many of the quotations above were taken from an October 8 lecture in Kōchi Prefecture.
22. Takashima openly opposed the Russo-Japanese War of 1904–1905, protested against the Ashio Copper Mine for polluting the environment in the postwar era, and had close ties to Meiji-era socialist Kōtoku Shūsui. See Stephen Large, "Buddhism, Socialism, and Protest in Prewar Japan: The Career of Seno'o Girō," *Journal of Japanese Studies* 21:1 (Winter 1987): 153–54.
23. Takashima, "Shin Tokyo no kensetsu to Tokyokko no iki," 56–57.
24. Haruno, "Shiren ni tōmen shite," 5–6.
25. Struber, "Divine Punishment or Object of Research?" 173–74.
26. Thomas Prentice, *Observations Moral and Religious on the Late Terrible Night of the Earthquake: A Sermon Preached at the Thursday Lecture in Boston, January 1st, 1756* (Boston: S. Kneeland, 1756), 357; quoted on Clark, "Science, Reason, and an Angry God," 357.
27. Other excellent studies that discuss divine interpretations of catastrophes not just confined to Europe are Clark, "Science, Reason, and an Angry God"; Russell Dynes and Daniel Yutzy, "The Religious Interpretation of Disaster," *Topic 10: A Journal of Liberal Arts* (Fall 1965): 34–48; Johns, "Introduction," xi–xxv; Kempe, "Noah's Flood," 151–71; Mulcahy, "A Tempestuous Spirit Called Hurri Cano," 11–38; Carl Smith, "Faith and Doubt: The Imaginative Dimensions of the Great Chicago Fire," in *American Disasters*, ed. Steven Biel, 129–69 (New York: New York University Press, 2001); Struber, "Divine Punishment or Object of Research?" 173–93; Maxine van de Wetering, "Moralizing in Puritan Natural Science: Mysteriousness in Earthquake Sermons," *Journal of the History of Ideas* 43:3 (July–September 1982): 417–38.
28. Suematsu, "Daishin ni kansuru shokan," 19–20.

29. Ugaki, *Ugaki Kazushige nikki*, 1:445.
30. Horie Ki'ichi, "Tokyoshi no saigai to keizaiteki fukkōan" (Tokyo's disaster and the economic restoration plan), *Chūō kōron* 38:11 (November 1923): 50–51.
31. Ichiki, "Fukkō shigi," 6–7.
32. Fukasaku Yasubumi, *Shakai sōsaku e no michi* (The path toward the creation of society), Tokyo: Kōbundō shoten, 1924, 162–72.
33. Ibid., 163–64.
34. Yamada Yoshio, *Kokumin seishin sakkō ni kansuru shōsho gikai* (A commentary on the imperial edict concerning the invigoration of the national spirit) (Tokyo: Hōbunkan, 1933), 89–90.
35. Sheldon Garon, *The State and Labor in Modern Japan* (Berkeley: University of California Press, 1987), 40–42.
36. Naimushō shakaikyoku shakaibu, eds., *Wakatsuki naimu daijin enjustu yōshi kinken shōrei ni tsuite* (A summary of Home Minister Wakatsuki Reijirō's speech regarding the encouragement of thrift and diligence) (Tokyo: Shakaikyoku, 1924), 3.
37. Ugaki, *Ugaki Kazushige nikki*, 1:445–447. Ugaki wrote in his diary that he believed Japan should move the capital.
38. Tawara, "Teito fukkō to jinshin," 18–19, 14.
39. Hoashi, "Tenrai no kyōkun," 26–27.
40. Fukasaku, *Shakai sōsaku e no michi*, 172–76.
41. Okutani, "Kantō no daisaigai wa ikanaru shin'i ka," 14–15.
42. Andre Sorensen, *The Making of Urban Japan* (London: Routledge, 2002), 62; Fujimori Terunobu, *Meiji no Tokyo keikaku* (Tokyo planning in the Meiji period) (Tokyo: Iwanami shoten, 2004), 15–27.
43. Fujimoto Taizō, *The Nightside of Japan* (London: T. Werner Laurie, 1915), 28–30.
44. Okutani, "Kantō no daisaigai wa ikanaru shin'i ka," 15–16.
45. These retailers were also located in the Ginza district. Horie, "Tokyoshi no saigai to keizaiteki fukkōan," 50–51.
46. Horie, "Tokyoshi no saigai to keizaiteki fukkōan, 51–52. The tale of the Heike documents a struggle between two clans for the control of Japan at the end of the Heian period. A central theme of this epic is the Buddhist law of impermanence (*mujō*). The tale suggests that the Taira clan sowed the seeds of their own demise through arrogance, hubris, and pride.
47. Fukasaku Yasubumi, "Kokumin seishin sakkō no goshōsho o haishite" (Having the honour of receiving the imperial rescript concerning the encouragement of national spirit), *Tōa no hikari* 19:11 (November 1924): 4–8.
48. Yamada, *Kokumin seishin sakkō ni kansuru shōsho gikai*, 88.
49. Ibid., 111–12.
50. Shimamoto, "Seishinteki fukkō saku," 105, 102–3, 107–8.
51. Haruno Ki'ichi, "Shiren ni tōmen shite" (Facing up to the ordeal), *Michi no tomo* 399 (October 5, 1923): 6.
52. Fukasaku "Kokumin seishin sakkō no goshōsho o haishite," 5–6.
53. Barbara Sato, *The New Japanese Woman: Modernity, Media, and Women in Interwar Japan* (Durham: Duke University Press, 2003), 125–26.

54. *Tokyo asahi shinbun*, July 8, 1923, 1.
55. Fukasaku, "Kokumin seishin sakkō no goshōsho o haishite," 6.
56. Nose Yoritoshi, "Fukkō seishin nahen ni ariya" (Where is the spiritual restoration), *Kyōiku* 502 (January 1925): 1–13. Quotes are from pp. 4–6.
57. *Japan Weekly Chronicle*, January 3, 1924, 25.
58. Ibid., October 4, 1923, 462, 464–65.
59. Garon, *Molding Japanese Minds*, 98–104.
60. Ichiki, "Fukkō shigi," 6.
61. Okutani, "Kantō no daisaigai wa ikanaru shin'i ka," 13–15.
62. Horie, "Tokyoshi no saigai to keizaiteki fukkōan," 52.
63. Fujimoto, *The Nightside of the City*, 1.
64. Ibid., 8. The stylistic errors and grammatical inconsistencies are repeated as they appear. The publishers claimed that this book, written by a Japanese, was a more authentic travel guide of Japan than those published by Westerners. They highlight in the preface that they had thought of giving the manuscript to an editor to correct but decided that revision would have destroyed much of its quaint charm.
65. *Tokyo nichi nichi shinbun*, September 24, 1923, 2; *Osaka mainichi shinbun*, September 27, 1923, 1.
66. Kawabata Yasunari, *The Scarlet Gang of Asakusa*, trans. Alisa Freedman (Berkeley: University of California Press, 2005), 59–61.
67. Nonomura Kaizō, "Kindai bunmei no hakai" (The destruction of modern civilization), *Kyōiku jiron* 1385 (December 1923): 2–8.
68. Okutani, "Kantō no daisaigai wa ikanaru shin'i ka," 15.
69. Sorensen, *The Making of Urban Japan*, 92–108; Yazaki Takeo, *Social Change and the City in Japan* (Tokyo: Japan Publications, 1968), 391–450.
70. Sorensen, *The Making of Urban Japan*, 92; Yazaki, *Social Change and the City in Japan*, 391.
71. Yazaki, *Social Change and the City in Japan*, 419, 390.
72. Ōkawa Kazushi, eds., *Chōki keizai tōkei* (Long-term economic statistics), 14 vols. (Tokyo: Tōyō keizai shinpōsha, 1967), 12: 228–29.
73. Mark Metzler, *Lever of Empire: The International Gold Standard and the Crisis of Liberalism in Prewar Japan* (Berkeley: University of California Press, 2006), 94–96.
74. Yazaki, *Social Change and the City in Japan*, 429, 425–26.
75. Ōkawa, *Chōki keizai tōkei*, 8:135–36.The figures are adjusted, with 1912 as a base of 100.
76. The amount of money spent on food in 1912 was ¥2.642 billion, while the 1920 figure stood at ¥7.299 billion. The amount for 1922 was ¥7.029 billion.
77. Ibid., 8:244–45. Yazaki listed Asakusa as an area known for home industries that specialized in *geta* making. See Yazaki, *Social Change and the City in Japan*, 457–59. One hundred *sen* equaled ¥1.
78. Ōkawa, *Chōki keizai tōkei*, 8:244–245.
79. Ibid., 6:250–51.
80. Sabine Früstrück, *Colonizing Sex: Sexology and Social Control in Modern Japan* (Berkeley: University of California Press, 2003); Ambaras, *Bad Youth*, 1–2.
81. Sato, *The New Japanese Woman*, 71.

82. Takenobu, ed., *The Japan Yearbook 1917*, 254–55; Takenobu, ed., *The Japan Yearbook, 1924–25*, 265–67.
83. Takashima Heizaburō, "Hensai o kikai ni" (The disaster as opportunity), *Jidō kenkyū* 27:2 (October 1923): 71.
84. Shimamoto, "Seishinteki fukkō saku," 105.

5. OPTIMISM

1. Wakatsuki Reijirō, "Teito fukkō to keizaisei keikaku" (The reconstruction of the imperial capital and financial planning), *Toshi kōron* 6:11 (November 1923): 18–19.
2. Fukuda, "Fukkō Nihon tōmen no mondai," 8. For information on Fukuda's background, see Inoue Takutoshi and Yagi Ki'ichirō, "Two Inquirers on the Divide: Tokuzō Fukuda and Hajime Kawakami," in *Economic Thought and Modernization in Japan*, ed. Sugihara Shiro and Tanaka Toshihiro, 60–77 (Cheltenham: Edward Elgar, 1998); Nishizawa Tamotsu, "Lujo Brentano, Alfred Marshall, and Tokuzō Fukuda," in *The German Historical School: The Historical and Ethical Approach to Economics*, ed. Shionoya Yuichi, 155–72 (London: Routledge, 2001).
3. Gotō Shinpei, "Toshi keikaku ni hitsuyō naru chishiki oyobi zaigen" (The financial resources and knowledge necessary for city planning), *Toshi kōron* 6:8 (August 1923): 2–14.
4. Abe, "Teito fukkō ni yōsuru ōnaru gisei," 19.
5. Tsurumi Yūsuke ed., *Gotō Shinpei* (Official biography of Gotō Shinpei), 4 vols. (Tokyo: Gotō Shinpei kaku denki hensankai, 1937–38): 4:586.
6. Abe, "Teito fukkō ni yōsuru ōnaru gisei," 17.
7. Nagai Tōru, "Daishinsai to shakai seisaku" (The great earthquake disaster and social policy), *Shakai seisaku jihō* 10:38 (November 1923): 13.
8. Sorensen, *The Making of Urban Japan*, 108.
9. Yazaki, *Social Change and the City in Japan*, 457–58.
10. Ōkawa Kazushi, ed., *Chōki keizai tōkei*, 8:243–51.
11. For an overview of the transformation in thinking about social problems as a manifestation of urban problems, see Koji Taira, "Urban Poverty, Ragpickers, and the Ants' Villa in Tokyo," *Economic Development and Cultural Change* 17:2 (January 1969): 156–60; Jeffrey Hanes, *The City as Subject: Seki Hajime and the Reinvention of Modern Osaka* (Berkeley: University of California Press, 2002), 64–189; Garon, *Molding Japanese Minds*, 42–57; and Sorensen, *The Making of Urban Japan*, 91–104. For contemporary accounts of poverty and the urban problem in Japan, see Tokyoshi, ed., *Tokyo no shinai no saimin ni kansuru chōsa* (An investigation into the poor in Tokyo), (Tokyo: Tokyoshi shakaikyoku, 1921); Inoue Teizō, *Hinminkutsu to Shōsū dōbō* (The slums and the minority brethren) (Tokyo: Ganshobō, 1923); Kusama Yasoo, "Dai Tokyo no saimingai to seikatsu no taiyō" (The poor-quarters and the living conditions in greater Tokyo), in *Nihon chiri taikei* (A compendium on Japanese geography), ed. Kaizōsha, 12 vols. (Tokyo: Kaizōsha, 1929–1931), 3: 370–80; Tokyofu gakumubu (Tokyo prefecture department of education), ed., *Tokyofu gunbu rinsetsu gogun shūdanteki furyō jūtaku chikuzushū* (Charts and diagrams on substandard hous-

ing in Tokyo prefecture and five neighboring counties) (Tokyo: Tokyofu gakumubu shakaika, 1928); and Tokyofu gakumubu (Tokyo prefecture department of education), ed., *Tokyofu gunbu ni okeru shūdanteki furyō jūtaku chiku jōkyō chōsa* (A survey of areas with a cluster of substandard housing) (Tokyo: Tokyofu gakumubu shakaika, 1930).

12. Yazaki, *Social Change and the City in Japan*, 449–51. The "most notorious indigent communities" were also named, including Kanda-misaki-chō, Shiba-shin'ami-chō, Asakusa-tamahime-chō, Fukagawa-tomioka-chō, Kyōbashi-hatchōbori-chō, and Yotsuya-samegahashi.
13. Ibid., 450; Tokyoshi, ed., *Tokyo no shinai no saimin ni kansuru chōsa*, 186.
14. For a discussion of the problems associated with Tokyo's slums, see Kikuchi Shinzō, "Teito fukkō to shakai seisaku" (The reconstruction project of the imperial capital and social policy), *Shakai seisaku jihō* 10:38 (October 1923): 211–12; and Abe Isoo, "Teito no kensetsu to sōzōteki seishin" (The construction of the imperial capital and the creative spirit), *Kaizō* 5:11 (November 1923): 65–67.
15. Fukuda, "Fukkō Nihon tōmen no mondai," 9–10.
16. Jeffrey Hanes examines this "discovery" of the urban problem in relation to Osaka. Hanes, *The City as Subject*, 174–91. For a related discussion on the discovery of "poverty" in Japan, both rural and urban, see W. Dean Kinzley, "Japan's Discovery of Poverty: Changing Views of Poverty and Social Welfare in the Nineteenth Century," *Journal of Asian History* 22:1 (January 1988): 1–24.
17. Hanes, *The City as Subject*, 185–86; Garon, *The State and Labor in Modern Japan*, 25–27.
18. Details on Gotō's early life can be found in Yukio Hayase, "The Career of Gotō Shinpei: Japan's Statesman of Research, 1857–1929," Ph.D. dissertation, Florida State University, 1974, 7–25.
19. Mizusawa shiritsu Gotō Shinpei kinenkan, eds., "Kenkō keisatsu ikan o mōkubeki kengen" (Proposal to establish health police officers), in *Gotō Shinpei monjo*, 3-1-b.
20. Poultney Bigelow, *Japan and Her Colonies* (London: Edward Arnold, 1923), 21.
21. Quoted in Frederick Dickinson, *War and National Reinvention: Japan in the Great War, 1914-1919* (Cambridge: Harvard University Press, 1999), 57.
22. Mochida Nobuki, "Gotō Shinpei to shinsai fukkō jigyō" (Gotō Shinpei and the earthquake reconstruction project), *Shakai kagaku kenkyū* 35:2 (1983): 4–7.
23. Hayase, "The Career of Gotō Shinpei," 118–20.
24. Koshizawa Akira, *Shokuminchi manshū no toshi keikaku* (The urban planning of colonial Manchuria) (Tokyo: Ajia keizai kenkyūjo, 1978), 34–48; Mochida, "Gotō Shinpei to shinsai fukkō jigyō," 6–7.
25. Tsurumi, *Gotō Shinpei*, 2:878–91.
26. Andre Sorensen, "Urban Planning and Civil Society in Japan: Japanese Urban Planning Development During the Taishō Democracy Period (1905–1931)," *Planning Perspectives* 16 (2001): 398–99.
27. Garon, *Molding Japanese Minds*, 50–52.
28. Ibid.; David Ambaras, *Bad Youth*, 4–8; Sally Hastings, *Neighborhood and Nation in Tokyo, 1905-1937* (Pittsburgh: University of Pittsburgh Press, 1995).
29. Tsurumi, *Gotō Shinpei*, 4:323–24.

30. Sorensen, "Urban Planning and Civil Society in Japan," 399.
31. A detailed budgetary chart of Gotō's ¥800 million plan is found in Nihon tōkei fukyūkai, ed., *Teito fukkō jigyō taikan*, chap. 9, inserted between pp. 14 and 15. A discussion of this plan can be found in Tsurumi, *Gotō Shinpei*, 4:245–52; and Murataka Mikihiro, *Tokyoshi no kaizō* (Reconstruction of Tokyo) (Tokyo: Minyūsha, 1922). Murataka's book is a detailed history of this ¥800 million plan and the responses to it. Published before the Great Kantō Earthquake hit, with the provocative title *Reconstruction of Tokyo*, the author laments that Gotō's plan was rejected and details how and why Tokyo must be renovated or reconstructed to become more modern and livable.
32. Murataka, *Tokyoshi no kaizō*, 128–29. Tokyo city officials were more precise. They claimed that only 1.7 percent of Tokyo was park space in 1923. Tokyo Municipal Office, *The Reconstruction of Tokyo*, 288.
33. Tsurumi, *Gotō Shinpei*, 4:252–60.
34. Ibid., 249–52, 267–85.
35. Hanes, *The City as Subject*, 174–93, 210–30; Jordan Sand, *House and Home in Modern Japan: Architecture, Domestic Space, and Bourgeois Culture, 1880–1930* (Cambridge: Harvard University Press, 2003), 203–21.
36. James C. Scott, *Seeing Like a State: How Certain Schemes to Improve the Human Condition Have Failed* (New Haven: Yale University Press, 1998), 88–102, quotes on pp. 90, 91.
37. Nagai, "Daishinsai to shakai seisaku," 23–24.
38. Ibid., 23–25.
39. Ibid., 16.
40. Tomoeda, "Shinsai to dōtokuteki fukkō," 67–69.
41. Ibid., 60, 67–70.
42. Kikuchi Shinzō, "Teito fukkō to shakai seisaku" (The reconstruction project of the imperial capital and social policy), *Shakai seisaku jihō* 10:38 (October 1923): 207–14; Kikuchi Shinzō, *Toshi keikaku to dōro gyōsei* (Urban planning and road administration) (Tokyo: Sūbundō shuppan, 1928).
43. Kikuchi, "Teito fukkō to shakai seisaku," 207–9.
44. Abe, "Teito fukkō ni yōsuru ōnaru gisei," 17–18.
45. Abe, "Teito no kensetsu to sōzōteki seishin," 59–60, 65–67.
46. Kobashi Ichita, "Teito fukkō to toshi keikaku no seishin" (Reconstruction of the imperial capital and the spirit of city planning), *Toshi kōron* 6:11 (November 1923): 30, 31.
47. Ibid.
48. Ibid., 32–33.
49. Mononobe, "Shinsai chihō fukkō ni taisuru kibō," 27–28.
50. Nagata, "Teito fukkō to shimin no danryokusei," 25.
51. Shitennō, "Kōkū hōmen yori kantaru teito no seishin teki fukkō ou ronzu," 34–35.
52. Ibid., 37, 33, 36.
53. Ibid., 36–37, 38.
54. Tsurumi, *Gotō Shinpei*, 4:625–47; Gotō Shinpei, "Teito fukkō ron" (Discussions concerning the imperial capital reconstruction), *Toshi kōron* 6:11 (November 1923): 5–8;

Gotō Shinpei, "Teito no daishinsai to jichiteki seishin no kanyō" (The great imperial capital earthquake disaster and the cultivation of an autonomous spirit), *Toshi kōron* 6:12 (December 1923): 12–14; Tokyo shisei chōsaikai, ed., *Teito fukkō hiroku*, 256–58; Ikeda Hiroshi, "Teito fukkō no yurai to hōsei" (The origins and legislation of the imperial capital reconstruction), *Toshi mondai* 10:4 (April 1930): 647–52.

55. Gotō Shinpei, "Teito fukkō keikaku no ōzuna" (The broad outline of the reconstruction plan for the imperial capital), *Toshi kōron* 6:12 (December 1923): 11–12; Tsurumi, *Gotō Shinpei*, 4:682–83.
56. Gotō, "Teito fukkō keikaku no ōzuna," 11.
57. Ibid., 14.
58. Anan, "Toshi seikon to daieidan," 81–82.
59. Horie, "Tokyoshi no saigai to keizaiteki fukkōan," 50–52.
60. Takashima, "Shin Tokyo no kensetsu to Tokyokko no iki," 53.
61. Ibid., 56–57.
62. Ugaki, *Ugaki Kazushige nikki*, 1:448, 445–46.
63. Charles Beard, *The Administration and Politics of Tokyo: A Survey and Opinions* (New York: Macmillan, 1923).
64. Charles Beard, *Biiado hakushi Tokyo fukkō ni kansuru iken* (Opinions on the reconstruction of Tokyo by Beard) (Tokyo: Tokyo shisei chōsakai, 1924), 1–2.
65. Tsurumi, *Gotō Shinpei*, 4:614–23.
66. Beard, *Biiado hakushi Tokyo fukkō ni kansuru iken*, 4–29.
67. Prime Minister Yamamoto Gonnohyōe was a retired admiral while Gotō was president of the Boy Scouts of Japan Association. The mural is discussed in Weisenfeld, "Imaging Calamity," 28.
68. Tsurumi, *Gotō Shinpei*, 4:724–27.
69. The *Jiji shinpō*, however, did ask its readers in 1924 to share their ideas and desires as to the style and type of city they hoped would be reconstructed. See Jiji shinpōsha, ed., *Atarashii Tokyo to kenchiku no hanshi* (Discussions on new Tokyo and architecture) (Tokyo: Jiji shinpōsha, 1924); cited in Sand, *House and Home in Modern Japan*, 207.

6. CONTESTATION

1. Mizuno Rentarō, "Gotō hakushaku tsuioku zadankai" (Round-table recollections of Count Gotō), *Toshi mondai* 8:6 (June 1926): 181. Part of this discussion is cited in Fukuoka Shunji, *Tokyo fukkō keikaku: Toshi saikaihatsu gyōsei no kōzō* (The reconstruction plan of Tokyo: The structure of urban redevelopment administration) (Tokyo: Nihon hyōronsha, 1991), 117–18.
2. Tsurumi, *Gotō Shinpei*, 4:737–42.
3. Ibid., 589–93.
4. Ibid., 593–94.
5. Ibid., 592, 597–98.
6. Tokyo shisei chōsakai, ed., *Teito fukkō hiroku*, 263–64.
7. Ibid., 264.

8. Yamada Hiroyoshi, "Fukkō gairo hi no kettei suru made" (Until we decide on the cost for road reconstruction) *Toshi mondai* 10:4 (April 1930): 193–94.
9. Aoki, *Inoue Junnosuke den*, 211–15.
10. Ikeda, "Teito fukkō no yurai to hōsei" 642–44; Tsurumi, *Gotō Shinpei*, 4:598–603.
11. Gotō's submission is contained in Tsurumi, *Gotō Shinpei*, 4:599–602.
12. Aoki, *Inoue Junnosuke den*, 211–12.
13. Matsuki Kan'ichirō, "Fukkōin no kaiko" (Reflections on the reconstruction institute), *Toshi mondai* 10:4 (April, 1930): 23–24; Tsurumi, *Gotō Shinpei*, 4:603–4; Matsuki Kan'ichirō denki hensankai hensan, eds., *Matsuki Kan'ichirō* (Official biography of Matsuki Kan'ichirō) (Tokyo: Matsuki Kan'ichirō denki hensankai hensan, 1941), 154–65.
14. Miyao Shunji, "Teito fukkō keikaku to sono yosan" (The imperial capital reconstruction plan and its budget), *Toshi mondai* 10:4 (April 1930): 606–7.
15. Tokyo shisei chōsakai, eds., *Teito fukkō hiroku*, 266.
16. Miyao, "Teito fukkō keikaku to sono yosan," 610.
17. Tokyo shisei chōsakai, ed., *Teito fukkō hiroku*, 266–67.
18. Miyao, "Teito fukkō keikaku to sono yosan," 606–8.
19. Aoki, *Inoue Junnosuke den*, 216–17.
20. Miyao, "Teito fukkō keikaku to sono yosan," 608.
21. Ibid., 612–13; Aoki, *Inoue Junnosuke den*, 212–15.
22. Miyao, "Teito fukkō keikaku to sono yosan," 612–13.
23. The plan submitted to Gotō and the cabinet from the Reconstruction Institute is discussed in Ikeda, "Teito fukkō no yurai to hōsei," 646–51; and Tsurumi, *Gotō Shinpei*, 4:625–36.
24. Details of this encounter are taken from Aoki, *Inoue Junnosuke den*, 221–24; Tokyo shisei chōsakai, eds., *Teito fukkō hiroku*, 296–98.
25. Aoki, *Inoue Junnosuke den*, 223.
26. Formal meetings were held on November 1 and 9. Teito fukkōin, ed., *Teito fukkōin sanyokai sokkiroku* (Stenographic record of the Imperial Capital Reconstruction Institute councilor's committee), 2 vols. (Tokyo: Teito fukkōin, 1924).
27. Ikeda, "Teito fukkō no yurai to hōsei," 650–52.
28. Fukkō chōsa kyōkai, ed., *Teito fukkōshi*, 1:111–12.
29. Metzler, *Lever of Empire*, 138–45.
30. Tokyo shisei chōsakai, ed., *Teito fukkō hiroku*, 280–81; Aoki, *Inoue Junnosuke den*, 213–15.
31. Tokyo shisei chōsakai, ed., *Teito fukkō hiroku*, 280–81; Tsurumi, *Gotō Shinpei*, 4:641–43; Miyao, "Teito fukkō keikaku to sono yosan," 607; Mochida, "Gotō Shinpei to shinsai fukkō jigyō," 20.
32. Aoki, *Inoue Junnosuke den*, 212–15.
33. Takakura, *Tanaka Giichi denki*, 2:339–43.
34. Tokyo shisei chōsakai, ed., *Teito fukkō hiroku*, 292–93; Tsurumi, *Gotō Shinpei*, 4:643.
35. Tokyo shisei chōsakai, ed., *Teito fukkō hiroku*, 282.
36. Ikeda, "Teito fukkō no yurai to hōsei," 641–42. This organization is also described in Nakamura Akira, "Shinsai fukkō no seijigaku" (The political science of earthquake reconstruction), *Seikei ronsō* 51:3–4 (April 1982): 64–75.

37. Gotō Shinpei wrote in November before Yamamoto convened the Shingikai that he viewed it as a "necessary organization" that could help "shape a nonpolitical party alliance when promoting the reconstruction of the central city of politics, economy, and culture." See Gotō Shinpei, "Teito fukkōron," *Toshi kōron* 6:11 (November 1923): 3.
38. The shorthand notes and minutes of the Shingikai's meetings can be found at the Tokyo Institute for Municipal Research (Tokyo shisei chōsakai) archives in Hibiya, Tokyo (hereafter abbreviated as TIMR). See TIMR Folder TIMR-OB447 *Teito fukkō shingikai sokkiroku*, November 24, a.m. meeting, 1–6. These handwritten minutes are not paginated but are broken down into sections: a.m. and p.m. meetings on November 24 and the final meeting on November 27. When the minutes are cited in the following section, I list page numbers for ease of referencing and use a.m. and p.m. to distinguish between the meetings. No formal minutes akin to these exist for the two special committee (*tokubetsu iinkai*) meetings of the Shingikai held on November 25 and 26. Detailed transcripts, however, are contained in Fukkō chōsa kyōkai, ed., *Teito fukkōshi* (A history of the imperial capital reconstruction), 3 vols. (Tokyo: Kōbundō, 1930), 1:183–203.
39. The draft budget is contained in Tsurumi, *Gotō Shinpei*, 4:653–54, while Gotō's and Miyao's orations are contained in TIMR-OB447, *Teito fukkō sokkiroku*, November 24, a.m. meeting, 1–6.
40. TIMR-OB447, *Teito fukkō sokkiroku*, November 24, a.m. meeting, 6–19.
41. Ibid., 20–66; Tsurumi, *Gotō Shinpei*, 4:657–63; Itō Miyoji's recollection in Tokyo shisei chōsakai, ed., *Teito fukkō hiroku*, 20–25; and Nakamura, "Shinsai fukkō no seijigaku," 69–72.
42. Tsurumi, *Gotō Shinpei*, 4:659
43. Tokyo shisei chōsakai, eds., *Teito fukkō hiroku*, 283–84; Tsurumi, *Gotō Shinpei*, 4:663–64.
44. TIMR-OB447, *Teito fukkō sokkiroku*, November 24, a.m. meeting, 67–70.
45. The text of Gotō's response can be found in ibid., p.m. meeting, 1–19; Tsurumi, *Gotō Shinpei*, 4: 667–76.
46. Tokyo shisei chōsakai, eds., *Teito fukkō hiroku*, 292.
47. TIMR-OB447, *Teito fukkō sokkiroku*, November 24, p.m. meeting, 3–4; Tsurumi, *Gotō Shinpei*, 4:667–68.
48. Tsurumi, *Gotō Shinpei*, 4:668–69.
49. Fukkō chōsa kyōkai, eds., *Teito fukkōshi*, 1:185.
50. Tsurumi, *Gotō Shinpei*, 4:674–75.
51. TIMR-OB447, *Teito fukkō sokkiroku*, November 24, p.m. meeting, 19–26.
52. Ibid., 39–40, for Itō; 40–61 for Egi; 75–76 for Aoki; 76–82 for Takahashi; and 82–83 for Katō.
53. Fukkō chōsa kyōkai, eds., *Teito fukkōshi*, 1:183–84. The breakdown of outstanding loans, their year of issue, the amount borrowed, interest payable, and final year of redemption is found in Takenobu, ed., *The Japan Year Book, 1924–25*, 488–90.
54. Fukkō chōsa kyōkai, ed., *Teito fukkōshi*, 1:184.
55. TIMR-OB447, *Teito fukkō sokkiroku*, November 24, p.m. meeting; Nakamura, "Shinsai fukkō no seijigaku," 64–67.

56. Anan Jōichi, "Teito fukkō shingikai" (The Imperial Capital Reconstruction deliberative committee), *Toshi kōron* 6:12 (December 1923): 96–98.
57. Fukuoka, *Tokyo no fukkō keikaku*, 166.
58. Nakamura, "Shinsai fukkō no seijigaku," 30–33.
59. TIMR-OB447, *Teito fukkō sokkiroku*, November, p.m. meeting, 83–87.
60. Fukkō chōsa kyōkai, ed., *Teito fukkōshi*, 1:193–94.
61. TIMR-OB447, *Teito fukkō sokkiroku*, November 17 meeting, 1–4, *Seiyū* 277 (January 1924): 3–4, Tsurumi, *Gotō Shinpei*, 4:687–89.
62. Fukkō chōsa kyōkai, ed., *Teito fukkōshi*, 1:200–201.
63. *Gotō Shinpei monjo*, R56: 21–26; Tsurumi, *Gotō Shinpei*, 4:684–88.
64. Tsurumi, *Gotō Shinpei*, 4:724–27.
65. Ibid., 4:682; Inoue Junnosuke's reflection in Tokyo shisei chōsakai, ed., *Teito fukkō hiroku*, 282.
66. Aoki, *Inoue Junnosuke den*, 225–26.
67. The tale of the forty-seven rōnin is an embellished historical account documenting the exploits of forty-seven samurai left masterless after their daimyō, Asano Naganori, was forced to commit ritual suicide for assaulting a court official named Kira Yoshinaka. For two years, the forty-seven rōnin plotted and eventually took revenge by killing Kira. Afterward, they committed suicide.
68. Nihon teikoku gikai, ed., *Dai Nihon teikoku gikaishi*, 14:136–37 (hereafter cited as *DNTG*); Naimushō fukkōkyoku, ed., *Teito fukkō jigyōshi* (Records of the imperial capital reconstruction project), 6 vols. (Tokyo: Fukkō jimukyoku, 1931–32), 4:157–59.
69. *DNTG*, 14: 1441–42; Aoki, *Inoue Junnosuke den*, 226–29.
70. *DNTG*, 14:1442–45.
71. Ibid., 1444.
72. Aoki, *Inoue Junnosuke den*, 193–197.
73. Seiyūkai, ed., *Seiyū* 277 (January 15, 1924): 11.
74. *DNTG*, 14:1441.
75. Ibid., 1492–96.
76. Ibid., 1478–79.
77. Naimushō fukkōkyoku, ed., *Teito fukkō jigyōshi*, 4:164–65.
78. *DNTG*, 14:1450–52.
79. Seiyūkai, ed., *Seiyū* 277 (January 15, 1924): 5.
80. Naimushō fukkōkyoku, ed., *Teito fukkō jigyōshi*, 4:171–72; Seiyūkai, ed., *Seiyū* 277 (January 15, 1924): 5.
81. Naimushō fukkōkyoku, ed., *Teito fukkō jigyōshi*, 4:168–80.
82. Tsurumi, Gotō Shinpei, 4:700–701.
83. Ibid., 701–8.
84. Naimushō fukkōkyoku, ed., *Teito fukkō jigyōshi*, 4:188–189.
85. Ibid., 200–201. Tagawa's plans for Tokyo are contained in Tagawa Daikichirō, *Tsukuraru beki Tokyo* (The Tokyo that should be built) (Tokyo: Hatano Jūtarō, 1923).
86. Seiyūkai, ed., *Seiyū* 277 (January 15, 1924): 3.
87. Miyao, "Teito fukkō keikaku to sono yosan," 613.

88. Tsurumi, *Gotō Shinpei*, 4:729–30.
89. Ibid., 730.
90. The draft dissolution proposal is quoted in ibid., 734–35.
91. *Japan Weekly Chronicle*, January 10, 1924, 49.
92. Tsurumi, *Gotō Shinpei*, 4:737–42.
93. *Japan Weekly Chronicle*, September 13, 1923, 359.
94. Mark A. Healy, "The Ruins of New Argentina: Peronism, Architecture and the Remaking of San Juan After the 1944 Earthquake," Ph.D. dissertation, Duke University, 1999.
95. Charles Beard, "Goto and the Rebuilding of Tokyo," *Our World* (April 1924): 21.

7. REGENERATION

1. Munakata Itsurō, "Saigo kokumin no kakugo sanken no uta" (The resolution of the nation after the disaster—A song of three ken), *Kyōiku* 20:48 (November 1923): 61–62.
2. Ambaras, *Bad Youth*, 4–5.
3. Takashima, "Shin Tokyo no kensetsu to Tokyokko no iki," 56–57.
4. Ichiki, "Fukkō shigi," 5–6.
5. For the text of the rescript and commentary on it, see Yamada, *Kokumin seishin sakkō ni kansuru shōshō gikai*, 80–86. This is a reprint of the first edition published in 1923 with an expanded section on the importance of the rescript given the state of emergency that the author believed existed in Japan in 1933.
6. Ibid., 80–84.
7. Ibid., 84–85.
8. Ibid., 48–49.
9. The text of these speeches can be found in ibid., 48–58. On November 17 Okano sent his message to every school principal and headmaster in Japan.
10. *Japan Weekly Mail*, November 22, 1923, 712–13.
11. Arima Sukemasa, "Seishin sakkō to Nogi taishō" (Spiritual promotion and general Nogi), *Tōa no hikari* 19:11 (November 1924): 12.
12. Ibid., 13–19.
13. Fukasaku, "Kokumin seishin sakkō no goshōsho o haishite," 7–8, 4.
14. Fukasaku, *Shakai sōsaku e no michi*, 172–83.
15. Ibid., 174–75.
16. Ibid., 175–76.
17. Ibid., 180–82.
18. Takashima, "Hensai o kikai ni," 71–72.
19. Takashima, "Daishinkasai to kodomo," 52–53.
20. Takashima, "Hensai o kikai ni," 71–72.
21. Miyazaki kenritsu Kobayashi chūgakkō, ed., "Kokumin seishin sakkō hōan" (Ideas for the promotion of national spirit), *Kyōiku* 495 (August 1924): 56–57.
22. Ibid., 57–59.
23. Ibid., 59–61.

24. Miyake Setsurei, *"Teito fukkō to seishin sakkō"* (Imperial capital reconstruction and spiritual renewal) (Tokyo: Nihon seinenkan, n.d), 9. Held in the Tokyo Metropolitan Archives under file 093.B MI, 8-9.
25. Abe Isoo, "Shinsai ni ataeta kokuminteki jikaku" (Self-awakening of the nation as a result of the earthquake), *Chihō gyōsei* 31:12 (December 1923): 20.
26. Abe, "Shinsai ni ataeta kokuminteki jikaku," 21.
27. Ōkawa, ed., *Chōki keizai tōkei*, 6: 222-223.
28. Abe, "Shinsai ni ataeta kokuminteki jikaku," 22.
29. Tawara, "Teito fukkō to jinshin," 18-19.
30. Abe, "Shinsai ni ataeta kokuminteki jikaku," 19-20.
31. Ibid., 22-23.
32. Yamakawa, "Teito fukkō to shakai shugiteki seisaku," 265-66.
33. Ugaki, *Ugaki Kazushige nikki*, 1:447-49, 451-52.
34. Nagai, *Kokumin seishin to shakai shisō* (National spirit and social ideology) (Tokyo: Ganshōdō shoten, 1924), 160-61.
35. Nagai, "Daishinsai to shakai seisaku," 17.
36. Fukuda, "Fukkō Nihon tōmen no mondai," 15.
37. Sheldon Garon, "Luxury Is the Enemy: Mobilizing Savings and Popularizing Thrift in Wartime Japan," *Journal of Japanese Studies* 26:1 (Winter 2000): 45-50; Sandra Wilson, *The Manchurian Crisis and Japanese Society, 1931-1933* (London: Routledge, 2002), 139-42.
38. *Yamato*, April 10, 1924; *Japan Weekly Chronicle*, April 14, 1924.
39. *Japan Weekly Chronicle*, July 10, 1924, 21.
40. The tariff bill was reproduced word for word in ibid., July 24, 1924, 122-23.
41. Ibid., July 10, 1924, 19-20.
42. Ibid., July 17, 1924, 29.
43. Ibid.
44. Ibid., 28.
45. Ibid., 25.
46. Ibid., 31.
47. *Japan Weekly Chronicle*, July 24, 1924, 41-43. His 1925 work is Sakaguchi Takenosuke, *Kōtō shōhingaku* (A study of high end merchandising) (Tokyo: Sanseidō, 1925).
48. Sheldon Garon, *Beyond Our Means: Why America Spends When the Rest of the World Saves* (Princeton: Princeton University Press, 2011), 223-33; Garon, "Saving for 'My Own Good and the Good of the Nation,'" 103-7; Sheldon Garon, "Fashioning a Culture of Diligence and Thrift," in *Japan's Competing Modernities: Issues in Culture and Democracy, 1900-1930*, ed. Sharon Minichiello, 312-34 (Honolulu: University of Hawaii Press, 1998).
49. Naimushō shakaikyoku shakaibu, ed., *Kinken shōrei undō gaikyō* (Outline of the campaign to encourage diligence and thrift) (Tokyo: Naimushō shakaikyoku shakaibu, 1927); Naimushō shakaikyoku shakaibu, ed., *Kinken shōrei eiga sujigakuishū* (Plot outlines from movies to encourage diligence and thrift) (Tokyo: Naimushō shakaikyoku shakaibu, 1926); Nihon kangyō ginkō, *Kinken shōrei ni kansuru kōen yōshi* (A summary of the essential points concerning the encouragement of diligence and thrift) (Tokyo: Nihon kangyō ginkō, 1924), 1-64; Naimushō shakaikyoku, ed., *Kinken kyōchō gaikan* (Outline on the importance of diligence and thrift) (Tokyo: Naimushō shakaikyoku, 1925), 1-12.

50. *Trans-Pacific*, August 28, 1926, 15.
51. Nose, "Fukkō seishin nahen ni ariya," 11–12. Katō's full speech was also translated and published in *Japan Weekly Chronicle*, September 4, 1924, 339.
52. Naimushō shakaikyoku, ed., *Wakatsuki naimu daijin enjutsu yōshi kinken shōrei ni tsuite*, 1, 9–10.
53. Ibid., 6–9.
54. Ibid., 10–14.
55. Ibid., 15.
56. Statistics Bureau, ed., *Historical Statistics of Japan*, 5 vols. (Tokyo: Japan Statistical Association, 1987), 3:239–41.
57. Horie Ki'ichi's statement was reprinted from the journal *Chūō kōron* in the *Japan Weekly Chronicle*, October 23, 1924, 561.
58. Ibid., November 20, 1924, 673.
59. Ibid., October 23, 1924, 561–62.
60. Statistics Bureau, ed., *Historical Statistics of Japan*, 3:182. All statistics in this paragraph related to postal savings come from this source.
61. *Trans-Pacific*, August 21, 1926, 20; Takenobu, ed., *The Japan Year Book 1929*, 485.
62. Statistics Bureau, ed., *Historical Statistics of Japan*, 3:239–41.
63. Ōkawa, *Chōki keizai tōkei*, 6:208–9.
64. Nose, "Fukkō seishin nahen ni ariya," 3–8.
65. *Jiji manga* 228, August 31, 1925; Weisenfeld, "Imaging Calamity," 125.
66. *Trans-Pacific*, February 20, 1926, 16.
67. Andrew Gordon, "Consumption, Leisure and the Middle Class in Transwar Japan," *Social Science Japan Journal* 10:1 (May 2007): 5–6.
68. Garon, *Molding Japanese Minds*, 52–56.
69. Statistics Bureau, ed., *Historical Statistics of Japan*, 5:52.
70. Uno, "San byaku nen no yume," 39.

8. READJUSTMENT

1. Nagata Hidejirō, *Kukaku seiri ni tsuite shimin shokun ni tsugu* (Notice to the people regarding land readjustment), May 10, 1924, Tokyo Metropolitan Archives, 093.22 NA, 1–4.
2. Koshizawa Akira, *Tokyo no toshi keikaku* (Urban planning of Tokyo) (Tokyo: Iwanami shoten, 1991), 59–86.
3. Nagata, *Kukaku seiri ni tsuite shimin shokun ni tsugu*, 3–9.
4. Egi Kazuyuki, *Egi hōjō no kenpō ihanron* (A commentary on Justice Minister Egi's discussion on constitutional law violations), Tokyo, n.d. Held at Tokyo Institute for Municipal Research Archives, TIMR OBZ-160.
5. Seidensticker, *Tokyo Rising*, 8–11.
6. Takebe Rokuzō, "Ni jū man tō no tatemono iten" (The transfer of 200,000 buildings), *Toshi mondai* 10:4 (April 1930): 139–40.
7. Ikeda Hiroshi, "Teito fukkō to tokubetsu toshi keikaku hō" (Imperial capital reconstruction and the special urban planning law), *Toshi kōron* 7:3 (March 1924): 2–10; a condensed English translation of this law is found in Bureau of Social Affairs, *The*

Great Earthquake of 1923 in Japan, 572–81. The law is discussed in Sorensen, *The Making of Urban Japan*, 125–28.
8. Sorensen, *The Making of Urban Japan*, 122–23.
9. The most thorough explanation of the land readjustment system can be found in Sano Toshikata, "Kukaku seiri an no seiritsu made" (Toward the establishment of plans for the land readjustment), in *Teito tochi kukaku seiri ni tsuite* (Concerning land readjustment in the imperial capital), ed. Gotō Shinpei, 27–36 (Tokyo: Kōseikai, 1924).
10. The source for this map is Takeuchi Rokuzō, "Kukaku seiri no sekkei narabi ni shikō" (The design and implementation of land readjustment), in ibid., 41.
11. Fukkō chōsa kyōkai, eds., *Teito fukkōshi*, 2:1477–78, 1418–19, 1420.
12. Ibid., 1477–78.
13. Ibid., 1480–82.
14. Ibid., 1484–85, 1488–89.
15. Tokyo Municipal Office, *The Reconstruction of Tokyo*, 248.
16. Fukkō chōsa kyōkai, ed., *Teito fukkōshi*, 2:1422–28.
17. Tokyo Municipal Office, *The Reconstruction of Tokyo*, 248.
18. Takebe, "Ni jū man tō no tatemono iten," 140–41.
19. Fukkō chōsa kyōkai, ed., *Teito fukkōshi*, 2:1431–37; Takebe, "Ni jū man tō no tatemono iten," 141.
20. Fukkō chōsa kyōkai, ed., *Teito fukkōshi*, 2:1429–30.
21. Takebe, "Ni jū man tō no tatemono iten," 138–39.
22. Fukkō chōsa kyōkai, ed., *Teito fukkōshi*, 2:1437.
23. Takebe, "Ni jū man tō no tatemono iten," 158.
24. Fukkō chōsa kyōkai, ed., *Teito fukkōshi*, 2:828, 1430–39.
25. Ibid., 828, 1437
26. Ibid., 883–84.
27. Ibid., 884.
28. Ibid., 889–91.
29. Ibid., 887–88.
30. Ibid., 888.
31. Ibid., 1288–89.
32. Ibid., 1422–29, 1318.
33. Ibid., 1318–19.
34. Ibid., 1321–22.
35. Ibid., 1323.
36. Ibid., 916–18.
37. Ibid., 928–29.
38. Ibid., 926–27.
39. Bureau of Social Affairs, *The Great Earthquake of 1923 in Japan*, 275–77. Statistics in this and the following paragraph are taken from this source.
40. Principle 5 of the "principles in the planning of roads and streets and the construction of bridges" adopted by the Home Ministry and the Tokyo municipal government stated that the government's aim was "to adopt, as far as possible, those road lines which were already determined and approved in the old scheme of town plan-

ning." Principle 10 suggested that, "in widening or constructing auxiliary roads with a width of 22 meters or less, little regard need be taken of slight irregularities or curves. As far as possible the existing roads and ways should be utilized for the purpose." Ibid., 201.
41. Nihon tōkei fukyūkai ed., *Teito fukkō jigyō taikan*, vol. 1, chap., charts on pp. 6–9.
42. Ibid., charts on p. 10.
43. Bureau of Social Affairs, *The Great Earthquake of 1923 in Japan*, 278–83; Nihon tōkei fukyūkai ed., *Teito fukkō jigyō taikan*, vol. 2, chap. 13, charts on pp. 7–18.
44. Two other projects funded entirely by the municipal government were also clearly successful in that they were vast improvements on what stood before 1923 and improved the quality of life for Tokyo's residents: primary schools and the city's sewage treatment system. For a detailed discussion of the reconstructed primary schools, see Borland, "Rebuilding Schools and Society after the Great Kantō Earthquake, 1923–1930."
45. Tokyo Municipal Office, *Tokyo, Capital of Japan Reconstruction Work, 1930* (Tokyo: Tokyo Municipal Office, 1933), 95–99.
46. Murataka, *Tokyo no kaizō*, 128–29.
47. Yasui Sei'ichirō, "Teito shakai jigyō no fukkō" (The reconstruction of imperial capital social welfare facilities), *Toshi mondai* 10:4 (April 1930): 210.
48. Ibid., 215–17.
49. Ibid., 217–18.
50. Tokyo shinsai kinen jigyō kyōkai, ed., *Hifukushōato*, 94–95.
51. Ibid., 95.
52. Ibid., 96–97.
53. *Japan Weekly Chronicle*, April 9, 1925, 450.
54. Tokyo shinsai kinen jigyō kyōkai, ed., *Hifukushōato*, 97–98.
55. *Trans-Pacific*, March 20, 1926, 13.
56. Ibid., April 3, 1926, 15.
57. Tokyo shinsai kinen jigyō kyōkai, ed., *Hifukushōato*, 102–4.
58. Ibid., 103–4.
59. Ibid., 104.
60. *Trans-Pacific*, November 12, 1927, 12.
61. Tokyo shinsai kinen jigyō kyōkai, ed., *Hifukushōato*, 104–5.
62. Ibid., 105–6, 109–12.

9. CONCLUSION

1. A detailed foldout map of the imperial inspection tour is included in Tokyoshi, ed., *Teito fukkōsaishi*.
2. *Trans-Pacific*, April 3, 1930, 1.
3. Tokyoshi, ed., *Teito fukkōsaishi*, 595.
4. Ronald Schaffer, *Wings of Judgment: American Bombing in World War II* (Oxford: Oxford University Press, 1988), 115–16; Conrad Crane, *Bombs, Cities, and Civilians*, (Lawrence: University Press of Kansas, 1993), 126–27.

5. Quoted in Koshizawa Akira, "Tokyo-to shikeikaku no shisō: Sono rekishi kōsatsu," 77.
6. Hastings, *Nation and Neighborhood in Tokyo*, 51.
7. Ian Buruma, "Japan's Next Transformation," in *Reimagining Japan: The Quest for a Future That Works*, ed. McKinsey & Company (San Francisco: VIZ Media LLC, 2011), 42.
8. Ōnishi Norimitsu, "In Tsunami Aftermath, 'Road to Future' Unsettles a Village," *New York Times*, December 31, 2011.
9. Hoffman, "After Atlas Shrugs," 311.
10. Ibid., 310.
11. Miyake, *Teito fukkō to seishin sakkō*, 12–13.
12. Nose, "Fukkō seishin nahen ni ariya," 3.
13. A larger discussion of Japan's emerging technocratic elites can be found in Janis Mimura, *Planning for Empire: Reform Bureaucrats and the Japanese Wartime State* (Ithaca: Cornell University Press, 2011), 29–40.
14. Ben-Ami Shillony, *Politics and Culture in Wartime Japan* (Oxford: Clarendon Press, 1981), 2–4.
15. Hastings, *Neighborhood and Nation in Tokyo*, 72–79.
16. Gregory Kasza, *The Conscription Society: Administered Mass Organizations* (New Haven: Yale University Press, 1995).
17. Thomas Havens, *Valley of Darkness: The Japanese People in World War Two* (New York: Norton, 1978), 38–40.
18. Matsushita Senkichi, "Kokuminteki kokubō no kyōiku wa masumasu hitsuyō nari" (The ever increasing need for national defence education), *Naigai kyōiku hyōron* 18:8 (August 1924): 6–7.
19. Ugaki, *Ugaki Kazushige nikki*, 1:445.
20. Tsuchida Hiroshige, "Kantō daishinsai go no shimin sōdoin mondai ni tsuite: Osaka no jirei o chūshin ni" (Concerning the problem of general civil mobilization following the Great Kantō Earthquake: The case of Osaka), *Shigaku zasshi* 106:12 (December 1997): 59–80.
21. Ibid., 74–75.
22. Terada Torahiko, *Tensai to kokubō* (Natural disasters and national defense) (Tokyo: Iwanami shoten, 1938), 137, 133–34.
23. Miura, *Planning for Empire*, 140–69.
24. Havens, *Valley of Darkness*, 18–19.
25. Garon, *Molding Japanese Minds*, 25–145; Elise Tipton, "Cleansing the Nation: Urban Entertainments and Moral Reform in Interwar Japan," *Modern Asian Studies* 42:4 (July 2008): 705–13.

BIBLIOGRAPHY

NEWSPAPERS

Fukuoka nichi nichi shinbun
Hōchi shinbun
Japan Weekly Chronicle
Kyoto hinode shinbun
Kyūshū nippō
New York Times
Osaka asahi shinbun
Osaka mainichi shinbun
Otaru shinbun
Tokyo nichi nichi shinbun
Yamato

SOURCES IN JAPANESE

Abe Isoo. "Shinsai ni ataeta kokuminteki jikaku" (Self-awakening of the nation as a result of the earthquake). *Chihō gyōsei* 31:12 (December 1923): 19–25.
———. "Teitō fukkō ni yōsuru ōnaru gisei" (The large sacrifice needed for imperial capital reconstruction). *Kaizō* 5:10 (October 1923): 17–24.

———. "Teito no kensetsu to sōzōteki seishin" (The construction of the imperial capital and the creative spirit). *Kaizō* 5:11 (November 1923): 49–71.
Anan Jōichi, "Teito fukkō shingikai," (The Imperial capital reconstruction deliberative committee). *Toshi kōron* 6:12 (December 1923): 96–100.
———. "Toshi seikon to daieidan" (The reconstruction of the city and a resolute decision). *Toshi kōron* 6:11 (November 1923): 75–84.
Anonymous. "Shikai no toshi o meguru: Saigai yokuyokujitsu no dai Tokyo" (Great Tokyo two days after calamity: Walking around a dead city). In *Taishō daishinsai daikasai* (The Taishō great earthquake and conflagration). Edited by Dainihon yūbenkai kōdansha, 33–46. Tokyo: Dainihon yūbenkai kōdansha, 1923.
Aoki Tokuzō. *Inoue Junnosuke den* (Official biography of Inoue Junnosuke). Tokyo: Inoue Junnosuke ronsō hensankai, 1935.
Arano Kōhei. *Shinsai romansu: Sanwa to bidan no maki* (Earthquake romance: A volume of tragic stories and beautiful tales). Tokyo: Seishindō shoten, 1923.
Arima Sukemasa. "Seishin sakkō to Nogi taishō" (Spiritual promotion and General Nogi). *Tōa no hikari* 19:11 (November 1924): 12–19.
Beard, Charles. *Biiado hakushi Tokyo fukkō ni kansuru iken* (Opinions on the reconstruction of Tokyo by Beard). Tokyo: Tokyo shisei chōsakai, 1924.
Den Kenjirō denki hensankai, ed. *Den Kenjirō den* (The official biography of Den Kenjirō). Tokyo: Den Kenjirō denki hensankai, 1932.
Egi Kazuyuki. *Egi hōjō no kenpō ihanron* (A commentary on Justice Minister Egi's discussion on constitutional law violations). Tokyo, n.d. Held at Tokyo Institute for Municipal Research Archives TIMR OBZ-160.
Fujimori Terunobu. *Meiji no Tokyo keikaku* (Tokyo planning in the Meiji period). Tokyo: Iwanami shoten, 2004.
Fukasaku Yasubumi. "Kokumin seishin sakkō no goshōsho o haishite" (Having the honor of receiving the imperial rescript concerning the encouragement of national spirit). *Tōa no hikari* 19:11 (November 1924): 4–8.
———. *Shakai sōsaku e no michi* (The path toward the creation of society). Tokyo: Kōbundō shoten, 1924.
Fukkō chōsa kyōkai, ed. *Teito fukkōshi* (A history of the imperial capital reconstruction). 3 vols. Tokyo: Kōbundō, 1930.
Fukuda Tokuzō. "Fukkō Nihon tōmen no mondai" (The urgent questions confronting Japanese reconstruction). *Kaizō* 5:10 (October 1923): 3–16.
Fukuoka Shunji. *Tokyo fukkō keikaku: Toshi saikaihatsu gyōsei no kōzō* (The reconstruction plan of Tokyo: The structure of urban redevelopment administration). Tokyo: Nihon hyōronsha, 1991.
Funaki Yoshie. "Hi de shinauka, mizu o erabauka" (Death by burning or death by drowning). In *Kantō daishinsai: Dokyumento*. (The Great Kantō Earthquake: Documents). Edited by Gendaishi no kai, 53–57. Tokyo: Sōfūkan, 1996.
Gotō Shinpei. "Teito fukkō keikaku no ōzuna" (The broad outline of the reconstruction plan for the imperial capital). *Toshi kōron* 6:12 (December 1923): 11–14.
———. "Teito fukkō ron" (Discussions concerning the imperial capital reconstruction). *Toshi kōron* 6:11 (November 1923): 2–8.

———. "Teito no daishinsai to jichiteki seishin no kanyō" (The great imperial capital earthquake disaster and the cultivation of an autonomous spirit). *Toshi kōron* 6:12 (December 1923): 4–10.

———. "Toshi keikaku ni hitsuyō naru chishiki oyobi zaigen" (The financial resources and knowledge necessary for city planning). *Toshi kōron* 6:8 (August 1923): 2–14.

Harada Katsumasa and Shiozaki Fumio, eds. *Tokyo Kantō daishinsai zengo* (Tokyo before and after the Great Kantō Earthquake). Tokyo: Nihon keizai hyōronsha, 1997.

Haruno Ki'ichi. "Shiren ni tōmen shite" (Facing up to the ordeal). *Michi no tomo* 399 (October 5, 1923): 2–6.

Hashimoto Manpei. *Jishingaku koto hajime* (Beginnings of earthquake studies). Tokyo: Asahi shinbunsa, 1983.

Hata Ikuhiko. *Nihon riku-kaigun sōgō jiten* (A comprehensive encyclopedia of Japanese army and navy personnel). Tokyo: Tokyo daigaku shuppankai, 1991.

Hatano Masaru, ed. *Kantō daishinsai to Nichi-Bei gaikō* (The Great Kantō Earthquake and Japanese-American foreign relations). Tokyo: Sōshisha, 1999.

Hirakata Chieko, Ōtake Yoneko, and Matsuo Shōichi, eds. *Kantō daishinsai seifu, rikukaigun kankei shiryō*. Vol. 1: *Seifu kaigenrei kankei shiryō* (Materials related to the administration of martial law). Tokyo: Nihon keizai hyōronsha, 1997.

Hiroi Osamu. *Saigai to Nihonjin: Kyodai jishin no shakai shinri* (Disasters and Japanese people: Social psychology of large-scale earthquakes). Tokyo: Jiji Tsūshinsha, 1995.

Hoashi Ri'ichirō. "Tenrai no kyōkun" (Heaven-sent lessons). *Chihō gyōsei* 31:10 (October 1923): 25–29.

Horie Ki'ichi. "Tokyoshi no saigai to keizaiteki fukkōan" (Tokyo's disaster and the economic restoration plan). *Chūō kōron* 38:11 (November 1923): 34–52.

Ichiki Kitokurō. "Fukkō shigi" (Personal opinion on reconstruction). *Chihō gyōsei* 31:11 (November 1923): 2–8.

Ikeda Hiroshi. "Teito fukkō no yurai to hōsei" (The origins and legislation of the imperial capital reconstruction). *Toshi mondai* 10:4 (April 1930): 627–77.

———. "Teito fukkō to tokubetsu toshi keikaku hō" (Imperial capital reconstruction and the special urban planning law). *Toshi kōron* 7:3 (March 1924): 2–10.

Imamura Akitsune. "Shigaichi ni okeru jishin no seimei oyobi zaisan ni taisuru songai o keigen suru kanpō" (Simple methods for reducing the damage to life and property caused by an earthquake in an urban area). *Taiyō* 11:12 (December 1905): 162–71.

Inoue Teizō. *Hinminkutsu to Shōsū dōbō* (The slums and the minority brethren). Tokyo: Ganshobō, 1923.

Ishida Yorifusa. *Nihon kindai toshi keikakushi kenkyū* (A study on the history of urban planning in modern Japan). Tokyo: Kashiwa shobō, 1992.

Ishizumi Harunosuke. *Asakusa ritan* (Little-known stories of Asakusa). Tokyo: Bungei shijō, 1927.

Jiji shinpōsha, ed. *Atarashii Tokyo to kenchiku no hanashi* (Discussion on new Tokyo and architecture). Tokyo: Jiji shinpōsha, 1924.

Kagawa Toyohiko. "Shinsai kyūgo undō o kaerimite" (Looking back on the earthquake disaster relief movement). In *Kantō daishinsai: Dokyumento* (The Great Kantō Earthquake: Documents). Edited by Gendaishi no kai, 278–85. Tokyo: Sōfūkan, 1996.

Kang Tok-sang and Kum Pyong-dong, eds. *Kantō daishinsai to Chōsenjin* (Koreans and the Great Kantō Earthquake). Tokyo: Misuzu shobō, 1963.

Kawatake Shigetoshi. "Sōnan ki" (Meeting with disaster). In *Taishō daishinkasai shi* (Documents of the Taishō great earthquake and conflagration). Edited by Kaizōsha, 22–31. Tokyo: Kaizōsha, 1924.

Kawatani Kenzaburō (Shōtei). *Boshin shōsho santaichō* (The substance of the Boshin rescript). Tokyo: Kōshi shodōkai, 1935.

Kiitsu kyōkai, ed. *Shinsai ni kansuru shūkyō dōtokuteki kansatsu* (Moral and religious observations related to the earthquake disaster). Tokyo: Kiitsu kyōkai, 1925.

Kikuchi Shinzō. "Teito fukkō to shakai seisaku" (The reconstruction project of the imperial capital and social policy). *Shakai seisaku jihō* 10:38 (October 1923): 207–14.

——. *Toshi keikaku to dōro gyōsei* (Urban planning and road administration). Tokyo: Sūbundō shuppan, 1928.

Kitō Shirō. "Shinda hito" (Dead people). In *Tokyo shiritsu shōgakkō jidō: Shinsai kinen bunshū* A collection of essays commemorating the earthquake by Tokyo's primary school children. Edited by Tokyo shiyakusho. 7 vols. Tokyo: Baifūkan, 1924.

Ko Hakushaku Yamamoto kaigun taishō denki hensankai, ed. *Yamamoto Gonnohyōe den* (Official biography of Yamamoto Gonnohyōe). 2 vols. Tokyo: Ko Hakushaku Yamamoto kaigun taishō denki hensankai, 1938.

Kobashi Ichita. "Teito fukkō to toshi keikaku no seishin" (Reconstruction of the imperial capital and the spirit of city planning). *Toshi kōron*. 6:11 (November 1923): 30–33.

Koizumi Tomi. "Hifukushō seki sōnan no ki" (Meeting with a disaster at the site of the clothing factory). In *Taishō daishinkasai shi* (Record of the Great Taishō earthquake calamity). Edited by Kaizōsha, 13–22. Tokyo: Kaizōsha, 1924.

Koshizawa Akira. *Shokuminchi manshū no toshi keikaku* (The urban planning of colonial Manchuria). Tokyo: Ajia keizai kenkyūjo, 1978.

——. *Tokyo no toshi keikaku* (Urban planning of Tokyo). Tokyo: Iwanami shoten, 1991.

——. "Tokyo-to shikeikaku no shisō: Sono rekishi kōsatsu (Ideology of Tokyo city planning: A historical inquiry). *Shisō* 181:3 (March 1991): 66–99.

Kusama Yasoo. "Dai Tokyo no saimingai to seikatsu no taiyō" (The poor-quarters and the living conditions in greater Tokyo). In *Nihon chiri taikei* (A compendium on Japanese geography). Edited by Kaizōsha. 12 vols. (Tokyo: Kaizōsha, 1929–1931).

Masuko Kikuyoshi and Sakai Genzō, eds. *Taru o tsukue toshite* (A barrel as a desk). Tokyo: Seibundō shoten, 1923.

Matsuda Genji. "Teito fukkyū no yōgi" (The important meaning of the restoration of the imperial capital). *Toshi kōron* 6:11 (November 1923): 27–29.

Matsui Shigeru. "Jikeidan no kaizō ni tsuite" (Regarding the reconstruction of self-defense groups). *Chihō gyōsei* 31:10 (October 1923): 10–14.

Matsuki Kan'ichirō. "Fukkōin no kaiko" (Reflections on the Reconstruction Institute). *Toshi mondai* 10:4 (April 1930): 23–28.

Matsuki Kan'ichirō denki hensankai hensan, ed. *Matsuki Kan'ichirō* (Official biography of Matsuki Kan'ichirō). Tokyo: Matsuki Kan'ichirō denki hensankai hensan, 1941.

Matsuo Shōichi. "Chōsenjin gyakusatsu to guntai" (The Korean massacre and the army). *Rekishi hyōron* 9 (September 1993): 29–41.

———. "Kantō daishinsai shi kenkyū no seika to kadai" (Results and research tasks of historical study on the Great Kantō Earthquake). *Hosei daigaku tama ronshū* 9 (March 1993): 79–148

———. *Kantō daishinsai to kaigenrei* (The Great Kantō Earthquake and martial law). Tokyo: Yoshikawa kōbunkan, 2003.

Matsuo Takayoshi. "Kantō daishinsai ka no Chōsenjin gyakusatsu jiken: 2" (Massacre of Koreans at the time of the Great Kantō Earthquake: 2). *Shisō* 476 (February 1964): 110–20.

———. "Kantō daishinsai ka no Chōsenjin gyakusatsu jiken: 1" (Massacre of Koreans at the time of the Great Kantō Earthquake: 1). *Shisō* 471 (September 1963): 44–61.

Matsushita Senkichi. "Kokuminteki kokubō no kyōiku wa masumasu hitsuyō nari" (The ever increasing need for national defense education). *Naigai kyōiku hyōron* 18:8 (August 1924): 5–7.

Miura Tōsaku. "Hijōji arawareru kyōiku no kōka ni tsuite" (Concerning the effectiveness of education in times of crisis). *Kyōiku jiron* 1384 (1923): 9–12.

Miyake Setsurei. *Teito fukkō to seishin sakkō* (Imperial capital reconstruction and spiritual renewal). Tokyo: Nihon seinenkan, n.d. Held in the Tokyo Metropolitan Archives Under file 093.B MI.

Miyao Shunji. "Teito fukkō keikaku to sono yosan" (The imperial capital reconstruction plan and its budget). *Toshi mondai* 10:4 (April 1930): 605–15.

Miyazaki kenritsu Kobayashi chūgakkō, eds. "Kokumin seishin sakkō hōan" (Ideas for the promotion of national spirit). *Kyōiku* 495 (August 1924): 56–63.

Mizuno Rentarō. "Gotō hakushaku tsuioku zadankai" (Roundtable recollections of Count Gotō). *Toshi mondai* 8:6 (June 1926): 181–85.

Mizusawa shiritsu Gotō Shinpei kinenkan, eds. *Gotō Shinpei monjo*. Reel 56, 21–11. Tokyo: Yūshōdō fuirumu shuppan, 1979.

Mochida Nobuki. "Gotō Shinpei to shinsai fukkō jigyō: Ka no toshi supendingu" (Gotō Shinpei and the earthquake reconstruction project: Urban spending under chronic recession). *Shakai kagaku kenkyū* 35:2 (1983): 1–60.

Mochizuki Masanori. "Kantō daishinsai: Gyakusatsu jiken to fukkōron o chūshin ni" (The Great Kantō Earthquake: The Korean massacre and reconstruction). *Rekishi hyōron* 9 (September 1993): 74–83.

Monbushō futsū gakumukyoku, ed. *Shinsai ni kansuru kyōiku shiryō* (Education materials related to the earthquake). 3 vols. Tokyo: Monbushō, 1923.

Mononobe Nagao. "Shinsai chihō fukkō ni taisuru kibō" (Hopes for the reconstruction of the earthquake disaster area). *Chihō gyōsei* 32:1 (January 1924): 23–33.

Munakata Itsurō. "Saigo kokumin no kakugo sanken no uta" (The resolution of the nation after the disaster—A song of three ken). *Kyōiku* 20:48 (November 1923): 61–62.

Murataka Mikihiro. *Tokyoshi no kaizō* (Reconstruction of Tokyo). Tokyo: Minyūsha, 1922.

Nagai Tōru. "Daishinsai to shakai seisaku" (The great earthquake disaster and social policy). *Shakai seisaku jihō* 10:38 (November 1923): 7–27.

———. *Kokumin seishin to shakai shisō* (National spirit and social ideology). Tokyo: Ganshōdō shoten, 1924.

Nagata Hidejirō. *Kukaku seiri ni tsuite shimin shokun ni tsugu* (Notice to the people regarding land readjustment). May 10, 1924. Tokyo Metropolitan Archives, 093.22.

———. "Teito fukkō to shimin no danryokusei" (The reconstruction of the imperial capital and resilience of the people). *Toshi kōron* 6:11 (November 1923): 23–26.

Naimushō fukkōkyoku. *Teito fukkō jigyōshi* (Records of the imperial capital reconstruction project). 6 vols. Tokyo: Fukkō jimukyoku, 1931–1932.

Naimushō keihokyoku, ed. *Shinsaigo ni okeru keikai keibi ippan* (Vigilance units after the earthquake). Tokyo: Naimushō, n.d. Written in September 1923. Held at Tokyo Metropolitan Archives 093.22 NA.

Naimushō shakaikyoku shakaibu, ed. *Kinken kyōchō gaikan* (Outline on the importance of diligence and thrift). Tokyo: Naimushō shakaikyoku, 1925.

———. *Kinken shōrei eiga sujigakuishū* (Plot outlines from movies to encourage diligence and thrift). Tokyo: Naimushō shakaikyoku shakaibu, 1926.

———. *Kinken shōrei undō gaikyō* (Outline of the campaign to encourage diligence and thrift). Tokyo: Naimushō shakaikyoku shakaibu, 1927.

———. *Wakatsuki naimu daijin enjutsu yōshi kinken shōrei ni tsuite* (A summary of Home Minister Wakatsuki's speech regarding the encouragement of diligence and thrift). Tokyo: Naimushō shakaikyoku, 1924.

Nakagome Motojirō. "Aa jishin no kuni Nihon" (Oh, Japan, the land of earthquakes). *Naigai kyōiku hyōron* 17:10 (October 1923): 22–29.

Nakamura Akira. "Shinsai fukkō no seijigaku" (The political science of earthquake reconstruction). *Seikei ronsō* 51:3–4 (April 1982): 1–94.

———. "Taishō hachi-nen toshi keikaku hō saikō" (The reevaluation of the city planning law of 1919). *Seikei ronsō* (Essays on politics and economics published by Meiji University) 69:6 (June 1980): 1–35.

Nakamura Shōichi. "Hifukushō ato" (The site of the clothing factory). In *Kantō daishinsai: Dokyumento*. (The Great Kantō Earthquake: Documents). Edited by Gendaishi no kai, 94–97. Tokyo: Sōfūkan, 1996).

Namae Takayuki. "Barakku mondai" (The barracks problem). In *Taishō daishinkasai shi* (Documents of the Taishō great earthquake and conflagration), 30–44. Edited by Kaizōsha. Tokyo: Kaizōsha, 1924.

Nihon kangyō ginkō. *Kinken shōrei ni kansuru kōen yōshi* (A summary of the essential points concerning the encouragement of diligence and thrift). Tokyo: Nihon kangyō ginkō, 1924.

Nihon teikoku gikai, ed. *Dai Nihon teikoku gikaishi* (A parliamentary record of imperial Japan). 18 vols. Tokyo: Dai Nihon teikoku gikai kankōkai, 1926–1930.

Nihon tōkei fukyūkai ed. *Teito fukkō jigyō taikan* (Survey of the imperial capital reconstruction project). 2 vols. Tokyo: Nihon tōkei fukyūkai. 1930.

Nonomura Kaizō. "Kindai bunmei no hakai" (The destruction of modern civilization). *Kyōiku jiron* 1385 (December 1923): 2–8.

Nose Yoritoshi. "Fukkō seishin nahen ni ariya" (Where is the spiritual restoration). *Kyōiku* 502 (January 1925): 1–13.

Ōkawa Kazushi, ed. *Chōki keizai tōkei* (Long-term economic statistics). 14 vols. Tokyo: Tōyō keizai shinpōsha, 1967.

Ōki Tōkichi. "Kokumin shisō zendō saku" (Policies for guiding the nation's ideology). *Chihō gyōsei* 32:2 (February 1924): 20–26.

Oku Hidesaburō. "Shinsai to dōtokushin no kaizō" (The earthquake and the renewal of a moral spirit). *Kyōiku* 490 (1924): 11–12.

Okutani Fumitomo. "Kantō no daisaigai wa ikanaru shin'i ka" (In what ways was the Great Kantō disaster divine will?). *Michi no tomo* 401 (November 5, 1923): 10–17.

Sakaguchi Takenosuke. *Kōtō shōhingaku* (A study of high-end merchandising). Tokyo: Sanseidō, 1925.

Sano Toshikata. *Teito tochi kukaku seiri ni oite* (Concerning imperial capital urban land readjustment). Tokyo Institute for Municipal Research Archives, TIMR OBZ-1080. Tokyo, 1924.

Shimaji Daitō. "Daijihen no dōtoku oyobi shinkō ni oyobosu eikyō" (The effect of the great disaster on morals and faith). *Nihon kyōiku* 2:11 (November 1923): 114–20.

Shimamoto Ainosuke. "Seishinteki fukkō saku" (Measures for spiritual restoration). *Teiyū rinrikai rinri kōenshū* 36:254 (1923): 102–11.

Shitennō Nobutaka. "Kōkū hōmen yori kantaru teito no seishin teki fukkō ou ronzu" (Discussing the spiritual reconstruction of the imperial capital from the viewpoint of aviation). *Chihō gyōsei* 32:1 (January 1924): 33–38.

Soeda Azenbō. *Asakusa teiryūki* (Record of the Asakusa underworld). Tokyo: Kindai seikatsusha, 1930.

———. "Taishō daishinsai no uta" (A song of the Taishō great earthquake). In *Kantō daishinsai: Dokyūmento* (The Great Kantō Earthquake: Documents). Edited by Gendaishi no kai, 210–12. Tokyo: Sōfūkan, 1996.

Suematsu Kai'ichirō. "Daishin ni kansuru shokan" (Impression related to the great earthquake disaster). *Chihō gyōsei* 31:10 (October 1923): 19–25.

Tagawa Daikichirō. *Tsukuraru beki Tokyo* (The Tokyo that should be built). Tokyo: Hatano jūtarō, 1923.

Takakura Tetsuichi, ed. *Tanaka Giichi denki* (Biography of Tanaka Giichi). 2 vols. Tokyo: Hara shobō, 1981.

Takashima Beihō. "Shin Tokyo no kensetsu to Tokyokko no iki" (Construction of new Tokyo and the spirit of Tokyoites). *Chūō kōron* 38:11 (October 1923): 52–57.

Takashima Heizaburō. "Daishinkasai to kodomo" (The great earthquake conflagration and children). *Jidō kenkyū* 27:2 (November 1923): 49–53.

———. "Hensai o kikai ni" (The disaster as opportunity). *Jidō kenkyū* 27:2 (October 1923): 71–72.

Takeuchi Rokuzō. "Kukaku seiri no sekkei narabi ni shikō" (The design and implementation of land readjustment). In *Teito tochi kukaku seiri ni tsuite* (Concerning land readjustment in the imperial capital). Edited by Gotō Shinpei, 36–45. Tokyo: Kōseikai, 1924.

Takebe Rokuzō. "Ni jū man tō no tatemono iten" (The transfer of 200,000 buildings). *Toshi mondai* 10:4 (April 1930): 135–160.

Tanaka Kōtarō. "Shitai no nioi" (The smell of dead bodies). In *Kantō daishinsai: Dokyumento* (The Great Kantō Earthquake: Documents). Edited by Gendaishi no kai, 71–83. Tokyo: Sōfūkan, 1993.

Tanaka Masataka and Ōsaka Hideaki, eds. Kantō daishinsai seifu rikukaigun kankei shiryō. Vol. 3: *Kaigun kankei shiryō* (Materials related to the navy). Tokyo: Nihon keizai hyōronsha, 1997.

Tasaki Kimitsukasa and Sakamoto Noboru, eds. *Kantō daishinsai seifu, riku, kaigun kankei shiryō*. Vol. 2: *Rikugun kaigenrei kankei shiryō* (Materials related to the army). Tokyo: Nihon keizai hyōronsha, 1997.

Tawara Magoichi. "Teito fukkō to jinshin" (The imperial capital reconstruction and the popular mind). *Chihō gyōsei* 31:10 (October 1923): 14–19.

Teito fukkōin, ed. *Kukaku seiri no jishi ni tsuite shimin shokun ni tsugu* (Announcement to the people regarding the implementation of land readjustment). June 5, 1925. Tokyo Institute of Municipal Research Archives, OBZ 75.

———. *Teito fukkōin hyōgikai sokkiroku* (Stenographic record of the imperial capital reconstruction institute consultative committee). 3 vols. Tokyo: Teito fukkōin, 1924.

———. *Teito fukkōin sanyokai sokkiroku* (Stenographic record of the imperial capital reconstruction institute councilor's committee). Tokyo: Teito fukkōin, 1924.

Terada Torahiko. *Tensai to kokubō* (Natural disasters and national defense). Tokyo: Iwanami shoten, 1938.

Tomoeda Takahiko. "Shinsai to dōtokuteki fukkō" (The earthquake disaster and moral reconstruction). *Shakai seisaku jihō* 10:38 (November 1923): 60–70.

Tokyofu, ed. *Taishō shinsai biseki* (Beautiful outcomes from the Taishō earthquake). Tokyo: Tokyofu, 1924.

Tokyofu gakumubu ed. *Tokyofu gunbu ni okeru shūdanteki furyō jūtaku chiku jōkyō chōsa* (A survey of areas with a cluster of substandard housing). Tokyo: Tokyofu gakumubu shakaika, 1930.

———. *Tokyofu gunbu rinsetsu gogun shūdanteki furyō jūtaku chikuzushū* (Charts and diagrams on substandard housing in Tokyo Prefecture and five neighboring counties). Tokyo: Tokyofu gakumubu shakaika, 1928.

Tokyoshi, ed. *Teito fukkō jigyō zuhyō* (Imperial capital reconstruction project diagrams). Tokyo: Tokyoshi, 1930.

———. *Teito fukkōsaishi* (Celebrating the completion of the imperial capital reconstruction). Tokyo: Tokyoshi, 1932.

———. *Tokyo no shinai no saimin ni kansuru chōsa* (An investigation into the poor in Tokyo). Tokyo: Tokyoshi shakaikyoku, 1921.

———. *Tokyo shinsairoku* (Record of the Tokyo earthquake). 5 vols. Tokyo: Tokyo shiyakusho, 1924.

Tokyo shinsai kinen jigyō kyōkai, ed. *Hifukushōato: Tokyo shinsai kinen jigyō kyōkai hōkoku* (Remains of the Honjō Clothing Depot: Report by the Tokyo Earthquake Commemorative Project Association). Tokyo: Tokyo shinsai kinen jigyō kyōkai, 1932.

Tokyo Shisei Chōsakai, ed. *Teito fukkō hiroku* (Secret notes on the imperial capital reconstruction). Tokyo: Hōbunkan, 1930.

Tsuchida Hiroshige. "Kantō daishinsai go no shimin sōdōin mondai ni tsuite: Osaka no jirei o chūshin ni" (Concerning the problems related to the mobilization of citizens after the Great Kantō Earthquake: The case of Osaka). *Shigaku zasshi* 106:12 (December 1997): 59–80.

Tsukamoto Seiji. "Teito fukkō no shinseishin" (The new spirit of imperial capital reconstruction). *Toshi kōron* 6:12 (December 1923): 15–18.
Tsurumi Yūsuke, ed. *Gotō Shinpei* (The official biography of Gotō Shinpei). 4 vols. Tokyo: Gotō Shinpei kaku denki hensankai, 1937–1938.
Uchida Shigebumi, ed. *Taishō daishin taika no kinen* (Commemorative album of the Taishō great earthquake and fire). Tokyo: Mainichi tsūshinsha, 1923.
Ugaki Kazushige. *Ugaki Kazushige nikki* (Diary of Ugaki Kazushige). 3 vols. Tokyo: Asahi shinbunsha, 1954.
Uno Kōji. "San byaku nen no yume" (Dream of three hundred years)." In *Taishō daishinkasai shi* (Documents of the Taishō great earthquake and conflagration). Edited by Kaizōsha, 38–42. Tokyo: Kaizōsha, 1924.
Wakatsuki Reijirō. "Teito fukkō to keizaisei keikaku" (The reconstruction of the imperial capital and financial planning). *Toshi kōron* 6:11 (November 1923): 16–22.
Yamada Hiroyoshi (Hakuai). "Fukkō gairo hi no kettei suru made" (Until we decide on the cost for road reconstruction). *Toshi mondai* 10:4 (April 1930): 193–201.
Yamada Taichi. *Tochi no kioku Asakusa* (Asakusa: Local memories). Tokyo: Iwanami shoten, 2000.
Yamada Yoshio. *Boshin shōsho gige* (Meaning of the Boshin imperial rescript). Tokyo: Hōbunkan, 1913.
———. *Kokumin seishin sakkō ni kansuru shōsho gikai* (A commentary on the imperial edict concerning the invigoration of the national spirit). Tokyo: Hōbunkan, 1933. (Expanded edition of 1923 edition).
Yamakawa Hitoshi. "Teito fukkō to shakai shugiteki seisaku" (Imperial capital reconstruction and socialist policy). In *Yamakawa Hitoshi zenshū*. Edited by Yamakawa Shinsaku, 264–75. Tokyo: Keisō shobō, 1968.
Yasui Sei'ichirō. "Teito shakai jigyō no fukkō" (The reconstruction of imperial capital social welfare facilities). *Toshi mondai* 10:4 (April 1930): 209–18.
Yasunari Jirō. "Barakku kara barakku e" (From one barrack to another). In *Kantō daishinsai: Dokyumento* (The Great Kantō Earthquake: Documents). Edited by Gendaishi no kai, 264–71. Tokyo: sōfūkan, 1996.
Yoshimi Shun'ya. *Toshi no doramaturugi: Tokyo sakariba no shakaishi* (The dramaturgy of the city: The social history of Tokyo entertainment districts). Tokyo: Kōbundō, 1987.

SOURCES IN ENGLISH

Allen, Michael. "The Price of Identity: The 1923 Earthquake and its Aftermath." *Korean Studies* 20 (1996): 64–93.
Ambaras, David. *Bad Youth: Juvenile Delinquency and the Politics of Everyday Life in Modern Japan*. Berkeley: University of California Press, 2006.
Anesaki, Masaharu. *History of Japanese Religion*. Tokyo: Charles Tuttle, 1963.
Bankoff, Gregory. *Cultures of Disaster: Society and Natural Hazard in the Philippines*. London: Routledge/Curzon, 2003.

———. "Living with Risk, Coping with Disasters: Hazard as a Frequent Life Experience in the Philippines." *Education About Asia* 12:2 (Fall 2007): 26–29.

Barry, John. *Rising Tide: The Great Mississippi Flood of 1927 and How it Changed America*. New York: Simon and Schuster, 1997.

Bates, Alex. "Catfish, Super Frog, and the End of the World: Earthquakes (and Natural Disasters) in the Japanese Cultural Imagination." *Education About Asia* 12:2 (Fall 2007): 13–19.

Beard, Charles A. *The Administration and Politics of Tokyo: A Survey and Opinions*. New York: Macmillan, 1923.

———. "Goto and the Rebuilding of Tokyo." *Our World* (April 1924): 20–25.

Bigelow, Poultney. *Japan and Her Colonies: Being Extracts from a Diary Made Whilst Visiting Formosa, Manchuria, Shantung, Korea and Saghalin in the Year 1921*. London: Edward Arnold, 1921.

Bloch, Marc. *Feudal Society*. 2 vols. Chicago: University of Chicago Press, 1961.

Borland, Janet. "Capitalising on Catastrophe: Reinvigorating the Japanese State with Moral Values Through Education Following the 1923 Great Kantō Earthquake." *Modern Asian Studies* 40:4 (October 2006): 875–908.

———. "Makeshift Schools and Education in the Ruins of Tokyo, 1923." *Japanese Studies* 29:1 (May 2009): 131–43.

———. "Rebuilding Schools and Society after the Great Kantō Earthquake, 1923–1930." Ph.D. dissertation, University of Melbourne, 2008.

———. "Stories of Ideal Subjects from the Great Kantō Earthquake." *Japanese Studies* 25:1 (May 2005): 21–34.

Boyer, Paul. *Urban Masses and Moral Order in America, 1820–1920*. Cambridge: Harvard University Press, 1978.

Bureau of Social Affairs, Home Ministry, Japan. *The Great Earthquake of 1923 in Japan*. 2 vols. Tokyo: Home Ministry, 1926.

Buruma, Ian. "Japan's Next Transformation." In *Reimagining Japan: The Quest for a Future that Works*. Edited by McKinsey & Company, 42–47. San Francisco: VIZ Media LLC, 2011.

Busch, Noel. *Two Minutes to Noon: The Story of the Great Tokyo Earthquake and Fire*. New York: Arthur Barker, 1962.

Brinkley, Douglas. *The Great Deluge: Hurricane Katrina, New Orleans, and the Mississippi Gulf Coast*. New York: Harper, 2006.

Chamberlain, Basil Hall. *Things Japanese: Being Notes on Various Subjects Connected with Japan for Use of Travellers and Others*. London: John Murray, 1905.

Clancey, Greg. *Earthquake Nation: The Cultural Politics of Japanese Seismicity, 1868–1930*. Berkeley: University of California Press, 2006.

———. "The Meiji Earthquake: Nature, Nation, and the Ambiguities of Catastrophe." *Modern Asian Studies* 40:4 (October 2006): 909–51.

Clark, Charles. "Science, Reason, and an Angry God: The Literature of an Earthquake." *New England Quarterly* 38:3 (September 1965): 340–62.

Coox, Alvin. *Nomonhan: Japan Against Russia, 1939*. 2 vols. Stanford: Stanford University Press, 1985.

Crane, Conrad. *Bombs, Cities, and Civilians: American Airpower Strategy in World War II*. Lawrence: University of Kansas Press, 1993.

Davison, Charles. "Fusakichi Omori." *Nature* 113:2830 (1924): 133–34.
Dickinson, Frederick R. *War and National Reinvention: Japan in the Great War, 1914–1919*. Cambridge: Harvard University Press, 1999.
Doering, Martin. "The Politics of Nature: Constructing German Reunification During the Great Oder Flood of 1997." *Environment and History* 9:2 (May 2003): 195–214.
Dynes, Russell, and Daniel Yutzy. "The Religious Interpretation of Disaster." *Topic 10: A Journal of the Liberal Arts* (Fall 1965): 34–48.
Früstrück, Sabine. *Colonizing Sex: Sexology and Social Control in Modern Japan*. Berkeley: University of California Press, 2003.
Fujimoto, Taizō. *The Nightside of Japan*. London: T. Werner Laurie, 1915.
Garcia-Acosta, Virgina. "Historical Disaster Research." In *Catastrophe and Culture: The Anthropology of Disaster*. Edited by Anthony Oliver-Smith and Susanna Hoffman, 49–66. Santa Fe: School of American Research Press, 2002.
Garon, Sheldon. *Beyond Our Means: Why America Spends When the Rest of the World Saves*. Princeton: Princeton University Press, 2011.
———. "Fashioning a Culture of Diligence and Thrift." In *Japan's Competing Modernities: Issues in Culture and Democracy, 1900–1930*. Edited by Sharon Minichiello, 312–34. Honolulu: University of Hawaii Press, 1998.
———. "Luxury Is the Enemy: Mobilizing Savings and Popularizing Thrift in Wartime Japan." *Journal of Japanese Studies* 26:1 (Winter 2000): 41–78.
———. *Molding Japanese Minds: The State in Everyday Life*. Princeton: Princeton University Press, 1997.
———. "Saving for 'My Own Good and the Good of the Nation': Economic Nationalism in Japan." In *Nation and Nationalism in Japan*. Edited by Sandra Wilson, 97–114. London: Routledge, 2002.
———. *The State and Labor in Modern Japan*. Berkeley: University of California Press, 1987.
Gordon, Andrew. "Consumption, Leisure and the Middle Class in Transwar Japan." *Social Science Japan Journal* 10:1 (May 2007): 1–21.
———. *Labor and Imperial Democracy in Prewar Japan*. Berkeley: University of California Press, 1991.
Griffis, William. *The Mikado's Empire*. New York: Harper & Brothers, 1886.
Hammer, Joshua. *Yokohama Burning: The Deadly 1923 Earthquake and Fire That Helped Forge the Path to World War II*. New York: Free Press, 2006.
Hanes, Jeffrey. *The City as Subject: Seki Hajime and the Reinvention of Modern Osaka*. Berkeley: University of California Press, 2002
———. "Urban Planning as an Urban Problem: The Reconstruction of Tokyo After the Great Kantō Earthquake." *Seisaku kagaku* 7:3 (March 2000): 123–37.
Harootunian, Harry. "Introduction: A Sense of Ending and the Problem of Taishō." In *Japan in Crisis: Essays on Taishō Democracy*. Edited by Bernard Silberman and Harry Harootunian, 3–28. Princeton: Princeton University Press, 1974.
Hastings, Sally. *Neighborhood and Nation in Tokyo, 1905–1937*. Pittsburgh: University of Pittsburgh Press, 1995.
Havens, Thomas. *Valley of Darkness: The Japanese People in World War Two*. New York: Norton, 1978.

Hayase, Yukio. "The Career of Gotō Shinpei: Japan's Statesman of Research, 1857–1929." Ph.D. dissertation, Florida State University, 1974.

Healy, Mark. "The Ruins of New Argentina: Peronism, Architecture and the Remaking of San Juan After the 1944 Earthquake." Ph.D. dissertation, Duke University, 1999.

Hein, Carola. "Visionary Plans and Planners: Japanese Traditions and Western Influences." In *Japanese Capitals in Historical Perspective*. Edited by N. Fieve and P. Waley 309–46. London: Routledge/Curzon, 2003.

Hewitt, Kenneth. "The Idea of Calamity in a Technocratic Age." In *Interpretation of Calamity: From the Viewpoint of Human Ecology*. Edited by Kenneth Hewitt, 3–32. London: Allen and Unwin, 1983.

Hoffman, Susanna. "After Atlas Shrugs: Cultural Change or Persistence After a Disaster." In *The Angry Earth: Disaster in Anthropological Perspective*. Edited by Anthony Oliver-Smith and Susanna Hoffman, 302–25. London: Routledge, 1999.

———. "The Monster and the Mother: The Symbolism of Disaster." In *Catastrophe and Culture: The Anthropology of Disaster*. Edited by Anthony Oliver-Smith and Susanna Hoffman, 113–41. Santa Fe: School of American Research Press, 2002.

Hoffman, Susanna, and Anthony Oliver-Smith. "Anthropology and the Angry Earth: An Overview." In *The Angry Earth: Disaster in Anthropological Perspectives*. Edited by Anthony Oliver-Smith and Susanna Hoffman, 1–16. London: Routledge, 1999.

Hones, Sheila. "Distant Disasters, Local Fears: Volcanoes, Earthquakes, Revolution, and Passion in The Atlantic Monthly, 1880–1884." In *American Disasters*. Edited by Steven Biel, 170–96. New York: New York University Press, 2001.

Humphreys, Leonard. *The Way of the Heavenly Sword: The Japanese Army in the 1920s*. Stanford: Stanford University Press, 1995.

Inoue, Takutoshi, and Yagi Ki'ichirō. "Two Inquirers on the Divide: Tokuzō Fukuda and Hajime Kawakami." In *Economic Thought and Modernization in Japan*. Edited by Sugihara Shiro and Tanaka Toshihiro, 60–77. Cheltenham: Edward Elgar, 1998.

Jackson, Jeffrey. *Paris Under Water: How the City of Light Survived the Great Flood of 1910*. New York: Palgrave Macmillan, 2010.

Johns, Alessa. "Introduction." In *Dreadful Visitations: Confronting Natural Catastrophe in the Age of Enlightenment*. Edited by Alessa Johns, xi–xxv. London: Routledge, 1999.

Kasza, Gregory. *The Conscription Society: Administered Mass Organizations*. New Haven: Yale University Press, 1995.

Kawabata, Yasunari. *The Scarlet Gang of Asakusa*. Translated by Alisa Freedman. Berkeley: University of California Press, 2005.

Kempe, Michael. "Noah's Flood: The Genesis Story and Natural Disasters in Early Modern Times." *Environment and History* 9:2 (May 2003): 151–71.

Kinzley, W. Dean. "Japan's Discovery of Poverty: Changing Views of Poverty and Social Welfare in the Nineteenth Century." *Journal of Asian History* 22:1 (January 1988): 1–24.

Koji, Taira. "Urban Poverty, Ragpickers, and the Ants' Villa in Tokyo." *Economic Development and Cultural Change* 17:2 (January 1969): 155–77.

Large, Stephen. "Buddhism, Socialism, and Protest in Prewar Japan: The Career of Seno'o Girō." *Journal of Japanese Studies* 21:1 (Winter 1987): 153–71.

Lees, Andrew. *Cities Perceived: Urban Society in European and American Thought, 1820–1940.* Manchester: Manchester University Press, 1985.
——. *Cities, Sin, and Social Reform in Imperial Germany.* Ann Arbor: University of Michigan Press, 2002.
Metzler, Mark. *Lever of Empire: The International Gold Standard and the Crisis of Liberalism in Prewar Japan.* Berkeley: University of California Press, 2006.
——. "Woman's Place in Japan's Great Depression: Reflections on the Moral Economy of Deflation." *Journal of Japanese Studies* 30:2 (Summer 2004): 315–52.
Milne, David. "Notices of Earthquake Shocks in Great Britain, and Especially in Scotland, with Inferences Suggested by These Notices as to the Nature and Causes of Such Shocks." *Edinburgh New Philosophical Journal* 31 (1841): 92–123.
Milne, John. *Seismology.* London: Kegan Paul, 1898.
Mimura, Janis. *Planning for Empire: Reform Bureaucrats and the Japanese Wartime State.* Ithaca: Cornell University Press, 2011.
Mulcahy, Matthew. *Hurricanes and Society in the British Greater Caribbean, 1624–1783.* Baltimore: Johns Hopkins University Press, 2008.
——. "A Tempestuous Spirit Called Hurri Cano: Hurricanes and Colonial Society in the British Greater Caribbean." In *American Disasters.* Edited by Steven Biel, 11–38. New York: New York University Press, 2001.
Mullin, John. "The Reconstruction of Lisbon Following the Earthquake of 1755: A Study in Despotic Planning." *Planning Perspectives* 7:2 (April 1992): 157–79.
Nakamura, Akira. "Cities Unplanned: The Failure of City Planning in Pre-War Japan: A Political Perspective." *Bulletin of the Institute of Social Sciences* 4:2 (1980): 1–16.
Oliver-Smith, Anthony. "Disaster Context and Causation: An Overview of Changing Perspectives in Disaster Research." In *Natural Disasters and Cultural Responses.* Edited by Anthony Oliver-Smith, 1–38. Williamsburg: College of William and Mary, 1986.
——. "Theorizing Disasters: Nature, Power, and Culture." In *Catastrophe and Culture: The Anthropology of Disaster.* Edited by Anthony Oliver-Smith and Susanna Hoffman, 23–47. Santa Fe: School of American Research Press, 2002.
——. "What Is a Disaster?: Anthropological Perspectives on a Persistent Question." In *The Angry Earth: Disaster in Anthropological Perspectives.* Edited by Anthony Oliver-Smith and Susanna Hoffman, 18–34. London: Routledge, 1999.
Oliver-Smith, Anthony, and Susanna Hoffman. "Introduction: Why Anthropologists Should Study Disasters." In *Catastrophe and Culture: The Anthropology of Disaster.* Edited by Anthony Oliver-Smith and Susanna Hoffman, 3–22. Santa Fe: School of American Research Press, 2002.
Prentice, Thomas. *Observations Moral and Religious on the Late Terrible Night of the Earthquake: A Sermon Preached at the Thursday Lecture in Boston, January 1st, 1756.* Boston: S. Kneeland, 1756.
Pyle, Kenneth. "The Technology of Japanese Nationalism: The Local Improvement Movement, 1900–1918." *Journal of Asian Studies* 33:1 (November 1973): 51–65.
Quarantelli, E.L. "The Basic Question, Its Importance, and How It Is Addressed in This Volume." In *What Is a Disaster? Perspectives on the Question.* Edited by E. L. Quarantelli, 1–7. London: Routledge, 1998.

Rodgers, Daniel. *Atlantic Crossings: Social Politics in a Progressive Age*. Cambridge: Belknap Press of Harvard University Press, 1998.

Rohr, Christian. "Man and Nature in the Late Middle Ages: The Earthquake of Carinthia and Northern Italy on 26 January 1348 and Its Perception." *Environment and History* 9:2 (May 2003): 127–49.

Sand, Jordan. *House and Home in Modern Japan: Architecture, Domestic Space, and Bourgeois Culture, 1880–1930*. Cambridge: Harvard University Press, 2003.

Sato, Barbara. *The New Japanese Woman: Modernity, Media, and Women in Interwar Japan*. Durham: Duke University Press, 2003.

Satō, Hiroshi, et al. "Earthquake Source Fault Beneath Tokyo." *Science* 309:5733 (July 15, 2005): 462–65.

Schaffer, Ronald. *Wings of Judgement: American Bombing in World War II*. Oxford: Oxford University Press, 1988.

Schencking, J. Charles. "Catastrophe, Opportunism, Contestation: The Fractured Politics of Reconstructing Tokyo Following the Great Kantō Earthquake of 1923." *Modern Asian Studies* 40:4 (October 2006): 833–74.

——. "The Great Kantō Earthquake and the Culture of Catastrophe and Reconstruction in 1920s Japan." *Journal of Japanese Studies* 34:2 (Summer 2008): 294–334.

——. "The Great Kantō Earthquake of 1923 and the Japanese Nation: Responding to an Urban Calamity of an Unprecedented Nature." *Education About Asia* 12:3 (Fall 2007): 20–25.

——. *Making Waves: Politics, Propaganda and the Emergence of the Imperial Japanese Navy, 1868–1922*. Stanford: Stanford University Press, 2005.

——. "1923 Tokyo as a Devastated War and Occupation Zone: The Catastrophe One Confronted in Post Earthquake Japan." *Japanese Studies* 29:1 (May 2009): 111–29.

Schwartz, Stuart B. "The Hurricane of San Ciriaco: Disaster, Politics, and Society in Puerto Rico, 1899–1901." *Hispanic American Historical Review* 72:3 (August 1992): 303–34.

Scott, James C. *Seeing Like a State: How Certain Schemes to Improve the Human Condition have Failed*. New Haven: Yale University Press, 1998.

Seidensticker, Edward. *Low City, High City: Tokyo from Edo to the Earthquake*. New York: Knopf, 1983.

——. *Tokyo Rising: The City Since the Great Earthquake*. New York: Knopf, 1990.

Shillony, Ben-Ami. *Politics and Culture in Wartime Japan*. Oxford: Clarendon Press, 1981.

Silverberg, Miriam. *Erotic Grotesque Nonsense: The Mass Culture of Japanese Modern Times*. Berkeley: University of California Press, 2006.

Smethurst, Richard. *From Foot Soldier to Finance Minister: Takahashi Korekiyo, Japan's Keynes*. Cambridge: Harvard University Press, 2007.

——. *A Social Basis for Prewar Japanese Militarism: The Army and the Rural Community*. Berkeley: University of California Press, 1974.

Smith, Carl. "Faith and Doubt: The Imaginative Dimensions of the Great Chicago Fire." In *American Disasters*. Edited by Steven Biel, 129–69. New York: New York University Press, 2001.

Smits, Gregory. "Shaking Up Japan: Edo Society and the 1855 Catfish Prints." *Journal of Social History* 39:4 (Summer 2006): 1045–77.

Sorensen, Andre. *The Making of Urban Japan: Cities and Planning from Edo to the Twenty-First Century*. London: Routledge, 2002.

———. "Urban Planning and Civil Society in Japan: Japanese Urban Planning Development During the Taishō Democracy Period (1905–1931)." *Planning Perspectives* 16 (2001): 383–406.

Statistics Bureau, ed. *Historical Statistics of Japan*. 5 vols. Tokyo: Japan Statistical Association, 1987.

Stein, Ross, Shinji Toda, Tom Parsons, and Elliot Grunewald. "A New Probabilistic Seismic Hazard Assessment for Greater Tokyo." *Philosophical Transactions of the Royal Society A* 364 (June 2006): 1965–88.

Steinberg, Ted. *Acts of God: The Unnatural History of Natural Disaster in America*. Oxford: Oxford University Press, 2000.

Struber, Martin. "Divine Punishment or Object of Research? The Resonance of Earthquakes, Floods, Epidemics and Famine in the Correspondence Network of Albrecht von Haller." *Environment and History* 9:2 (May 2003): 171–93.

Takenobu, Yoshitarō, ed. *The Japan Year Book, 1924–25*. Tokyo: Foreign Affairs Association of Japan, 1925.

———. *The Japan Year Book, 1929*. Tokyo: Foreign Affairs Association of Japan, 1929.

Taub, Liba. *Ancient Meteorology*. London: Routledge, 2003.

Taylor, John Gates Jr. "Eighteenth-Century Earthquake Theories: A Case-History Investigation into the Character of the Study of the Earth in the Enlightenment." Ph.D. dissertation, University of Oklahoma, 1975.

Tipton, Elise. "Cleansing the Nation: Urban Entertainments and Moral Reform in Interwar Japan." *Modern Asian Studies* 42:4 (July 2008): 705–31.

———. "In a House Divided: Japanese Christian Socialist Abe Isoo." In *Nation and Nationalism in Japan*. Edited by Sandra Wilson, 80–96. London: Routledge/Curzon, 2002.

Tokyo Municipal Office, Tokyo. *The Reconstruction of Tokyo*. Tokyo: Tokyo Municipal Office, 1933.

Tsu, Timothy. "Making Virtues of Disaster: 'Beautiful Tales' from the Kobe Flood." *Asian Studies Review* 32 (June 2008): 197–214.

Udias, Agustin, and Alfonso Lopez Arroyo. "The Lisbon Earthquake of 1755 in Spanish Contemporary Authors." In *The 1755 Lisbon Earthquake: Revisited*. Edited by L. A. Mendes-Victor et al., 7–24. New York: Springer-Verlag, 2008.

Van de Wetering, Maxine. "Moralizing in Puritan Natural Science: Mysteriousness in Earthquake Sermons." *Journal of the History of Ideas* 43:3 (July–September 1982): 417–38.

Watanabe Shun'ichi. "Planning History in Japan." *Urban History* 7 (May 1980): 63–75.

Weiner, Michael A. "Koreans in the Aftermath of the Kantō Earthquake of 1923." *Immigrants and Minorities* 2:1 (April 1983): 5–32.

———. *The Origins of the Korean Community in Japan, 1910–1923*. Manchester: Manchester University Press, 1989.

Weiner, Susan B. "Bureaucracy and Politics in the 1930s: The Career of Gotō Fumio." Ph.D dissertation, Harvard University, 1984.

Weisenfeld, Gennifer. "Designing After Disaster: Barrack Decoration and the Great Kantō Earthquake." *Japanese Studies* 18:3 (December 1998): 229–46.

———. "Imaging Calamity: Artists in the Capital After the Great Kantō Earthquake." In *Modern Boy, Modern Girl: Modernity in Japanese Art, 1910–1935*. Edited by Jackie Menzies, 25–29. Sydney: Art Gallery of New South Wales, 1998.

Wilson, Sandra. *The Manchurian Crisis and Japanese Society, 1931–1933*. London: Routledge, 2002.

———. "The Past in the Present: War in Narratives of Modernity in the 1920s and 1930s." In *Being Modern in Japan: Culture and Society from the 1910s to the 1930s*. Edited by Elise Tipton and John Clark, 170–84. Honolulu: University of Hawaii Press, 2000.

Winchester, Simon. *Krakatoa: The Day the World Exploded, 27th August 1883*. New York: Viking Press, 2003.

Woodward, William P. "Interfaith Communication and the Confrontation of Religions in Japan." *Journal of the American Academy of Religion* 32:3 (September 1964): 231–38.

Worster, Donald. *Dust Bowl: The Southern Plains in the 1930s*. New York: Oxford University Press, 1979.

———. *The Wealth of Nature: Environmental History and the Ecological Imagination*. New York: Oxford University Press, 1993.

Yazaki, Takeo. *Social Change and the City in Japan*. Tokyo: Japan Publications, 1968.

Zeilinga de Boer, Jelle, and Donald Theodore Sanders. *Earthquakes in Human History: The Far-Reaching Effects of Seismic Disruptions*. Princeton: Princeton University Press, 2005.

INDEX

Abe Isoo, 1, 6, 137–138, 154–155, 170–173, 238–241, 292, 312; compares the Great Kantō Earthquake and the challenges of reconstruction to war, 113; describes the entertainment quarters of Tokyo as the "dark side of civilization," and a "monster" 117, 154; visions for new Tokyo that emphasized pedestrian promenades, 171–172; prescriptions for spiritual reconstruction, 238–240; program of diligence, thrift, and frugality, 239–240; advocates the abolition of licensed prostitution as part of a program for spiritual renewal, 240–241

Air raids, fear of future occurrence heightened by experience of the Great Kantō Earthquake, 48–49, 72–77, 311–314

Akaike Atsushi, 49–50

Anan Jōichi, 6, 153, 180–181; opinion that a successful reconstruction of Tokyo would illustrate Japan's strength and ingenuity to the international community, 180

Anderson, Edwin Alexander, xvi

Ansei Earthquake of 1855, 16, 56

Arima Sukemasa, 232–234; program of spiritual reconstruction following the Great Kantō Earthquake, 232–233; suggests that General Nogi Maresuke become a symbol of spiritual reconstruction, 233–234

Arishima Incident (1923), 135–137; interpretation of incident as an act symptomatic of the moral and sexual degeneracy of post–World War I Japan, 136–137

Asakusa Ward, 13, 20, 26, 34, 40, 43, 58, 68, 93, 104, 289–292, 302, 308; described

Asakusa Ward (*continued*)
as a place of hedonism, debauchery, and licentious entertainment, 116–117, 138–143; picture postcards of, before the Great Kantō Earthquake, 140; lithograph print depicting fire in, 142–143; picture postcard of the twelve-story tower of, before, and after the Great Kantō Earthquake, 144, 148; number of factories located in, and population density of, in 1919, 156–157, 261; petitions submitted against land readjustment by residents of, 269–270

Barrack housing, 57, 64, 67, 69–71, 86, 95, 234; constructed at sites of destroyed primary schools across Tokyo, 70; descriptions of, 70–71
Beard, Charles, 182–184; visit to Tokyo following the Great Kantō Earthquake, 182–183; recommendations for the reconstruction of Tokyo, 183–184; warnings concerning possible contestation associated with the reconstruction of Tokyo issued to Home Minister Gotō Shinpei, 212–213; critique of Japan's political system in light of reconstruction contestation, 224–225

Campaign for the encouragement of diligence and thrift (*kinken shōrei undō*), 250–254; employment of Great Kantō Earthquake in support of, 250–252
Chihō gyōsei, 27–28, 73–74, 120, 122, 174, 238–239
Chūō kōron, 71, 121, 123, 130, 181, 227–228
Consumer spending in Japan, 147–152, 246–248, 254–258; on phonographs and records, 149, 255; on cosmetics, 149, 256–257; on sake and other alcohol, 150, 255–256; on items deemed "luxuries" by the 1924 luxury tariff, 246

Dead bodies in Tokyo as a result of Great Kantō Earthquake, 13–14, 68, 90, 95, 97–99, 101–110, 294, 297; descriptions of, as charred sticks or logs, 29; descriptions of endless rows of, stacked at the site of the Honjo Clothing Depot, 29, 30–35; descriptions of, as charred, dried, sardines, 33; picture postcards depicting, 30–33, 99, 101
Degeneration, view of Japanese society in a state of, following World War I, xvii, 123–152, 227, 231, 308–310, 314; economic, 126–134; social, moral, and sexual, 135–143; perceptions vs. realities, 143–152
Deliberative council. *See* Imperial capital reconstruction deliberative council (*Teito fukkō shingikai*)
Divine punishment (*tenken* or *tenbatsu*), interpretation of the Great Kantō Earthquake as an act of, xvii, 116–128, 132–135, 143, 152–154, 227, 277; opinions expressed by: Okutani Fumitomo, 116–117, 121, 128–130; Shitennō Nobutaka, 120; Takashima Beihō, 121–122; Haruno Ki'ichi, 122–123; Ugaki Kazushige, 123, 125–126; Fukasaku Yasubumi, 123–124; Yamada Yoshio, 124–125, 132–133; Horie Ki'ichi, 130; Shimamoto Ainosuke, 133; Ichiki Kitokurō, 123; Tawara Magoichi, 127; Nose Yoritoshi, 136–137

Earthquake memorial hall design competition. *See* Taishō earthquake memorial hall design competition
Earthquake refugees, 24–25, 43, 52–55, 67–72, 95, 97; relocation to various parts of Japan by, 68; relocation to various parts of Japan's empire by, 68; establishment of temporary housing in the open spaces of Tokyo by, 68–70; picture postcards of: leaving Tokyo by train, 67; living in Hibiya Park, 69

Egi Kazuyuki, 204–213, 264–265, 269; extended oration delivered at the imperial capital reconstruction deliberative council meeting, 206–208; criticism of land readjustment, 264

Emergency earthquake relief bureau (*Rinji shinsai kyūgo jimukyoku*), 57–72, 83, 104; providing medical relief, 57–59; providing food and water, 59–67; providing relocation assistance and temporary shelter, 67–72; securing and distributing rice donated from across Japan, 61–62

First World War. *See* World War I

Fukagawa Ward, 20, 37, 40, 43–44, 66, 68, 70, 95, 104, 116, 138, 148, 156–158, 167–169, 290–292, 302, 308–309; annihilation in 1923, 59–61; photograph of, burning, as seen from air, 60; number of factories in, and population density of, in 1919, 156–157; Land Readjustment District No. 52 located in, 278–280; petitions submitted against land readjustment by residents of, 278–279

Fukasaku Yasubumi, 47, 127–128, 132, 233–235; interpretation of the Great Kantō Earthquake as an act of divine punishment expressed by, 123–124; belief that modern urban life fostered increased promiscuity and adultery, 135–136; prescriptions for spiritual reconstruction, 233–234

Fukkō kyoku. *See* Reconstruction bureau

Fukuda Tokuzō, 4–6, 154–155, 242–243; critiques of pre-1923 Tokyo made by, 4–5, 159; belief that Great Kantō Earthquake and fires cleansed Tokyo, 5–6, 154; prescriptions for spiritual reconstruction, 242

Fukuoka nichi nichi shinbun, 38, 83

Funaki Yoshie, 14, 24–26; tale of harrowing escape from burning Tokyo via boat on the congested Sumida River, 25–26, 90

Genroku Earthquake of 1703, 16

Ginza, 20, 71, 138–139, 196, 261, 304; described as center of up-market consumer spending, 128–130, 147, 170

Gotō Fumio, 27, 50

Gotō Shinpei, xvi, 109, 154–155, 159–165, 178–180, 183–184, 187–215, 215–225, 228–233, 294, 305–307; opinion that the Great Kantō Earthquake disaster and the challenges of reconstruction were similar to war, 112; early career: as doctor, 160; as colonial bureaucrat in Taiwan, 161–162; as president of the South Manchurian Railway Company, 161; compares the city to an organic body, 162; activities undertaken as mayor of Tokyo in relation to urban betterment, 163–164; visions for new Tokyo that emphasized projecting power through architecture, 178–179; belief espoused that a golden opportunity existed to rebuild Tokyo and reconstruct Japan after the Great Kantō Earthquake, 187–192; presents initial reconstruction ideas to cabinet colleagues, 189–190; cabinet-level brawl with Finance Minister Inoue Junnosuke over reconstruction budget, 198–201; responses to critiques leveled against the government's reconstruction plan at the imperial capital reconstruction deliberative council meeting, 208–209, 212–213; responses to critiques leveled against the government's reconstruction plan at the 47th session of parliament, 215–222; reflections on the failure of the government's reconstruction plan expressed by, 221–222; influence over the drafting of the imperial rescript

Gotō Shinpei (*continued*)
regarding the invigoration of the national spirit, 228–230
Great Kantō Earthquake, descriptions of, 18–27, 29–35; conflagrations, 13–14, 20–27; sounds, 18–19, 25–26, 29–30; seismic upheaval, 18; devastation wrought on Tokyo, 31–35; pandemonium, 72–77
Great Kantō Earthquake, scientific explanations, 15–18; map of tectonic plates that produce earthquakes in Japan, 17

Hamaguchi Osachi, 244–246, 249; opening speech before 49th session of parliament delivered by, 244–245; plan to "replenish the country's economic vitality," 245
Haruno Ki'ichi, 122; belief that Tokyo exemplified the "total spiritual corruption of mankind," 135
Hell, Tokyo compared to following the Great Kantō Earthquake, 1–3, 13, 20, 23–24, 26, 29–30, 32, 37, 47, 61, 81, 102–103, 227
Hibiya Park, 58, 68–69; picture postcard of refugees in, 69
Hifukushō. See Honjo Clothing Depot
Hirohito, the Shōwa Emperor, 93–96; inspection tour of Tokyo on September 15, 1923, as crown prince and prince regent, 94–95; lithograph print (*sekiban*) of, 94; picture postcard, 96; imperial inspection tour of reconstructed Tokyo as emperor, in 1930, 301–307; reflections on Gotō Shinpei's grandiose reconstruction plan and its failure made public in 1983, 307
Hoashi Ri'ichirō, 28, 78, 113, 153; reflections on, interpretation and condemnation of, Korean massacre following the Great Kantō Earthquake, 28; belief that degenerate state of affairs existed in post-World War I Japan, 116, 127

Hōchi shinbun, 36–37
Honjo Clothing Depot (*Hifukushō*), 14, 25–26, 29–35, 37, 58, 78–80, 90–91, 97–110, 114, 293–300; packed with refugees, 26; as a site populated by "hungry ghosts," 29; picture postcards of dead bodies at, 33, 101; photographs of bodies being cremated at, 34; memorial service of October 19, 1923, 78–80, 107–110; lithograph print (*sekiban*) of whirlwind firestorm approaching the site of, 91; postcards of refugees being directed to, as fires approached, 100; selection of the site for the cremation of bodies, 103–104; earmarked for memorial hall, 107–108; construction of earthquake memorial hall at site of, 295–300; emperor's visit to, as part of reconstruction celebrations held in 1930, 302–303
Honjo Ward, 18, 20, 23, 25–26, 29–34, 37, 40, 43, 44, 58, 66, 70, 78, 90–91, 95, 97, 100–101, 103–108, 116, 138, 148, 167–169, 272, 274, 289–292, 295–300, 308; number of factories in, and population density of, in 1919, 156–157; petitions submitted against land readjustment by residents of, 270, 278–279; Land Readjustment District No. 52 located in, 278–281
Horie Ki'ichi, 130–132, 138–139, 181, 253; interpretation of the Great Kantō Earthquake as an act of divine punishment, 123; critiques of consumerism exhibited at Mitsukoshi Department Store, 130, 132; claim that Japan had reached the international "summit" of decadence in the post-World War I era made by, 138
Horie Shōzaburō, 79–80
Horikiri Zenjirō, 192, 301, 303–306; work carried out as mayor of Tokyo in conjunction with festivities held to celebrate the completion of reconstruction in 1930, 304–306

Ichiki Kitokurō, 6, 137–138, 227–228; critiques against Japanese society as a result of the anarchy and murders in Tokyo following the Great Kantō Earthquake, 74; interpretation of the Great Kantō Earthquake as an act of divine punishment, 123; prescriptions for spiritual reconstruction, 228

Ikeda Hiroshi, 191, 200–201, 223

Imamura Akitsune, 56–57, 119–120; scientific and popular controversy, as a result of Tokyo earthquake prediction published in 1905, 57

Imperial capital reconstruction councilor's committee (*Teito fukkō sanyokai*), 200–201, 204–205

Imperial capital reconstruction deliberative council (*Teito fukkō shingikai*), 204–213, 221–222; creation and membership of, and powers held by, 204–207; reduction of the government's reconstruction budget agreed to, by members of, 211–212

Imperial capital reconstruction institute (*Teito fukkōin*), 193–201, 204, 207; creation and membership of, 194–195; disputes between planners of, 195–201; creation of ¥1.3 billion reconstruction plan, 197–198

Imperial Japanese army, 15, 25, 38, 45, 47, 49–57, 59–67, 77, 94, 125, 142, 201–204, 241, 254, 262, 281, 313; troop deployments to Tokyo by, 51, 54, 56; reconnaissance flights flown over the disaster zone by, 15, 53, 60; use of carrier pigeons to facilitate communication by, 53, reports from field commanders deployed to the disaster zone, 54–56; photograph of Tokyo burning taken from reconnaissance flight operated by, 60; transportation and distribution of food rations by, 61–62; repair of infrastructure to facilitate transportation and distribution of food in disaster zone by, 63–65; photographs of personnel repairing infrastructure and distributing food in disaster zone from, 64

Imperial Japanese Diet, lower house of parliament (House of Representatives), 29, 213–222, 244–250; critiques leveled against the reconstruction plan in, by: Yoshiue Shōichirō, 214–215; Tsuhara Takeshi, 215–216; Miwa Ichitarō, 216; Tagi Kumejirō, 216–217; Hata Toyosuke, 217; Mito Tadazo, 217; Shimada Toshio, 217; Suzuki Jōzō, 218; Inoue Takaya, 218; reduction of the government's reconstruction budget from ¥598 million to ¥468 million in the 47th session of, 219–220; debates concerning the adoption of a luxury tariff during the 49th session of, 244–249

Imperial Japanese navy, 38, 45, 61–65, 88, 112, 202, 220, 254, 313; transportation of rice and other foodstuffs to Tokyo, 63–64; repair of docks and harbor works to facilitate the transportation and distribution of food in the disaster zone, 64

Imperial rescript regarding the invigoration of the national spirit (*Kokumin seishin sakkō ni kansuru shōsho*), 228–233; background to, 229–230; announcement by Prime Minister Yamamoto Gonnohyōe of, 229–231; responses to, published in newspapers, 231–232

Inoue Junnosuke, 189–190, 193, 196–222, 229–231; secret budget meetings with vice president of the imperial capital reconstruction institute (*Teito fukkōin*), Miyao Shunji, 196–197; cabinet level brawl with Home Minister Gotō Shinpei over the reconstruction budget, 198–201; classical economic financial outlook held by, 201–204; announcement of the government's ¥703 million reconstruction budget

Inoue Junnosuke (*continued*)
by, 204; response to critiques leveled against the government's reconstruction plan at the imperial capital reconstruction deliberative council, 209; discussions held with finance ministry subordinates on the eve of the 47th session of parliament, 213–214; response to critiques leveled by parliamentarians against the government's reconstruction budget plan introduced by, 215–218
Insurance, issuance of payments following the Great Kantō Earthquake, 41–43; total amount paid, 43
Inukai Tsuyoshi, 204–213, 220–221
Ishimitsu Maomi, 49–50
Itō Chūta, 276, 295–300; design of Taishō earthquake memorial hall, 299–300
Itō Miyoji, 204–213, 269; extended oration criticizing the government's reconstruction plan at the imperial capital reconstruction deliberative council meeting, 207–208; concerns about the financial position of Japan, 209–210

Japan Weekly Chronicle, xvii, 70, 95, 101, 222, 248–250, 253–254, 297–298; editorial published in, comparing devastation meted out by the Great Kantō Earthquake to that caused by combat on the Western Front in World War I, 72; editorial criticizing the luxury tariff of 1924, 248–249
Jikeidan. *See* Neighborhood vigilance groups

Kaizō, 5, 113, 170–171
Kanda Ward, 20–21, 31–32, 40, 58, 104, 129, 148, 167; photograph of the remains of, 44; number of factories in, and population density of, in 1919, 156–157; petitions submitted against land readjustment by residents of, 269;

274–277; Land Readjustment District No. 8 located in, 274–278
Kasuya Gizō, 79–80, 113; speech delivered at the 19 October memorial service held at the site of the Honjo Clothing Depot, 79, 113
Katō Takaaki, 204–213, 223, 232, 250–254; suggestions to disband the imperial capital reconstruction deliberative committee, 210–211; employment of the Great Kantō Earthquake in support of the campaign for the encouragement of diligence and thrift, 251–253; policy of fiscal retrenchment introduced by, as prime minister, 253–254
Kawatake Shigetoshi, 18, 23–25, 32, 90, 101–103; harrowing story of escape from burning Tokyo, 23–25; paralysis experienced at the sight of so many dead bodies in Tokyo, 32; graphic tale of journey through "dead Tokyo," 32; vivid description of searching piles of dead bodies, 102–103
Kiitsu kyōkai, 120–122; research into the interpretation of the Great Kantō Earthquake, 121–122
Kikuchi Shinzō, 169–170; visions for new Tokyo that emphasized transportation infrastructure, 170
Kinken shōrei undō. *See* Campaign for the encouragement of diligence and thrift
Kitazawa Rakuten, xxi, 133–134, 247–248, 259–260; political cartoons created by, and published in *Jiji manga*, 134, 247, 260
Kiyosu Bridge, 286–287, 302; picture postcards of, 287, 304
Kobashi Ichita, 173–174; visions for new Tokyo that emphasized parks, community gardens, and green space, 174
Koizumi Tomi, 25–32; graphic description of fires that burned Tokyo, 26; harrowing survivor account of the tragedy that took place at the site of the Honjo

Clothing Depot, 26–27; first impression upon waking up at the site of the Honjo Clothing Depot on September 2, 1923, 29–31
Korean massacre following the Great Kantō Earthquake, 26–29, 52–55, 73–76; reflections, interpretations, and condemnation of, 28–29, 73–75; condemnation leveled on the floor of the Imperial Japanese Diet, 29; Koreans placed in "protective custody," 55
Kurushima Takehiko, 88–89, 235
Kyōbashi Ward, 20, 40, 104, 138, 167, 290–292; number of factories in, and population density of, in 1919, 156–157
Kyōiku, 28, 226, 236
Kyōiku jiron, 9, 145
Kyoto hinode shinbun, 83
Kyoto nichi nichi shinbun, 83
Kyūshū nippō, 37–38; description of disaster zone as a "sea of fire" (zenshi hi no umi), 37

Land readjustment (*kukaku seiri*), xviii, 164, 192–194, 197–198, 217–219, 259, 263–264, 266–284, 286–288, 294, 303–306, 310; implementation of in Tokyo as defined by Imperial Ordinance No. 49, 267–268; map of land readjustment districts in Tokyo, 268; statistics associated with implementation across Tokyo of, 272–274, 304–306; carried out in Land Readjustment District No. 8, Kanda Ward, 274–277; carried out in Land Readjustment District No. 52, Honjo and Fukagawa wards, 278–280; carried out in Land Readjustment District No. 10, Nihonbashi Ward, 280, 282
Lithographic prints (*sekiban*) depicting scenes from the Great Kantō Earthquake, 79–80, 88, 90–95, 141–142; of Nihonbashi Bridge on fire, 22; of the whirlwind firestorm that swept through the site of the Honjo Clothing Depot, 91; of the neighborhood around Matsuzakaya Department Store ravaged by fires, 92; of the imperial inspection tour at a lookout in Ueno Park, 94; of the spread of fires around the Twelve-Story Tower of Asakusa, 142
Luxury tariff of 1924, 148, 244–250, 256–258, 311; introduction of, in the 49th session of the Imperial Japanese Diet, 244–245; definition of "luxury items," 245–246

Maeda Kenjirō, 295–300; winning design for the Taishō earthquake memorial hall created by, 295–297; controversy over winning design, 297–299; line drawing of winning design, 296; scrapping of winning design of, 298–299
Martial law, 29, 48, 51–55, 93, 142; powers granted to authorities, 51–53; expansion of geographic jurisdiction, on 3 and 4 September, 1923, 53–55
Manseibashi Train Station, 97, 102; photograph of, 31
Matsuki Kan'ichirō, 194–198; appointment as a vice president of the imperial capital reconstruction institute (*Teito fukkōin*), 194
Matsuzakaya Department Store, 91–93, 258
Meiji Constitution of 1889, 189, 206, 208, 224
Meiji Restoration of 1868, 78, 112–113, 133, 178, 182, 227, 311
Memorial hall design competition. *See* Taishō earthquake memorial hall design competition
Michi no tomo, 113, 121–122
Ministry of Education (*Monbushō*), 87–89, 195; publication of materials on the Great Kantō Earthquake for students, 87–88; sponsorship of catastrophe road show, 88–89

Mitsukoshi Department Store, 129–133, 258, 281; before-and-after postcards of, 131; destruction of, interpreted as an act of divine punishment, 129–130

Miura Tōsaku, 9

Miyake Setsurei, 237–238, 311; critique of post–World War I Japanese society leveled by, 238

Miyako shinbun, 36, 231

Miyao Shunji, 62, 193–198, 206–207, 219–220; appointment as a vice president of the imperial capital reconstruction institute (*Teito fukkōin*) of, 194; secret meetings with Finance Minister Inoue Junnosuke arranged by, 196–197

Modernity, anxieties about in post–World War I Japan, xvi, xviii, 11–12, 48, 118, 125–143, 155, 311–315; as expressed by concerned commentators in relation to the Great Kantō Earthquake, 126–143

Modernity, desire to embrace technocratic, 12, 172, 180, 184, 286, 292, 311–315

Mononobe Nagao, 18, 174–176; visions for new Tokyo that emphasized disaster preparedness shared by, 175–176

Munakata Itsurō, 226–228; song written by, selected as an anthem for national reconstruction by the journal *Kyōiku* (Education), 226–227

Nagai Tōru, 71, 155, 242–243, 292; visions for new Tokyo that emphasized social welfare infrastructure shared by, 167–168; prescriptions for spiritual reconstruction following the Great Kantō Earthquake, 242–243

Nagata Hidejirō, 59–63, 70, 85, 103, 107, 200, 269, 294–295, 301; policies implemented by, as mayor of Tokyo in September 1923, 59–60; shock and disbelief expressed by, upon learning of Fukagawa Ward's annihilation, 60–61; comparisons of the Great Kantō Earthquake and the challenges of reconstruction to war, 111–112; visions for new Tokyo that emphasized disaster preparedness shared by, 175–176; public campaign to sell the merits of land readjustment launched by, 263–265

Neighborhood associations (*tonarigumi*), 11, 237–238, 261, 281, 312–313

Neighborhood vigilance groups (*jikeidan*), 27–28, 48, 52–55, 58, 74–75, 261; photograph of, 75

Nihonbashi Bridge, 22–23

Nihonbashi Ward, 14, 20, 22–25, 40–43, 66, 70, 99, 104, 138, 167, 269, 274, 280–283, 289, 302–304; photograph of, following the Great Kantō Earthquake, 41; picture postcard of dead bodies in, 99; number of factories in, and population density of, in 1919, 156–157; petitions submitted against land readjustment by residents of, 269, 280–282; Land Readjustment District No. 10 located in, 280–282

Nonomura Kaizō, 143, 145–147, 151; comparisons of 1920s Japan to Rome in the final stages of its decline, 146–147

Nose Yoritoshi, 136–138, 258–261, 311; critique of sexual degeneracy in post–World War I Japan leveled by, 136–137; claims that the government's program for spiritual reconstruction had failed, 258–260

Okano Keijirō, 87–88, 192, 205, 211, 228–233; activities undertaken to promote spiritual reconstruction as education minister, 88, 228–230; influence in drafting the imperial rescript regarding the invigoration of the national spirit, 228–230; call to all principals across Japan to embrace the imperial rescript regarding the invigoration of the national spirit, 230–231

Oku Hidesaburō, 28–30; condemnation of Korean massacre following the Great Kantō Earthquake, 28–29
Okutani Fumitomo, 128–130, 138, 145–146; claim that the loss suffered as a result of the disaster equaled the total loss of all previous wars, 113–114; interpretation of the Great Kantō Earthquake as an act of divine punishment, 116–117, 121, 128–130; critique of Ginza district, 128–129; suggestion that the destruction of Mitsukoshi Department Store was a targeted act of heavenly displeasure, 129–130
Osaka asahi shinbun, 35, 81
Osaka mainichi shinbun, 13–14, 35, 69, 81–83, 103–105; earthquake motion picture news reels made by, and shown to audiences in Shanghai, China, 82–83
Otaru shinbun, 37, 83

Picture postcards from the Great Kantō Earthquake, 19, 22, 42, 66–69, 79, 88, 90–95, 105–110, 130–131, 144, 245, 285, 286–289, 298–299; of dead bodies, 30, 33, 96–99, 101; arrests made in relation to selling images of dead bodies, 95–96; of Tokyo before the Great Kantō Earthquake, 131, 140, 143; of reconstructed Tokyo, 285, 287, 289, 299, 302, 304
Popular mind (*jinshin* or *minshin*), 126–143; concern about the decline of, expressed in Japan following the Great Kantō Earthquake, 126–143, 312; policies advocated for strengthening of, expressed in Japan following the Great Kantō Earthquake, 227–230, 234, 236, 238, 258
Postal saving in Japan, 254–255

Reconstruction bureau (*Fukkō kyoku*), 267–270; announcement of land readjustment regulations by, 269–270; consideration of petitions received from angry landowners in Tokyo, 273–282
Reconstruction of Tokyo, initiatives undertaken, 283–306; on roads, bridges, and canals, 284–288; on social welfare facilities, 288–293, 304–306; on parks, 288–290, 304–306; on land readjustment, 304–306; total money spent on, 304–306
Relief bureau. *See* Emergency earthquake relief bureau
Rumors, spread following the Great Kantō Earthquake, 26–29, 48, 52–55, 73–74, 107; of armed bands of Koreans marching on Tokyo, 52; of Korean uprisings across eastern Japan, 53–54
Russo-Japanese War, 1904–05, 47, 73–74, 88, 95, 97, 124, 159, 161, 180–182, 189, 199, 227, 233, 237; view of the Great Kantō Earthquake as a national test akin to, 112–113, 220

Sagami Bay, 2, 15–17, 54; as epicenter of the Great Kantō Earthquake, 16–17
Saimin (people of scant means), 156–159; average living conditions of, 158–159; distribution of, across Tokyo, 157–158
San Francisco Earthquake of 1906, 4, 180–182
Sano Toshikata, 191, 295
Seiyūkai political party, 79, 189, 205, 231, 249; challenges against the government's reconstruction bill in the 47th session of the Imperial Japanese Diet launched by members of, 213–225
Shakai seisaku jihō, 167–168
Shibusawa Eiichi, 204–213, 294
Shimamoto Ainosuke, 2, 114, 133, 152; description of the Great Kantō Earthquake as the "mother of all happiness," 6, 152; critique of murder and anarchy that erupted following the Great Kantō Earthquake, 73–74;

Shimamoto Ainosuke (*continued*)
critique of rampant individualism in Japan, 133
Shingikai. See Imperial capital reconstruction deliberative council (*Teito fukkō shingikai*)
Shitennō Nobutaka, 120, 176–178; interpretation of the Great Kantō Earthquake as an act of divine punishment, 120
Sino-Japanese War, 1894–1895, 47, 112–113, 180–182, 189, 199
Smell, or stench, of the dead in Tokyo, 1, 13–15, 29–31, 34–35, 45, 49, 105–106
Spiritual reconstruction (*seishin fukkō*), 74, 87, 120, 136, 152, 226–262; as advocated in the imperial rescript regarding invigoration of the national spirit (*Kokumin seishin sakkō ni kansuru shōsho*), 228–233; suggestions for securing, made by: Fukasaku Yasubumi, 233–235; Takashima Heizaburō, 235–236; Kobayashi Junior High School, 236–237; Miyake Setsurei, 237–238; Abe Isoo, 238–241; Yamakawa Hitoshi, 241; Ugaki Kazushige, 241–242; Nagai Tōru, 242–243; Fukuda Tokuzō, 242–243; Hamaguchi Osachi, 244–246; Tokutomi Sohō, 246–247; Wakatsuki Reijirō, 251–253
Soeda Azenbō, 2–4, 102–106; Song of the Great Kantō Earthquake, 3–4; description of the cremation of bodies at the site of the Honjo Clothing Depot, 105–106
Suematsu Kai'ichirō, 27–28, 74; discussion of "absurd rumors" heard in relation to Koreans, 27; interpretation of the Great Kantō Earthquake as an act of divine punishment, 122–123
Sumida River, 14, 24–25, 63, 78, 90–91, 102, 156, 270, 286–287, 301

Tabuchi Toyokichi, 28–29
Takahashi Korekiyo, 164–165, 205, 209–211, 223

Taishō earthquake disaster memorial project association (*Taishō shinsai kinen jigyō kyōkai*), 294–300; formation of, 294–295; sponsorship of memorial hall design competition, 295–297; controversy over winning entry of Maeda Kenjirō, 297–299; decision to change design of memorial hall, 298–300
Taishō earthquake memorial hall design competition, xviii, 89, 183, 264, 293–300; competition announced and judges appointed for, 295; entry of Maeda Kenjirō selected as winner, 295; winning design of, displayed to the public in 1925, 295–296; public outcry over winning design, 296–299; winning entry of Maeda Kenjirō scrapped, 298–299
Taishō Restoration, 6, 114, 146, 227, 232–233
Takarabe Takeshi, 49, 63–64, 202–203, 205, 220–222; pragmatic compromise over dispute associated with reconstruction budget, 220–221
Takashima Beihō, 45–46, 227–228; description of damage caused to Tokyo, 2, 13; comparisons of Great Kantō Earthquake and the challenges of reconstruction to war, 112; interpretation of the Great Kantō Earthquake as an act of divine punishment or the "Buddha's punishment," 121–122; belief that a successful reconstruction project would be viewed by the international community as a sign of Japan's strength, 181–183
Takashima Heizaburō, 14, 152, 226; calls for spiritual reconstruction that emphasized children and families, 235–236
Tanaka Giichi, 50–53, 203–205; deployment of army units to Tokyo directed by, 51–53; concerns expressed to Finance Minister Inoue Junnosuke that a large reconstruction budget would

INDEX - 373

result in a reduction of the army's budget, 203–204

Tanaka Kōtarō, 14, 18, 20–23, 27, 32–34, 101; harrowing description of his escape from fires, 20–21; description of pandemonium that engulfed Tokyo following the Great Kantō Earthquake, 20–22; recounting posters that warned Japanese to be on guard against illegal activities of XXX [Koreans], 27–28; journey to the Honjo Clothing Depot to view dead bodies, 32–33

Tawara Magoichi, 127, 133, 187, 222, 240; suggestion that the "Korean incident" exposed a "major defect of national spirit," 28, 73; reflections at the sight of dead bodies in Tokyo recounted by, 107

Teito fukkō shingikai. *See* Imperial capital reconstruction deliberative council

Teito fukkōin. *See* Imperial capital reconstruction institute

Terada Torahiko, 314–315

Tōa no hikari, 232–234

Tōhoku Earthquake of 2011, xv, 8, 11, 16, 309; amount of energy released by, 119

Tokutomi Sohō, 231–233, 246–248; praise given to the government of Katō Takaaki on passage of the luxury tariff of 1924, 246

Tokyo asahi shinbun, 81, 136, 210, 248; analysis of the imperial rescript regarding the invigoration of national spirit, 231–232

Tokyo, physical reconstruction visions for, 156–186; that emphasized social welfare infrastructure, 167–171; that emphasized public housing, 168–169; that emphasized parks and green space, 173–175; that emphasized disaster resilience, 174–178; that emphasized a visually impressive appearance, 178–182

Tokyo nichi nichi shinbun, 35–36, 80, 83, 142, 187; use of motion picture (*katsudō shashin*) to relay news of the Great Kantō Earthquake to people across Japan, 80–81

Tomoeda Takahiko, 114, 168–169

Toshi kōron, 153, 162, 173, 175, 180

Tsukamoto Seiji, 45

Twelve-story tower of Asakusa (Ryōunkaku), 13, 34, 139–143; viewed as symbol of hedonism, decadence, and frivolity by social commentators prior to the Great Kantō Earthquake, 139–142; picture postcard of before Great Kantō Earthquake, 140; picture postcards of, before, and after the Great Kantō Earthquake, 143–144

Ueno Park, 13, 34, 58, 68–69, 91–94, 105, 295, 303; lithograph print of the imperial inspection tour that viewed Tokyo from, 94

Ugaki Kazushige, 75–77, 118, 123, 181–183, 200–201, 312–313; concerns about the vulnerability of Japan in relation to future wars or aerial bombardment expressed by, 76–77; critique of consumerism and urban life in post–World War I Japan, 76–77, 126–127; interpretation of the Great Kantō Earthquake as an act of divine punishment, 123, 126–127; concerns over how a "failed reconstruction" project would be viewed by international observers, 182–183; prescriptions for spiritual reconstruction following the Great Kantō Earthquake, 241–242

Uno Kōji, 13–14, 34, 261–262; reflections on the "shroud of cremation" that hung over Tokyo following the Great Kantō Earthquake, 105–106; longing for a return of Tokyo's bright lights and bustling night life, 261

Urbanization in Japan, 146, 155–157; population growth of Tokyo between 1900 and 1920, 156

Wakatsuki Reijirō, 125–126, 153–154, 251–254; concerns about the state of economic degeneration in post–World War I Japan, 125–126; belief that the Great Kantō Earthquake provided a "golden opportunity" to fashion a new Tokyo, 153; employment of Great Kantō Earthquake in support of campaign for the encouragement of diligence and thrift by, 251–253; desire for Japan to embrace austerity as Britain had done in the aftermath of World War I expressed by, 252–253

World War I, 6, 11, 35, 47, 72, 146–147, 151, 157, 160, 176, 193; conflation of Europe's suffering in, to that experienced by Japan from the Great Kantō Earthquake, 110–115; era afterwards in Japan described as one of hedonism, decadence, consumerism, frivolity, and immorality, 117–118, 124, 128, 138–139, 147, 258

Yamada Hiroyoshi (Hakuai), 191–193, 195–199, 223; initial Tokyo reconstruction plans drawn up by, 191–192; ordered subordinates to devise four reconstruction plans simultaneously, 192; describes debates over, and contestation associated with reconstruction planning as a "war," 196

Yamada Yoshio, 124–125, 132–133; interpretation of the Great Kantō Earthquake as an act of divine punishment, 132–133

Yamakawa Hitoshi, 45, 241–242, 312; prescriptions for spiritual reconstruction following the Great Kantō Earthquake, 241

Yamamoto Gonnohyōe, 49, 51–52, 57–58, 183, 191, 200–202, 205–206, 208–215, 218–222, 229–232, 265; description of tortured journey from Akasaka to Nagatachō that was hindered by vigilantes, 51–52; reflections on rumors that circulated in Tokyo following the Great Kantō Earthquake, 52; proclamation, urging calm mindedness in response to anarchy and murder in Tokyo, 55; cartoon image of, shaking hands with a well-dressed catfish, 165; opening speech before 47th session of the Imperial Japanese Diet, 214; resignation as prime minster, 221–222

Yamanashi Hanzō, 47–50, 55, 72–77; personal background and military experience prior to 1923, 47–48; desire to use example of the Great Kantō Earthquake to gain support for the implementation of civil and air defense training programs in Japan, 76–77

Yanagisawa Yasue, 78, 109–110

Yokohama, xvi, 1, 4–5, 14–17, 37–39, 43, 45, 54–56, 58, 61, 63, 71–72, 82, 87, 101, 114, 146, 197, 204, 207–208, 211, 219, 222, 250, 265; rumors claiming that Yokohama had been washed away by tsunami, 26; army field reports on conditions in, following the Great Kantō Earthquake, 54–55; reconstruction budget, 197

Yoshiwara, 97–99, 137–138, 241, 308; picture postcard of the bodies of dead prostitutes from, 99